Stealth Adapted Viruses; Alternative Cellular Energy (ACE) & KELEA™ Activated Water

---

A New Paradigm of Healthcare

W. John Martin, MD, PhD.
Institute of Progressive Medicine
South Pasadena CA 91030

AuthorHouse™ LLC
1663 Liberty Drive
Bloomington, IN 47403
www.authorhouse.com
Phone: 1-800-839-8640

Cover: A Figure used at a Lecture given at the University of Southern California School of Medicine on September 24, 1994. The Figure depicts the range of neuropsychiatric illnesses attributed at the time to infection with stealth viruses. The term stealth was chosen to emphasize the lack of effective immune recognition of the viruses, as well as an understanding of a generic process seemingly applying to various viruses, rather than a single virus type. Similarly, the term "encephalopathy" is used in distinction to" encephalitis" to emphasize the lack of any accompanying inflammation. The following illnesses were abbreviated: CFS - chronic fatigue syndrome; GWS – Gulf war syndrome; FM – fibromyalgia; MCS – multiple chemical sensitivity; Dep – depression; ADD – attention deficit disorder. Psychosis and autism are also included, with psychosis somewhat distanced from those illnesses, such as CFS, in which the patients are defensive as being labeled as psychiatric patients. SVE – stealth virus encephalopathy.

This is Volume 1 of an intended series of publications indexed as issues of the Journal of Progressive Medicine.

The author can be contacted at wjohnmartin@ccid.org

Published by AuthorHouse 06/06/2014

ISBN: 978-1-4969-0496-6 (sc)
         978-1-4969-0497-3 (e)

authorHOUSE®

Stealth Adapted Viruses;
Alternative Cellular Energy (ACE)
& KELEA™ Activated Water

"Dedicated to those who place kindness and compassion above the sense of entitlement and financial gain. The book is also in partial fulfillment of a promise I made to my mother before she died. I am also very thankful to my wife and daughters who have made social and financial sacrifices so that the work could continue and to my sister, Liz, for her unwavering trust and confidence."

CONTENTS

Preface to Section 1

The reason for compiling this book is to introduce a new paradigm for preventing and treating many common neuropsychiatric illnesses by enhancing the alternative cellular energy (ACE) pathway. The groundwork for this paradigm has been studies conducted for more than 25 years on viruses, which are not effectively recognized by the cellular immune system and, therefore, do not evoke a typical inflammatory response. Fortunately, these "stealth adapted" as well as the conventional viruses from which they are derived can be suppressed through a non-immunological mechanism, which involves the ACE pathway. Simple methods for monitoring and enhancing the ACE pathway have been devised and are ready for large scale testing in the prevention and, if necessary, the treatment of many infectious diseases. More recent advances have been i) understanding of the ACE pathway in terms of a natural physical force, tentatively termed "KELEA" and ii) broadening of illnesses attributed to an insufficiency of cellular energy (ICE) and potentially treatable via the ACE pathway.

KELEA refers to "kinetic energy limiting electrostaic attraction" and can be shown by the increased vaporization of liquids, including water and ethanol. Seemingly, this is a result of diminished hydrogen bonding due to a fundamental force counteracting the natural attraction between opposing electrical charges. Water with KELEA absorbing properties has potential benefit well beyond its immediate clinical applications to a wide range of human and animal illnesses. As supported by ongoing studies, it has the potential of greatly enhancing the quality and quantity of agricultural crops as well as having important industrial applications.

The more academic information in Section 1 of the book is in the form of 6 research articles, listed as Chapters 1 through 6. Chapter 1 is a brief history of the discovery of stealth adapted viruses, along with an updated sequence analysis of DNA extracted from cultures of the first of these viruses to be isolated. Chapter 2 provides extensive evidence leading to identification of the ACE pathway. Chapters 3-5 document three different clinical applications of ACE pathway based therapies. They are i) the successful treatment of herpes simplex virus (HSV) infections, as performed in conjunction with Dr. Jon Stoneburner; ii) the therapy of children with autism; and iii) a collaborative study treating children in El Salvador with severe diarrhea. Chapter 6 is a comprehensive review on the various means of energizing (activating) water, which led to the KELEA hypothesis. These 6 Chapters have not previously been published as research articles, although they do make reference to numerous previously published articles.

A major purpose of the first six Chapters is to give scientific credibility to the research as can be assessed by experts on the respective topics. The task then becomes how quickly the research can be applied to the betterment of mankind. Political barriers to the research need to be removed and replaced with a growing enthusiasm to educate and enable others around the world to benefit from the research. This challenge is addressed in Section 2.

CHAPTER 1

Stealth Adaptation of Viruses: Review of Earlier Studies and Updated Molecular Analysis on a Stealth Adapted African Green Monkey Simian Cytomegalovirus (SCMV)

W. John Martin

Institute of Progressive Medicine, South Pasadena CA 91030

Abstract

The available DNA sequence data on an African green monkey simian cytomegalovirus (SCMV)-derived stealth adapted virus are summarized. The data provide important insight into a generic mechanism by which viruses avoid effective immunological recognition by the cellular immune system. This process is termed stealth adaptation and comprises the deletion or mutation of the relatively few virus components, which are normally directly targeted on virus infected cells by T lymphocytes. The sequence data also reveal the potential complexity of stealth adapted virus genomes resulting from genetic instability and also from the apparent involvement with replicating cellular and bacterial genes. Stealth adapted viruses, including those which presumptively originated from SCMV contaminated poliovirus vaccines, pose a serious threat to public health and can readily explain the increasing prevalence of neuropsychiatric illnesses, including autism, mental diseases and the chronic fatigue syndrome (CFS).

Keywords: Stealth, SCMV, Cytomegalovirus, T lymphocytes, Cell Mediated Immunity, Vaccines, Viteria, Ochrobactrum, Mycoplasma, Encephalopathy, Encephalitis, Chronic Fatigue Syndrome, CFS, autism, immune evasion, genetic instability, PCR, Virus Culture, Brucella, Rhizobium, Lyme Disease.

Address: 1634 Spruce St., South Pasadena, CA 91030: 626-616-2868 wjohnmrtin@hotmail.com

Abbreviations: CFS – chronic fatigue syndrome; CTL – cytotoxic T lymphocytes; CPE – cytopathic effect; HTLV- human T lymphotrophic virus; MHC – major histocompatibility complex; MRI – magnetic resonance imaging; HCMV – human cytomegalovirus; RhCMV – rhesus monkey cytomegalovirus; SCMV – African green monkey simian cytomegalovirus.

Introduction

While inflammation is the expected hallmark of infectious diseases, situations have been described in which the body fails to mount an inflammatory response to an ongoing virus infection. One example is infection occurring in an individual without a functioning cellular immune system, as can be seen with JC virus infections in AIDS patients (1,2). Prenatal virus infections can potentially induce immunological tolerance to virus antigens, as occurs with congenitally acquired hepatitis B virus (3,4). Changes in a virus can also explain a lack of an accompanying inflammatory reaction. Although still not widely appreciated among virologists, relatively few components of most viruses are displayed on the surface of virus-infected cells in a manner that allows for effective immunological recognition by T lymphocytes. The restricted diversity of virus antigen presentation is a corollary of the Clonal Selection Theory of Acquired Immunity (5). Each lymphocyte can only engage with cells expressing multiple copies of the actual antigenic specificity for which that particular lymphocyte is genetically preprogrammed to recognize. Virus antigen presentation is a complex process requiring partial degradation of selected virus proteins to small peptides, which then bind to a specific region of newly synthesized major histocompatibility complex (MHC) proteins, prior to these proteins being transported to and lodging within the cell membrane (6).

The restriction on virus antigenic recognition is particularly striking for human cytomegalovirus (HCMV). Although, this virus codes for more than 200 proteins, the majority of anti-HCMV cytotoxic T cells (CTL) are directed against the protein coded by virus gene UL83, where UL refers to the unique long segment of the virus genome and US refers to an adjoining unique short segment (7-10). Additional smaller antigenic contributions are made by UL55 and UL123 coded proteins, such that in aggregate these three components comprise over 90% of targeted antigens for the CTL response against HCMV (11-15). Deletion or mutation in the relatively few genes encoding critical immunogenic virus antigens provides a relatively simple explanation for how active virus infections may persist without evoking an inflammatory reaction.

The possibility of this novel immune evasion mechanism was supported by an examination of a brain biopsy obtained in 1990 from an immunocompetent patient with an unexplained neurological illness (16). She had been experiencing social difficulties as a primary school teacher and sought psychological counseling in an unsuccessful effort to avoid being dismissed. Upon reemployment as a kindergarten teacher, she found it difficult to express herself, either verbally or in writing. Near confluent, bilateral periventricular opacities were seen on MRI (magnetic resonance imaging), justifying a stereotactic needle biopsy of this region. The biopsy showed no inflammation, yet was weakly positive when tested using the polymerase chain reaction (PCR) method of virus detection.

The PCR assay uses small synthetic oligonucleotides as primers for recycling DNA replication. Ideally the DNA sequences of the synthetic primers match exactly to relatively closely spaced regions on opposing strands of the targeted DNA virus, allowing for the assay to be performed under high stringency conditions and to be uniquely specific for the targeted virus. In the PCR study employed on the brain biopsy, however, the primers used were designed to be more broadly reactive with multiple human herpesviruses when tested using low stringency conditions. It was not possible at the time to further characterize the actual virus presumptively responsible for the weak, but decidedly positive PCR. Cellular damage comprising vacuolization with intracellular inclusions was clearly present on histological examination of the brain tissue and confirmed by electron microscopy (16,17). The indications of cellular damage, together with the positive PCR, were highly suggestive of a virus infection, in spite of the lack of an inflammatory response. The putative virus causing the cellular brain damage was characterized as "stealth" in its apparent ability to bypass the cellular immune system.

Earlier support for this basic premise was provided by positive PCR assays performed on blood and/or cerebrospinal fluid (CSF) obtained from several patients with unexplained or atypical neurological illnesses. A memorable strikingly positive PCR assay was repeatedly obtained on an infant born with hepatomegaly and thrombocytopenia. The infant remained in neonatal intensive care because of choroid plexus hemorrhage and an overall failure to thrive. While he was suspected of having a virus illness, routine commercial laboratory attempts at virus cultures were negative as was IgM serology for common viruses. A ventriculovenous shunt allowed for repeated sampling of CSF. Consistently positive PCR were obtained on CSF samples using the herpesvirus-reactive PCR primers. While not indicative of regular HCMV, the results strongly supported a viral cause for the infant's illness. Another example was an adolescent with residual brain damage for which his mother was seeking legal compensation. She alleged his clinician had been negligent in not starting Acyclovir therapy soon after her son's hospital admission for headache and cognitive confusion. The early possibility of herpes simplex virus (HSV) encephalitis was considered but not supported due to the absence of a cellular reaction in the patient's CSF and the rather slight lowering of mental status (consciousness). Only after the patient clinically deteriorated and developed diplopia, did the physician arrive at the diagnosis of HSV encephalitis and begin to prescribe Acyclovir. While this therapy appeared to improve his medical condition, he remained

cognitively impaired with severe learning and visual disorders. A blood sample yielded clearly positive PCR findings with the broadly herpesvirus-reactive primer set, but was negative when tested using a primer set specific for HSV or primer sets specific for other known human herpesviruses.

PCR Studies in Patients with the Chronic Fatigue Syndrome (CFS)

The capacity to perform PCR assays on the brain biopsy and on patients with complex neurological illnesses, along with the high level of suspicion that atypical viruses were able to inflict brain damage without an accompanying inflammatory reaction, were outcomes of ongoing efforts to find the cause of the chronic fatigue syndrome (CFS). The major impetus to these studies was the report in 1986 of a possible epidemic virus illness at Lake Tahoe, Nevada (18). Affected patients were experiencing persisting fatigue and were tentatively being diagnosed as having CFS. A new herpesvirus, initially called HBLV and later HHV-6, had been described in the same year (19), as had the PCR assay (20). An obvious line of inquiry was to use the PCR assay to search for HHV-6 in CFS patients.

Zaki Salahuddin, working in Dr. Robert Gallo's laboratory at the National Institutes of Health, kindly provided sufficient DNA sequence data on HHV-6 to prepare sets of primers specifically reactive with the newly described herpesvirus. Experience using the PCR assay was gained in other studies aimed at identifying mutant *ras* gene in some human cancers (21); human papillomaviruses in cervical (22) and ocular tissues (23); and HCMV in HIV infected patients (24), kidney transplant recipients and tissue samples of salivary gland tumors. Two sets of HCMV primers were generally used, one of which was directed against the gene coding the immediate early (IE) gene (coded by UL123) and the other against the UL83 coded gene (24). Interestingly, in a small minority of patients, there was a lack of concordance using the two sets of primers, suggesting that variant forms of HCMV may exist (24). When applied to patients with complex neurological diseases, weak positive PCR responses were not uncommonly observed using the UL83 directed primers. Unlike in the positive HCMV assays, however, quite low amounts of multiple products were being generated, none of which corresponded to the expected size based on the HCMV genome. By reducing the stringency conditions of the PCR assay and employing a sensitive dot blot hybridization detection method for the PCR generated DNA products, positive PCR assays were obtainable using these primers with the other human herpesviruses; HSV-1, HSV-2, Epstein Barr virus (EBV), varicella-zoster virus (VZV) and HHV-6. The assays were conducted such that no discernable hybridizable products were generated in PCR assays performed on DNA samples from laboratory personnel and other normal volunteers. As noted above, however, positive results were obtainable using this assay in several patients with severe neurological illnesses.

Dr. Jay Goldstein, an Orange County physician specializing in CFS, kindly provided blood samples from many of his patients for both virus culture and PCR assays. While the cultures were not definitive, the PCR assays using the broadly reactive herpesvirus primers under low stringency conditions, were clearly yielding weak positive reactions in about a third of the CFS patients (16). Even more efficient was a primer set originally designed to screen for possible retrovirus sequences. The primers corresponded to regions of the *tax* gene of human T lymphotrophic viruses (HTLV) I and II, respectively (25), but such

that they should not specifically amplify either of the conventional forms of these two retroviruses. Using these SK43 (HTLV-I) and SK44 (HTLV-II) primers, unmistakable, clearly positive findings were obtainable in various patients with complex neurological illnesses, as well as in more than a third of Dr. Goldstein's CFS patients.

The varying patterns of PCR reactivity argued against their being a single, molecularly stable virus common to different patients. The PCR data also distinguished the provisional viruses from known human herpesviruses and from HTLV. Dr. Goldstein provided the clinical insight that his patients were manifesting clinical symptoms more consistent with limbic encephalopathy than encephalitis (26). Similar, therefore, to the PCR positive patients with more severe neurological illnesses, the CFS patients were apparently not responding with an inflammatory reaction; as would be expected from overt activation of their cellular immune system. This conclusion further justified the term stealth adapted in referring to the putative viruses causing CFS and potentially also causing a wide spectrum of non-inflammatory neurological, psychiatric and other illnesses (27).

Culture of Stealth Adapted Viruses

Earlier attempts at culturing viruses from CFS patients were yielding equivocal and essentially unconvincing evidence for a progressive cytopathic effect (CPE). A determined effort was made in 1990 to closely follow the blood cultures of a PCR positive 43-year-old healthcare provider. She had been in her usual state of good health before an acute onset illness in August of that year. Her illness was characterized by intense headaches, generalized myalgia, and fever, developing one week after a sore throat. She was hospitalized with a provisional diagnosis of encephalitis/meningitis. Her CSF had normal protein and glucose levels with only a single white blood cell per cu millimeter. She improved somewhat without therapy and was discharged 7 days after admission. A blood sample was subsequently obtained because of her persisting fatigue and cognitive impairment. It clearly tested positive with both the retrovirus and herpesvirus-based PCR primer sets. Additional blood samples were, therefore, requested for virus cultures, which were performed on human foreskin fibroblasts. After 4-6 weeks, the cultures began showing subtle cellular changes, which were initially thought to possibly reflect a nutritional deficiency. Following several re-feedings of the cultures with fresh medium, the fibroblast cells began to exhibit a CPE with signs of cellular damage somewhat comparable to that seen previously in the brain biopsy. Re-culturing additional blood samples from the patient on MRC-5 human fibroblasts and on rhesus monkey kidney cells also yielded a CPE, which occurred more rapidly than previously because of early and frequent re-feeding of the cultures (28, 29). The striking CPE was characterized by the formation of foamy vacuolated cells that formed large syncytia. The cultures were negative in specific antibody and/ or PCR testing for HCMV, HHV-6, HSV, EBV, HTLV and enteroviruses. Yet the cultures reacted strongly with the HTLV-related primers, which had previously and were subsequently shown to test positive on the patient's blood. Using the culture generated PCR products as a probe, the size of the viral DNA isolated from the culture supernatant and from cell pellets was noted to be significantly smaller than that of an intact herpesvirus based on agarose gel electrophoresis (29). The foamy appearance of the CPE, along with reactivity using HTLV-based primers, led to an early consideration of a possible spumavirus (spuma being the Latin word for foamy). Yet, numerous herpesvirus-like virus particles were seen by electron microscopy (29). A single gene cluster, referred to as bel (between the envelope gene and the long terminal repeat) distinguishes spumaviruses from simple retroviruses and can cause illness independent of the other retrovirus genes (30, 31). A reasonable suggestion, therefore, was that a bel-like gene might have been acquired by an atypical herpesviruses, essentially allowing it to create the observed foamy CPE.

The distinctive syncytial foamy cell CPE was easily transferable to secondary and subsequent cultures of many long-term human cell lines of epithelial, glial and lymphoid origin. Even more striking was the wide species host range of the virus. Thus, CPE was demonstrable on murine, feline, rabbit, hamster, duck and chicken fibroblast cell lines (29). The virus, designated stealth adapted virus-1, was subsequently noted to be also cytopathic for an insect cell line.

Fifteen of 18 additional blood samples from the patient obtained over a three-year period produced the same characteristic CPE within several days of culturing (29). Moreover, a stored CSF sample collected at the time of her initial illness also yielded a positive culture. The primary development and expression of CPE by the patient's blood samples was shown to be improved by using frozen thawed extracts of the patient's mononuclear cells cultured in a serum free medium.

A male non-intimate social contact of the patient had concurrent symptoms of fatigue and cognitive impairments. His blood was cultured and yielded very similar CPE to that of the female patient. He was shown to be HIV positive and his clinical condition deteriorated rather rapidly leading to his death within 6 months. The female patient has been unable to work and has remained disabled for more than 20 years with a diagnosis of CFS.

Recovery from the CPE in Cultures of Stealth Adapted Viruses

Under routine virus culture conditions, the early CPE caused by cultured stealth adapted viruses can be mistaken for non-specific toxicity, especially since it tends not to progress and commonly disappears in infrequently fed cultures. Reversal of the CPE can even occur in well established, strongly positive cultures and is associated with the production of materials that can seemingly supply a non-mitochondria-based source of energy to the virus infected cells. These materials are referred to as alternative cellular energy (ACE) pigments and have been described elsewhere (32).

The unappreciated need for repeated re-feeding of the cultures as a means of reducing levels of ACE pigments explained the many suggestive, but never definitive, virus culture results, which were previously obtained in many of Dr. Goldstein's patients. Once the stealth adapted virus culture methodology was standardized, virtually all of the blood samples subsequently received from Dr. Goldstein and other clinicians specializing in CFS yielded positive cultures. Moreover, positive cultures were regularly being obtained on blood samples on autistic children (33) and their mothers and on the majority of tested patients in a psychiatric facility. Other illnesses yielding consistently positive cultures included multiple myeloma, amyotrophic lateral sclerosis, Gulf War syndrome and so-called chronic Lyme disease. In blinded control studies established and read independently by different technologists,

patients with severe CFS, multiple myeloma and autism showed consistently positive cultures in contrast to the 10% of selected healthy controls and up to 20% of randomly selected blood samples. Strongly positive cultures of rapid onset, were not seen in control cultures, but were not infrequently observed in the cultures from patients, especially those with more severe illnesses. It should be noted that not all culture positive patients report fatigue as their primary symptom or even as a major component of their present illness. Moreover, some individuals will report being healthy, but on closer questioning reveal either prior or ongoing episodes of sub-optimal cognitive functioning, with accompanying mood and/or sleep disorders. These considerations limit the usefulness of trying to validate a CFS-specific diagnostic assay and basically challenge the many efforts at defining CFS as a discrete illness.

Animal Inoculation Studies

The functional lack of immunogenic antigens recognizable by the cellular immune system was confirmed by intravenously inoculating cats with a frozen-thawed extract of cultured stealth adapted virus-1 infected cells. (34) Cats were chosen because several CFS patients had reported behavioral illnesses and even unexplained deaths within their household pets. The animal studies were done in a university setting with full institutional approval. The inoculated animals remained asymptomatic for 48 hrs, but then developed a severe neurological illness, which peaked at between 2-4 weeks. During this period, the animals lost the playfulness that was present prior to injection. Rather, they became reclusive and irritable; yet the altered behavior was not accompanied by a marked reduction in consciousness. The body temperature of the animals dropped $0.6° – 0.8°$ F below normal, which is characteristic of ill cats. They had dilated pupils and were clearly bothered by the light. Several animals had balding areas on their head from repeated rubbing against the cage. Another animal had torn part of his face from scratching; while two others had bloody ocular and nasal discharges, also probably from scratching. During this initial period, the animals' gums were swollen. Peripheral enlarged lymph nodes could also be easily palpated and many muscle groups were clearly painful when squeezed. The cats were euthanized at 1, 2, 4, 6 and 15 weeks after virus inoculation. Although only a few animals were observed, the severity of the illness began to wane after 4 weeks with definite improvement noted in the cats at 6 weeks. Indeed, the cat on which the necropsy occurred at 15 weeks had seemingly resumed normal activities by week 10.

Histological examination of brain tissue at each of the time points showed foci of cells with cytoplasmic vacuolization, which occasionally formed syncytia. These changes occurred in the complete absence of any inflammatory reaction. Marked intracellular inclusions and extracellular deposited materials were noted upon staining with periodic acid Schiff (PAS) and Stains-all dyes. Electron microscopy confirmed the presence of foamy vacuolated cells with structured intracellular inclusions. Cellular damage was still apparent in the cat on whom the necropsy was performed at week 15. Vacuolating cytopathology without discernable inflammation (except for a suggested slight increase of tissue eosinophils), was observed within other organs, indicating a widespread systemic infection.

Illness did not occur in a cat inoculated with heat inactivated virus infected cells. Interestingly, two follow up inoculation of non-inactivated virus failed to induce any apparent disease in this animal. The protection provided by the heat-inactivated virus may well have been mediated by antibody, since the failure to generate a cellular immune response does not necessarily preclude antibody-mediated protective immunity. Consistent with this interpretation and with the differences among stealth adapted viruses isolated from different patients, the previously protected cat developed a severe, but recoverable illness, when inoculated with a different stealth adapted virus isolate (34). The other isolate was obtained from a patient with systemic lupus erythematosus. Interestingly, this patient described her own cats as having aberrant behaviors and that some of her earlier cats had died from undiagnosed neurological illnesses.

Positive stealth adapted virus cultures were subsequently obtained using blood samples from symptomatic cats and dogs of several CFS patients. Moreover, direct inoculation of the blood from sick into healthy cats caused a similar illness as when using CFS patient infected cultured cells. Necropsy of a newborn kitten of a virus inoculated pregnant cat showed widespread histological changes within its various tissues, including the brain. Moreover, milk collected from the stomach of the kitten tested positive by culture for stealth adapted virus. Similar events are predicted for infants born of stealth adapted virus infected humans.

Initial Sequencing Studies on Stealth Virus-1

DNA sequencing of PCR products amplified from the stealth virus-1 culture provided direct support for an atypical herpesvirus. As with the patient's blood sample, a strong PCR response was seen using HTLV I directed primer SK43 in conjunction with the HTLV II directed primer

SK44. The PCR reactions yielded two distinct bands on agarose gel electrophoresis (29). One band migrated into the agarose gel with an estimated size of approximately 1.5 kilobase (kb) and the other band migrated further into the gel with an estimated size of approximately 0.5 kb. The PCR products were cloned into pBluescript plasmids and sequenced. The sequencing was performed at the City of Hope Molecular Core Facility (Duarte, CA).

The results showed that the larger band comprised two separate products (15-5-2 and 15-5-4) with the SK44 primer flanking both sequences. Using BlastN analysis (discussed later), the 15-5-4 sequence showed significant DNA matching to a region within the UL36 gene of HCMV (29). The sequence was not, however, identical to HCMV. Moreover, the 15-5-2 product could not be matched to HCMV, or indeed to any sequences then available within GenBank; a repository of DNA sequences maintained by the National Center for Biotechnology Information (NCBI). Primer SK43 flanked the smaller PCR product, comprising 507 nucleotides. Its sequence matched to a cellular gene, although there was no amplification of this sequence in performing the PCR assay on uninfected cells.

In experiments using two different cultures obtained from the CFS patient, DNA was extracted, cut with a restriction enzyme and cloned into pBluescript plasmids. EcoR1 was used on DNA obtained from the ultracentrifuged pellet of Millipore filtered culture supernatant. The resulting clones were designated as the 3B series. DNA pellet from a later culture was further purified by agarose gel electrophoresis as an approximate 20 kb band, cut with SacI restriction enzyme and cloned into pBluescript plasmids. These clones were designated

as the C16 series. Sequencing of the majority of the clones was performed either at the City of Hope Molecular Core Facility or Lark Technologies, a major DNA sequencing corporation in Houston Tx. Other commercial sequencing facilities occasionally used were US Biochemical (Cleveland, OH) and BioServe Biotechnology (Weaton, MD). Complete definitive sequencing was performed on most of the clones referenced in this paper. For a few clones, however, only the initial T3 and T7 readouts are available, with occasional nucleotide uncertainties occurring beyond 500-600 bases. These extended regions were not included in the sequence analysis presented in this or in prior publications.

Many of the newly sequenced clones confirmed the relatedness, but non-identity, to HCMV. One of the clones corresponded to a region of HCMV for which there was also a known sequence for the Colburn strain (35) of African green monkey simian cytomegalovirus (SCMV). Although, not 100% identical, the clone matched far more closely to SCMV than to HCMV or other sequenced animal herpesviruses (36). The translated amino acid sequence matched even more closely to that of SCMV. By using different primer sets, similar but still distinguishable reactivity could be shown between the patient's virus and the Colburn strain of SCMV (36). It was concluded that the patient's virus had unequivocally arisen from SCMV, but had undergone further genetic changes. As more sequence data of the Colburn and other strains of SCMV (37) became available, it was also possible to assign the sequence of PCR product 15-5-2 to SCMV. It clearly matched to the region of SCMV, which corresponds to the UL20 gene of HCMV.

The nucleotide alignments were performed using BLAST (Basic Local Alignment Search Tool) of the National Center for Biotechnology Information (NCBI). The blastN program provides a statistical measure of the probability of a sequence alignment occurring simply by chance. It is expressed as a "Score," which relate to the percentage of identical nucleotides, size of the aligned regions and the requirement to place "gaps" for missing nucleotides within one or both sequences to obtain the optimal alignment. The higher the score, the more significant is the alignment, with limited meaning for Scores below 100. Very high Score values are recorded as an exponential to base "e." The statistics of the sequence alignment is also expressed as a function of the total number of available sequences within the database being searched, which is the entire non-redundant collection of all sequences within GenBank. This probability value is recorded as an "Expect" value and is stated as 0.0 when it is beyond the upper limit of 1 in 10^{179} probability of the alignment occurring by chance (38, 39). The alignments of clones 15-5-4 and 15-5-2 (minus the primers) with Colburn SCMV sequences yielded very high Scores of 2,379 and 2,140 respectively; (95% and 93% nucleotide identity), and both alignments had an Expect value of 0.0. No comparable matching occurred with any virus other than SCMV, including rhesus cytomegalovirus (RhCMV).

Evidence for Genetic Instability and Recombination of Stealth Virus-1

Further sequencing of the clones revealed some remarkable findings. First, individual clones corresponding to essentially the same region of the putative SCMV derived stealth adapted virus, would not uncommonly show minor nucleotide differences, including deletions, substitutions and apparent duplications (40). Examples of matching occurring

at only one of the two ends, were also found within sets of both the EcoRI and SacI derived plasmids. An artifact of the rejoining of different restriction fragments in the cloning procedure was excluded by ensuring that there were no internal sequences corresponding to the restriction enzyme being used. Rather the heterogeneity was explained by mutations that affected restriction sites. Occasional examples were also seen of major recombination. For example, in three of the C16 series clones, a region matching to SCMV nucleotides 202,128 to 201,415 (coding for US18) was contiguous with a region matching to SCMV nucleotides 220,153 to 223,156 (coding for US32 and beyond). The sequence at the point of recombination was unrelated to the specificity of SacI restriction enzyme. Other clones of the C16 series shared one but not both ends of the clones containing the apparent recombination. The deduced coded proteins of these clones were as expected and corresponded to US18 adjoining to US19 in some of the clones and US31 adjoining to US32 in the other clones. Within the 3B series of clones, one showed recombination linkage between the UL57 and UL69 regions, while other 3B clones contained regions corresponding to portions of the intervening sequences deleted by the apparent recombination.

As more partial sequence data were obtained from unrelated clones, it became clear that the entire genome far exceeded the predicted ~20kb size as suggested by the agarose gel electrophoresis of the purified, uncut, viral DNA. It was concluded, therefore, that the virus genome existed as fragments, as well as being genetically unstable (40).

Also consistent with a fragmented genome, was the finding that the clones were non-randomly distributed when matched against corresponding genes of HCMV (41). Five HCMV genes (UL 36, 47, 52, 86, and US28) were each matched by 10 or more of the initially sequenced clones. Ten additional HCMV genes were each matched by 5-9 clones. The majority of the represented HCMV genes were matched by only 1-4 of the partially sequenced clones. No clones were identified with sequences matching to two of the three major immunogenic coding regions of HCMV (UL83 and UL55). The inability to identify clones corresponding to a region within the virus could arise if it was contained within a portion of the virus genome without flanking EcoRi and SacI sites closer than ~9 kb, the practical upper size limit of cloned fragments within the pBluescript plasmid. This was not the case with UL55, since clones containing sequences corresponding to the UL54 and UL56 genes were found. It may still, however, be the case with the UL83 gene. Thus, there is a span of 11,080 and 16,726 nucleotides, respectively, between identified EcoRI and SacI sites, in the cloned sequences corresponding to the UL76 and UL84 coding genes, respectively. The distance between the cutting SacI sites is particularly long because a SacI cutting site present in the UL84 coding gene of SCMV was mutated in stealth adapted virus-1. The SCMV genome actually shows two additional SacI cutting sites within the unidentified region of stealth adapted-1 virus between UL76 and UL84. Again, however, mutation in one or both of these sites cannot be excluded. If the UL83 related gene is present in stealth adapted virus-1, however, it's sequence is presumably mutated in order to explain the lack of an immunogenic product.

A clone (3B546) of stealth adapted virus-1 was identified, which corresponded to the UL123 gene, which in HCMV codes for the major immediate antigen-1, one of the three major immunogenic targets for anti-CMV CTL (13, 14). Sequencing of the clone from stealth adapted virus-1 revealed several major mutations when compared with the sequence of the Colburn and other strains of SCMV (41). The major immunogenic regions of the UL123 coded protein in HCMV is in the third exon of the major immediate antigen 1 (MIE-1). The deduced amino acid sequence of stealth adapted virus-1 shows only 82% identity with SCMV and interestingly lacks a stretch of 10 amino acids, within the third exon, just distal to a relatively highly conserved region in primate and human cytomegaloviruses (42). The major deletion was confirmed in a second clone (C16139). While specific T cell epitopes of the MIE-1 gene in HCMV have been defined (13, 14), no comparable data are available for SCMV. Nevertheless, it can be concluded that the apparent MIE-1 coding sequence in stealth adapted virus-1 is not coding for an antigen capable of evoking an inflammatory response in either the patient from which the virus was isolated or in inoculated cats.

Long stretches of contiguous sequences could be assembled using representative examples of the cloned genes, e.g. UL28-48; UL48-54; UL84-102; UL115-132; UL141-146 and US19-28. Overall, the sequences showed statistically far greater homology to SCMV than to any other cytomegalovirus of human or primate origin. Still, throughout the assembled sequences, there were widespread deviations from SCMV, again confirming the remarkable genetic instability of the stealth adapted virus. For all but one of the stretches of sequences from stealth adapted virus-1, the level of nucleotide identity was only 94-95% when compared with the fully sequenced GR2715 strain of SCMV. Optimal alignments also required the insertion of numerous gaps into one or other of the sequences. Moreover, some of alignments did not extend throughout the entire stretch of the stealth adapted virus-1 sequence. In part, this reflected significant differences among some of the fully and partially sequenced SCMV isolates, but in no case did the alignment exceed 95% or not require the insertion of numerous gaps. For one stretch of the stealth adapted virus-1 sequence, the nucleotide identity with SCMV was only 80%, again reflecting widespread mutation. Yet, even with this sequence, the alignment with SCMV is still highly significant with a Score values well beyond 100 and an Expect values of 0.0. While, it is unequivocal that stealth adapted virus-1 originated from SCMV, it is clearly distinguishable from any of the known isolates of SCMV and should not be disregarded as a possible laboratory contaminant (43).

Are Cellular Sequences Involved in Stealth Adapted Virus Replication?

As noted earlier, the sequence of the smaller PCR product generated from the culture of stealth adapted virus-1, matched to cellular DNA and not to known SCMV sequences. The matching was unique to an intron within the Rho guanine neuleotide exchange factor 10 on chromosome 8. Still the matching was not 100%. Indeed, optimal alignment showed only 89% nucleotide identity and required 18 gaps. Still the Score was highly significant at 568 and an Expect value of 2e-158. Moreover, partially matching products were generated in PCR assays on stealth adapted virus cultures from two other patients. No such product was ever generated using the PCR assay on uninfected cells. Upon cloning of the DNA isolated from stealth adapted virus-1, approximately 5% of the clones contained sequences, which showed significant homologies to cellular sequences (44). While these clones could have been dismissed as simply being contaminating cellular DNA, consideration was given to the possibility of

cellular DNA being somehow involved in the virus replication process. Supporting this possibility was that cellular DNA related sequences were present in both the 3B and C16 series of clones; the latter being derived from agarose banded virus DNA. Furthermore, the actual sequences of the cloned DNA were not always an exact match to the apparent corresponding normal cellular genes. While, the lack of exact matching could presumably be because the actual cellular sequence of origin had not yet been identified, it is also consistent with genetic instability if the DNA had become incorporated into the virus replication process. It is quite possible that some of the DNA is of monkey origin.

An originally reported stealth adapted-1 DNA sequence was mistakenly identified as newly acquired chemokine-related cellular genes. As additional sequence data of intact SCMV isolates became available, it became apparent that these viruses too had essentially the same cellular sequences as reported for stealth adapted virus-1. The acquisition of this particular sequence was not, therefore, involved in the stealth adaptation process. Specifically, it was reported that sequences adjacent to the gene corresponding to the UL145 gene of HCMV matched closely three copies of an alpha CXC chemokine coding gene (45). The gene had been loosely implicated as a possible cause of melanoma and is, therefore, designated melanoma growth stimulatory activity (MGSA). The third copy of the gene contains an intron, indicating that this gene had probably been incorporated from cellular DNA rather than as reverse transcribed RNA. As mentioned above, it was incorrectly assumed in an earlier publication that the cellular-derived gene had been incorporated into the virus as part of the stealth adaptation process. Instead, it is now apparent that the incorporation of MGSA-related cellular genes has occurred much earlier and even before the divergence of SCMV and RhCMV.

Interestingly, different strains of SCMV vary in the retention of the incorporated MGSA-related cellular genes (43). The pattern in the stealth adapted virus was similar to that in a partially sequenced SCMV, but differed from that in the completely sequenced SCMV strains (Colburn, 2715 and GR2715). It is also worth noting that slight differences exist between the MGSA related sequences in the stealth adapted virus from those of the sequenced SCMV, consistent with the genetic instability of the stealth adapted virus.

The complete sequencing of SCMV isolates has also led to a reinterpretation of the presence of five copies of the gene coding US28 chemokine receptor-related genes in stealth adapted virus-1. HCMV and RhCMV have a single copy of the US28 gene. In an earlier published paper (46) it was incorrectly suggested that US28 gene amplification had occurred in the stealth adapted virus. This is not so since some other SCMV isolates also contain five copies of this gene (43, 47). Other isolates have fewer copies, indicating once again the unexpected heterogeneity among SCMV isolates. The US28 gene of human and primate cytomegaloviruses is particularly noteworthy since it has been implicated as a potential oncogene (48); a major co-factor for HIV infection (49, 50); and a stimulus for aberrant vascular proliferation (51).

The issue still persists, however, that among the apparently amplified and potentially mutated cellular gene sequences there are potentially genes with cell growth regulatory and even oncogenic activity. Furthermore, for some herpesviruses, oncogenic potential may not normally be seen because of the cell killing caused by the virus (52). It may emerge, however, if the cell killing activity is lost due to gene deletion or mutation, as can seemingly occur in stealth adapted viruses. Sequencing and biological studies are, therefore, indicated for stealth adapted viruses isolated from cancer patients to help better define the occurrence of such possibilities.

Are Bacteria-Derived Sequences Involved in Stealth Adapted Virus Replication?

The biggest surprise with sequencing of the clones derived from DNA extracted from the stealth adapted virus cultures came with the identification of occasional clones that matched not to virus or to cellular DNA sequences, but to sequences of bacteria (53). Rather than being dismissed as coming from contaminating bacteria, the sequences were considered part of the patient's disease process for several compelling reasons. First, cells from parallel cultures to those used for DNA extraction were subsequently maintained for several months in antibiotic free media with no evidence of bacterial overgrowth. Second, fecal cultures from some virus positive patients have grown bacteria with unusual metabolic activities. For example, the API and Vitek typing systems (54), which rely on slightly different metabolic profiles, would sometimes yield discordant identifications of the same fecal-derived bacterial colony. These typing systems rely on somewhat different metabolic profiles and can be misled when bacteria either gain or lose a metabolic activity. Furthermore, in some culture plates, a particular metabolic activity, such as hydrogen sulfide formation would seemingly occur sequentially along a line of non-touching bacterial colonies. It appeared as if an infectious agent was progressively transforming the bacteria and, thereby, inducing changes in their metabolic activities. Finally, filtered extracts from fecal cultures of some stealth adapted virus infected patients were able to produce the same vacuolating CPE as seen with their blood.

The bacteria-related sequences from the stealth adapted virus-1 cloned DNA were entered into GenBank to try to identify their bacterial origins. Both BlastN (nucleotide comparisons) and BlastX (comparison of deduced amino acid sequences) were employed. The preliminary analysis indicated that the sequences were clearly not derived from a single type of bacteria. Some sequences initially appeared to be derived from brucella, an alpha proteobacterium (55). Some other clones matched to mycoplasma, while several remained unmatchable against the existing databases. Some of the sequences were also suggestive of possible genetic instability. For example, significant differences were noted between a cloned ribosomal sequence and highly conserved sequence in a broad family of matching bacteria. This observation also argued against the cloned bacterial DNA coming from contaminating bacteria. The possibility of genetic recombination was further suggested by the finding of discrete stretches of sequences matching to either different types of bacteria or to widely disparate regions within a known type of bacteria. The term viteria was introduced to describe the possible inclusion of bacteria sequences in the virus replicating process (53).

It has now been more than a decade since the preliminary findings were published. Many more bacterial sequences are now available in GenBank allowing for more extensive sequence comparisons. Still the basic conclusions still hold. Essentially, the bacteria matching sequences can be divided into three groups (Tables 2-4). Clearest is the extremely close matching of seven of the 3B series clones to

regions of mycoplasma fermentans. The nucleotide matching covered the entire sequences of the clones with from 97 - 99% identity and only occasional gaps. None of the C16 series of clones matched to mycoplasma, raising the possibility that the organism was a secondary contaminant of only one of the cultures from which the DNA was cloned. Upon review, however, mycoplasma fermentans has very few Sac1 restriction sites and those that are present are widely dispersed over the nearly million nucleotide base genome.

The second grouping of bacteria related sequences comprise those, which originally matched best to brucella. As sequences of the closely related ochrobactum bacterial species became available, many but not all of the brucella matching clones showed a significantly greater homology to ochrobactrum (taxid 528) and most specifically to ochrobactrum anthropi. Ochrobactrum belongs to the family Brucellaceae within the alpha proteobacteria order Rhizobales (56). In contrast to the mycoplasma related sequences, the nucleotide identities of the ochrobactrum-matching sequences were generally in the range of 80-90% homology with many gaps required for optimal alignments. Moreover, the pairing of the sequences of these clones with ochrobactrum/brucella species did not, with one exception (clone 3B629 of only 165 nucleotides), extend over the entire clone. Regions of statistically significant homology were separated by regions, which could not be reasonably aligned to ochrobactrum. For example, clone 3B23 contained a 1,201 nucleotide sequence that best matched to the chromosomal region of Agrobacterium radiobacter and clone C1616 contained a 884 nucleotide sequence, which best matched to a plasmid identified in Agrobacterim rhizogenes.

With several clones, e.g. C16134, portions of the sequences still statistically matched slightly better to brucella rather than ochrobactum, but were sufficiently close to the latter to probably reflect this species. For clones 3B47 however, the matching was clearly much more to a plasmid associated with a more distant rhizobia, than to members of the Brucellaceae family (57). Similarly, the T3 and T7 readouts from clone C16125 matched to a plasmid of Chelativorans sp., (Mesorhizobium), rather than to an ochrobactum plasmid.

These data are suggestive of acquired genetic recombinations and/or insertions, rather than indicative of an existing novel, single bacterial species. In further support of this suggestion and consistent with the marked genetic instability observed with the SCMV-matching sequences, are the many discrete intervening regions within several of the ochrobactun matching clones, which are unrelated to any known bacteria, cellular or viral sequences. The nucleotide homologies of the clones with ochrobactrum bacteria related sequences are summarized in Table 3. Some of these clones had intervening sequences, which matched to rhizobium (Ribobiaceae family) or to plasmids of these bacteria, while other regions could not be aligned to any of the available sequences in GenBank. These intervening sequences, which did match to Rhizobia and some other clones with rhizobium plasmid-related sequences are listed in Table 4. Included is clone 3B513 in which widely separated 596 and 202 nucleotide regions within the entire length of 8106 nucleotides roughly matched to sequences within distinct plasmids of a methylobacterium and a rhizobium species, respectively..

Of special interest with clone 3B513, is the matching of a region of clone 3B513 with the T3 sequence of clone 3B525. The continuation of the T3 readout beyond the 3B513 matching region, as well as the T7 readout of this clone, matched to sequences of SCMV. The T7 sequence also matched to several other clones, none of which overlapped with clone 3B513.

There is a third grouping of the cloned bacteria-related sequences, which cannot yet be even partially typed with any confidence to any specific bacteria. The BlastN searching was extended beyond the "nr/nt nucleotide collection" of NCBI to include various shotgun genomic sequences, microbial genomes and environmental samples. The BlastX program was also employed to see if the possible translations of the DNA sequence resulted in recognizable amino acid sequences. In doing so, only marginal and limited partial homology was seen to some bacterial proteins of certain actinomyces (gamma proteobacteria). Interpretation of these sequences may become easier as GenBank continues to acquire more data.

As will be discussed elsewhere, several of the matching bacterial genes code for proteins with rather interesting metabolic activities. If actually expressed in infected cells, these proteins might facilitate further passage of the virus within bacteria. The presumed ability of stealth adapted viruses to infect and to be transmitted by bacteria has enormous public health implication. Moreover, the bacterial coded proteins may also contribute to the CPE observed when stealth adapted viruses are grown in eukaryotic cells and also to the production of extraneous materials, as previously described.

The bacterial sequence data are relevant to possible over interpretation of positive PCR and serology-based assays for putative bacteria pathogens in various chronic illnesses. Mycoplasma fermentans has been reported as commonly positive in CFS patients (58-61), as well as AIDS patients (62), often in the absence of confirmatory serology. Some CFS patients have also tested positive in assays for brucella reactive antibodies (63). Little consideration is usually given to the potential antibody cross reactivity among bacteria. Certainly, the reports of finding antibodies reactive with Borrelia bundgoferi in patients with chronic illness similar to CFS (64), is not a convincing argument that the patients have Lyme disease, especially since large numbers of patients said to have chronic Lyme disease have tested positive in cultures for stealth adapted viruses. Similar considerations should be given to whether bacteria are the actual cause of PANDAS (Pediatric Autoimmune Neuropsychiatric Disorder Associated with Streptococcal Infection) (65).

SCMV Origin of Stealth Adapted Viruses Cultured From Other Patients

Although stealth adaptation is regarded as a generic process potentially applicable to all forms of cytopathic viruses, it is apparent that several are derived from SCMV. For instance, a SCMV-derived stealth adapted virus, similar but not identical to stealth virus-1, was also isolated in 1991 from the CSF of a 23 year old woman with a 4 year history of bipolar psychosis (66). Her clinical condition had recently deteriorated with gross hallucinations and delusions. She had repeated seizures during an ambulance trip to the Los Angeles County Hospital and a brief cardiac arrest while in the Emergency Room. Acyclovir was administered from day 2-13 after admission. She remained comatose for several days and when she did awake showed only vegetative activity. Her illness was tentatively assumed to have resulted from a drug overdose although no supporting toxicology data were ever obtained. The

patient's vegetative state persisted till her death 6 years later. Six repeat blood samples obtained over this period and a repeat CSF sample one year prior to her death all showed the same CPE. Numerous virus particles were seen on electron microscopy of virus-infected cultures. Her culture and a subsequently CSF sample collected 4 years later yielded positive PCR using primer sets based on stealth virus-1. Partial sequencing of a PCR product showed 94% and 87% nucleotide identity, with clone 15-5-2 of stealth adapted virus-1 and the SCMV genome, respectively (66).

Another gentleman had his blood analyzed for SCMV by PCR at a primate laboratory testing facility. He had returned from a business trip that involved some partying. He became ill, as did his son and visiting father. Learning of the published work on SCMV derived stealth adapted viruses; he sent his and his father's blood to a primate PCR testing facility under the guise that they were monkey-derived specimens. Both blood samples tested positive for SCMV using primers designed to amplify part of the DNA polymerase coding gene. The laboratory provided assurance on the validity of the results and that the specificity of the assay excluded human, rhesus or baboon CMV. The Centers for Disease Control and Prevention (CDC) was informed of the finding. Although a blood sample was requested for culture, the CDC essentially opted to disregard the PCR-based report, as they have disregarded other communications regarding the culturing of SCMV-derived pathogenic viruses. Interestingly, the gentleman's blood also tested positive for anti-brucellosis antibodies in a commercial laboratory.

PCR positivity on blood samples using stealth adapted virus-1 reactive primers was also noted among patients affected by a stealth adapted virus community outbreak occurring in the Mohave Valley region of Western Arizona and involving the town of Needles, California (67). Among these patients was a boy whose brain biopsy showed a severe vacuolating encephalopathy (68). His illness presented as a behavioral/learning disorder for many months prior to his showing objective neurological signs. A second brain biopsy confirmed the non-inflammatory histological findings and on electron microscopy, showed markedly vacuolated cells with structured intracellular inclusions (69). The child responded somewhat to ganciclovir but subsequently died in the second year of his severe, yet fluctuating illness.

Several other patients, including children with autism and a patient whose illness was complicated by the development of cerebral vasculitis (70), have shown positive PCR on blood and/or CSF samples using primers reactive with stealth adapted virus-1. Many other isolates, however, have not shown reactivity with these primers. Yet some may possibly still be derived from SCMV, but may have undergone far more deletions and other changes than stealth adapted virus-1. Consistent with multiple origins of stealth adapted viruses, weak positive PCR have occasionally been observed with primers designed to react more specifically with other DNA viruses, including HSV, HHV-6, EBV and adenoviruses.

A limitation of the regular PCR assay for stealth adapted DNA viruses was revealed during PCR assays on a positive culture obtained from the CSF of a patient who became ill when working as an Emergency Room nurse. Her PCR assay was negative using three stealth virus-1 reactive primer sets directed against portions of the SCMV genome. A positive PCR was obtained using a primer set if the PCR was preceded by reverse transcribing RNA

sequences into DNA (71). While the sizes of the PCR products were slightly different from SCMV, the strong reactivity using the SCMV-reactive primers strongly supported the presence of SCMV, but mainly in the form of RNA rather than DNA. This finding is consistent with RNA forms of SCMV, undergoing replication by endogenous retroviruses, which can potentially be induced by various herpesviruses (72). Alternatively, it is possible that some stealth adapted viruses may remain as RNA and utilize their own RNA polymerases or a polymerase incorporated from other RNA viruses.

RNA replication lacks the editing functions, which generally accompany DNA replication. RNA replication via reverse transcriptases of endogenous retroviruses or by RNA virus polymerases could well account for the observed heterogeneity of viral, cellular and bacterial DNA sequences as seen in stealth adapted virus-1. Even with DNA replication, the polymerase and the editing functions involve different regions of the enzyme, such that error prone synthesis of DNA can proceed because of a lack or defectiveness in the editing component (73, 74). Whatever the cause, the genetic instability of DNA sequences isolated from cultures of stealth adapted virus-1, is truly remarkable.

Because of genetic variability, within related stealth adapted viruses and even more so because of the multiple potential origins from different DNA and RNA viruses, the PCR assay, even with prior reverse transcription, is an unreliable initial diagnostic test for stealth adapted viruses. Virus cultures, while less easily quantified, provide a more useful screening method. Moreover, as shown in this paper, virus DNA or RNA can be extracted from the positive cultures and used to identify portions of the virus on which sensitive molecular probes can be developed. Infected cells or cell extracts can also be useful in developing individual patient's antibody based assays since, as noted above, antibody responses can occur even without direct T cell recognition of virus-infected cells.

Confirmation of SCMV Contamination of Earlier Batches of Poliovirus Vaccines

FDA officials have now examined prior batches of poliovirus vaccines for DNA of SCMV using PCR. Three of 8 batches released in the mid 1970's tested positive, but they were unable to retrieve infectious virus (75). Investigators from the United Kingdom Bureau of Standards also performed PCR testing for SCMV on its collection of earlier poliovirus vaccines. (76). Of over 90 vaccines tested, nearly half showed the presence of SCMV, Again, the investigators were unsuccessful in attempts at culturing live virus from the retrieved poliovaccine lots. Neither FDA nor CDC has been forthright in attributing polio vaccines as a probable major source of stealth adapted viruses affecting humans and animals. Some FDA officials are mindful of a 1972 joint study with Lederle (part of Wyeth, which is now a component of Pfizer). In this study, kidney cell cultures from 11 African green monkeys, which would otherwise have been used to produce polio vaccines, were instead tested for SCMV. All 11 cultures were SCMV positive, confirming an earlier finding of Smith and his colleagues (77). No public notification was provided on the Lederle/ FDA results.

An incidental finding included in the United Kingdom Bureau of Standards study was the continued use of rhesus monkeys by one poliovaccine manufacture, as evidenced by the detection of DNA of RhCMV. Even more relevant was the confirmation of RhCMV DNA in the CHAT poliovirus vaccine (76). This vaccine was developed by Dr. Hilary Koprowski and extensively tested in Africa. Dr. Albert Sabin had expressed his concern regarding an unexplained virus CPE, which was still detectable in anti-polio virus antibody neutralized CHAT vaccine (78). Firsthand accounts indicated that the CHAT vaccine was commonly inoculated into chimpanzees and also used to infect kidney tissues removed from chimpanzees. Reportedly, many of the chimpanzees became ill, as did some of the animal handlers. The human illness was characterized by wasting (Thin Man Syndrome) and bacterial infections. These events have led to a reasonable suggestion that RhCMV contaminated CHAT vaccine induced the transformation of chimpanzee SIV to HIV. This topic has yet to be publicly addressed by public health officials.

Summary

Stealth adaptation is envisioned to be a generic process, which allows viruses to bypass effective recognition by the cellular immune system. It requires deletion or mutation of the relatively few virus antigens, which are normally processed and presented at the surface of virus-infected cells for functional engagement by antigen specific T lymphocytes. While stealth adapted viruses have presumably been in existence since the inception of the cellular immune system, the development of live vaccines has provided additional opportunities for their transmission to humans. Specifically, poliovirus vaccines produced in kidney cell cultures of rhesus and African green monkeys have allowed for the introduction of stealth adapted CMV from these species to humans. The infecting, vaccine-derived, stealth adapted viruses can subsequently spread among humans and along with other stealth adapted viruses, can explain the growing incidence of many illnesses, including CFS, autism and psychiatric disorders. The unique susceptibility of the brain to exhibit clinical signs from even limited and localized cellular damage is consistent with the propensity of neuropsychiatric manifestations in many stealth adapted virus infected individuals. A prototype stealth adapted virus unequivocally derived from African green monkey simian cytomegalovirus (SCMV) seemingly exists as genomic fragments of DNA. It has an extraordinary wide range of infectivity for cells of different species and is pathogenic when inoculated into cats. The cats developed a severe encephalopathy, from which clinical recovery ensures in spite of the lack of an inflammatory response. A striking feature of the genetic analysis of the prototype stealth adapted virus is instability of the virus replication process, with widespread minor nucleotide differences between this virus and conventional SCMV isolates. The data are consistent with replication occurring via an RNA intermediate; a suggestion supported by the need to use reverse transcriptase prior to DNA based PCR detection of virus in one of the tested cultures. Stealth adapted virus replication appears also to involve the assimilation of cellular genes and also genes of bacterial origin. The dual prospects of stealth adapted viruses acquiring oncogenic cellular genes and of being transferrable within bacteria potentially pose a very serious public health challenge, which will be difficult to address. Overall, the process of stealth adaptation greatly extends the scope of viral illnesses and places an important emphasis on better understanding of the non-immunological anti-virus defense mechanism operative against these viruses in cultures and in inoculated animals.

References

1. Cinque P, Koralnik IJ, Gerevini S, Miro JM, Price RW (2009) Progressive multifocal leukoencephalopathy in HIV-1 infection. Lancet Infect Dis. 9:625-36.

2. Kedar S, Berger JR (2011) The changing landscape of progressive multifocal leukoencephalopathy. Curr Infect Dis Rep. 13:380-6.

3. Pungpapong S, Kim WR, Poterucha JJ (2007) Natural history of hepatitis B virus infection: An update for clinicians. Mayo Clin Proc. 82: 967-75.

4. Chang MH (2007) Hepatitis B virus infection. Semin Fetal Neonatal Med.12: 160-7.

5. Burnet, FM (1959) The Clonal Selection Theory of Acquired Immunity. Nashville: Vanderbilt University Press.

6. Hombach J, Pircher H, Tonegawa S, Zinkernagel RM (1995) Strictly transporter of antigen presentation (TAP)-dependent presentation of an immunodominant cytotoxic T lymphocyte epitope in the signal sequence of a virus protein. J Exp Med. 182: 1615-9.

7. Chee MS, Bankier AT, Beck S, Bohni R, Brown CM et al. (1990) Analysis of the protein-coding content of the sequence of human cytomegalovirus strain AD169. Curr. Top. Microbiol. Immunol.154: 126–169.

8. Gibson W (2008) Structure and formation of the cytomegalovirus virion. Curr Top Microbiol Immunol. 325: 187-204.

9. Wills MR, Carmichael AJ, Mynard K, Jin X, Weekes MP et al. (1996) The human cytotoxic T-lymphocyte (CTL) response to cytomegalovirus is dominated by structural protein pp65: frequency, specificity, and T-cell receptor usage of pp65-specific CTL. J Virol. 70: 7569-79.

10. Kern F, Bunde T, Faulhaber N, Kiecker F, Khatamzas E et al. (2002) Cytomegalovirus (CMV) phosphoprotein 65 makes a large contribution to shaping the cell repertoire in CMV-exposed individuals. J Infect Dis. 185: 1709-16.

11. Khan N, Best D, Bruton R, Nayak L, Rickinson AB et al. (2007) T cell recognition patterns of immunodominant cytomegalovirus antigens in primary and persistent infection. J Immunol. 178: 4455-4465.

12. Babel N, Brestrich G, Gondek LP, Sattler A , Wlodarski MW et al. (2009) Clonotype analysis of cytomegalovirus-specific cytotoxic T lymphocytes. J Am Soc Nephrol. 20: 344-352.

13. Gyulai Z, Endresz V, Burian K, Pincus S, Toldy J et al. (2000) Cytotoxic T lymphocyte (CTL) responses to human cytomegalovirus pp65, IE1-Exon4, gB, pp150, and pp28 in healthy individuals: reevaluation of prevalence of IE1-specific CTLs. J Infect Dis. 181: 1537-46.

14. Khan N, Cobbold M, Keenan R, Moss PA (2002) Comparative analysis of CD8+ T cell responses against human cytomegalovirus proteins pp65 and immediate early 1 shows similarities in precursor frequency, oligoclonality, and phenotype. J Infect Dis.185: 1025-34.

15. Borysiewicz K, Hickling JK, Graham S, Sinclair J, Cranage MP et al. (1988) Human cytomegalovirus-specific cytotoxic T-cells, Relative frequency of stage-specific CTL recognizing the 72 kD immediate early protein and glycoprotein B expressed by recombinant vaccinia viruses. J. Exp. Med. 168: 919–931.

16. Martin WJ (1992) Detection of viral related sequences in CFS patients using the polymerase chain reaction. In: Hyde BM, editor. The Clinical and Scientific Basis of Myalgic Encephalomyelitis Chronic Fatigue Syndrome. Ottawa. Nightingale Research Foundation Press. pp 278283.

17. Martin WJ (1996) Severe stealth virus encephalopathy following chronic fatigue syndrome-like illness: Clinical and histopathological features. Pathobiology, 64: 1-8.

18. Barnes DM (1986) Mystery disease at Lake Tahoe challenges virologists and clinicians. Science. 234:541-2.

19. Salahuddin SZ, Ablashi DV, Markham PD, Josephs SF, Sturzenegger S et al. (1986) Isolation of a new virus, HBLV, in patients with lymphoproliferative disorders. Science. 234: 596-601.

20. Mullis K, Faloona F, Scharf S, Saiki R, Horn G et al. (1986) Specific enzymatic amplification of DNA in vitro: the polymerase chain reaction. Cold Spring Harb Symp Quant Biol. 51: 263-73.

21. Almoguera C, Shibata D, Forrester K, Martin WJ, Arnheim N et al. (1988) Most human carcinomas of the exocrine pancreas contain mutant cKras genes. Cell 53: 549554.

22. Shibata DK, Arnheim N, Martin WJ (1988) Detection of human papilloma virus in paraffin-embedded tissue using the polymerase chain reaction. J Exp Med. 167: 225-30.

23. McDonnel JM, Mayr A, Martin WJ (1989) Detection of human paplllomavlrus type 16 in dysplasias and squamous carcinomas of the conjunctive. N Eng J Med. 320; 14421446.

24. Shibata D, Martin WJ, Appleman MD, Causey DM, Leedom JM et al. (1988) Detection of cytomegalovirus DNA in peripheral blood of patients infected with human immunodeficiency virus. J Infect Dis.158: 1185-92.

25. Ehrlich GD, Greenberg S, Abbott MA. (1990) Detection of human T-cell lymphoma/leukemia viruses. In: Innis Ml, Gelfand DH, Sninsky JJ, White TJ, editors. PCR Protocols: A Guide to Methods and Applications. New York: Academic Press, pp 325-336.

26. Goldstein JA (2003) Chronic Fatigue Syndrome: The Limbic Hypothesis. Binghamton, NY Haworth Medical Press,.

27. Martin WJ (1994). Stealth viruses as neuropathogens. CAP Today 9: 67-70.

28. Martin WJ: Viral infection in CFS patients. In: Hyde BM, editor. The Clinical and Scientific Basis of Myalgic Encephalomyelitis Chronic Fatigue Syndrome. Ottawa. Nightingale Research Foundation Press pp 325-327.

29. Martin WJ, Zeng LC, Ahmed K, Roy M (1994) Cytomegalovirus-related sequences in an atypical cytopathic virus repeatedly isolated from a patient with the chronic fatigue syndrome. Am J Pathol. 145: 44-51.

30. Flügel RM, Rethwilm A, Maurer B, Darai G (1987) Nucleotide sequence analysis of the env gene and its flanking regions of the human spumaretrovirus reveals two novel genes. EMBO J. 6: 2077-84.

31. Aguzzi A (1994) Neurotoxicity of human foamy virus in transgenic mice. Verh Dtsch Ges Pathol. 78: 180-8.

32. Martin WJ (2003) Stealth virus culture pigments: A potential source of cellular energy. Exp Mol Path. 74: 210-223.

33. Martin WJ (1995) Stealth virus isolated from an autistic child. J Autism Dev Disord. 25: 223-224.

34. Martin WJ, Glass RT (1995) Acute encephalopathy induced in cats with a stealth virus isolated from a patient with chronic fatigue syndrome. Pathobiology 63: 115-118.

35. Huang ES, Kilpatrick B, Lakeman A, Alford CA (1978) Genetic analysis of a cytomegalovirus-like agent isolated from human brain. J Virol. 26: 718–723.

36. Martin WJ, Ahmed EN, Zeng LC, Olsen JC, Seward JG et al. (1995) African green monkey origin of the atypical cytopathic

'stealth virus' isolated from a patient with chronic fatigue syndrome. Clin Diagn Virol. 4: 93-103.

37. Barry PA, Chang W (2007) Primate betaherpesviruses. In: *Human Herpesviruses: Biology, Therapy, and Immunoprophylaxis.* Arvin A, Campadelli-Fiume G, Mocarski E, Moore PS, Roizman B, Whitley R, Yamanishi K, editors. Cambridge University Press; Chapter 59.

38. Altschul SF, Gish W, Miller W, Myers EW, Lipman DJ (1990) Basic local alignment search tool. J Mol Biol. 215: 403-410.

39. Reich JG, Drabsch H, Däumler A (1984) On the statistical assessment of similarities in DNA sequences. Nucleic Acids Res. 12: 5529-43.

40. Martin WJ (1996) Genetic instability and fragmentation of a stealth viral genome. Pathobiology 64: 917.

41. Martin WJ (1999) Stealth adaptation of an African green monkey simian cytomegalovirus. Exp Mol Path. 66:3-7.

42. Du G, Dutta N, Lashmit P, Stinski MF (2011) Alternative splicing of the human cytomegalovirus major immediate-early genes affects infectious-virus replication and control of cellular cyclin-dependent kinase. J Virol. 85: 804-17.

43. Alcendor DJ, Zong J, Dolan A, Gatherer D, Davison AJ et al. (2009) Patterns of divergence in the vCXCL and vGPCR gene clusters in primate cytomegalovirus genomes. Virology. 395: 21-32.

44. Martin WJ (1998) Cellular sequences in stealth viruses. Patobiology 66: 53-58.

45. Martin WJ (1999) Melanoma Growth stimulatory activity (MGSA/GRO-alpha) chemokine genes incorporated into an African green monkey simian cytomegalovirus (SCMV)-derived stealth virus. Exp Mol Path. 66: 15-18.

46. Martin WJ (2000) Chemokine receptor-related sequences in an African green monkey simian cytomegalovirus (SCMV)-derived stealth virus. Exp Mol Path. 69: 10-16.

47. Sahagun-Ruiz A, Sierra-Honigmann AM, Krause P, Murphy PM (2003) Simian cytomegalovirus encodes five rapidly evolving chemokine receptor homologues. Biologicals. 31: 63-73.

48. Beisser PS, Lavreysen H, Bruggeman CA, Vink C (2008) Chemokines and chemokine receptors encoded by cytomegaloviruses. Curr Top Microbiol Immunol. 325: 221-42.

49. Sequar G, Britt WJ, Lakeman FD, Lockridge KM, Tarara RP et al. (2002). Experimental coinfection of rhesus macaques with rhesus cytomegalovirus and simian immunodeficiency virus: pathogenesis. J Virol. 76: 7661-71.

50. Baroncelli S, Barry PA, Capitanio JP, Lerche NW, Otsyula M et al. (1997) Cytomegalovirus and simian immunodeficiency virus coinfection: longitudinal study of antibody responses and disease progression. J Acquir Immune Defic Syndr Hum Retrovirol. 15: 5-15.

51. Streblow DN, Soderberg-Naucler C, Vieira J, Smith P, Wakabayashi E et al. (1999) Cell 99: 511-20.

52. Duff R, Rapp P (1973) Oncogenic transformation of hamster embryo cells after exposure to inactivated herpes simplex virus type 1. J Virol 12: 209-17.

53. Martin WJ (1999) Bacteria-related sequences in a simian cytomegalovirus-derived stealth virus culture. Exp Mol Path 66: 8-14.

54. Sadar HS, Biedenbach D, Jones RN (1995) Evaluation of Vitek and API 20S for species identification of enterococci. Diag microbial Infect dis. 22: 315-9.

55. Bergey DH, Holt JG. (1999). Bergey's Manual of Determinative Bacteriology 9. Baltimore. Williams and Wilkins.

56. Patrick SG, Chain DM, Lang DJ, Comerci SA, Malfatti LM et al. (2011) Genome of Ochrobactrum anthropi ATCC 49188^T, a versatile opportunistic pathogen and symbiont of several eukaryotic hosts. J Bacteriol. 193: 4274-4275.

57. Teyssier C, Marchandin H, Jumas-Bilak E (2004) The genome of alpha-proteobacteria: Complexity, reduction, diversity and fluidity. Can J Microbial 50: 383-96.

58. Endresen GK (2003) Mycoplasma blood infection in chronic fatigue and fibromyalgia syndromes. Rheumatol Int. 23: 211-5.

59. Choppa PC, Vojdani A, Tagle C, Andrin R, Magtoto L (1998) Multiplex PCR for the detection of mycoplasma fermentans, m. hominis and m. penetrans in cell cultures and blood samples of patients with chronic fatigue syndrome. Mol Cell Probes. 12: 301-8.

60. Vojdani A, Choppa PC, Tagle C, Andrin R, Samimi B et al (1998) Detection of mycoplasma genus and mycoplasma fermentans by PCR in patients with chronic fatigue syndrome. FEMS Immunol Med Microbiol. 22: 355 - 65.

61. Nasralla M, Haier J, Nicolson GL (1999) Multiple mycoplasmal infections detected in blood of patients with chronic fatigue syndrome and/or fibromyalgia syndrome. Eur J Clin Microbiol Infect Dis. 18: 859-65.

62. Dawson MS, Hayes MM, Wang RY, Armstrong D, Kundsin RB et al. (1993). Detection and isolation of Mycoplasma fermentans from urine of human immunodeficiency virus type 1-infected patients. Arch Pathol Lab Med. 117: 511-514.

63. Nicolson GL, Gan R, Haier J (2004) Evidence for brucella spp. and mycoplasma spp. co-infections in blood of chronic fatigue syndrome patients. J Chronic Fatigue Synd. 12: 5-17.

64. Treib J, Grauer MT, Haass A, Langenbach J, Holzer G et al. (2002) Chronic fatigue syndrome in patients with Lyme borreliosis. Eur Neurol. 43: 107-9.

65. Swedo SE, Leonard HL, Garvey M, Mittleman B, Allen AJ et al. (1998) Pediatric autoimmune neuropsychiatric disorders associated with streptococcal infections:

clinical description of the first 50 cases. Am J Psychiatry. 155: 264-71.

66. Martin WJ (1996) Simian cytomegalovirus-related stealth virus isolated from the cerebrospinal fluid of a patient with bipolar psychosis and acute encephalopathy. Pathobiology. 64: 64-6.

67. Martin WJ, Anderson D (1997) Stealth virus epidemic in the Mohave Valley. I. Initial report of virus isolation. Pathobiology. 65: 51-6.

68. Martin WJ, Anderson D (1999). Stealth virus epidemic in the Mohave Valley: severe vacuolating encephalopathy in a child presenting with a behavioral disorder. Exp Mol Pathol. 66: 19-30.

69. Martin WJ (2003) Complex intracellular inclusions in the brain of a child with a stealth virus encephalopathy. Exp Mol Pathol. 74: 197-209.

70. Martin WJ Stealth virus encephalopathy: Report of a case complicated by vasculitis. Pathobiology. 64: 59-63.

71. Martin WJ (1997) Detection of RNA sequences in cultures of a stealth virus isolated from the cerebrospinal fluid of a health care worker with chronic fatigue syndrome. Case report. Pathobiology. 65: 57-60.

72. Brudek T, Lühdorf P, Christensen T, Hansen HJ, Møller-Larsen A (2007) Activation of endogenous retrovirus reverse transcriptase in multiple sclerosis patient lymphocytes by inactivated HSV-1, HHV-6 and VZV. J Neuroimmunol. 187: 147-55.

73. Franklin MC, Wang J, Steitz TA (2001) Structure of the Replicating Complex of a Pol Family DNA Polymerase. Cell. 105: 657–667.

74. Jayaraman R (2011) Hypermutation and stress adaptation in bacteria. J Genet. 90: 383-391.

75. Sierra-Honigmann AM, Krause PR (2002) Live oral poliovirus vaccines and simian cytomegalovirus. Biologicals. 30: 167-74.

76. Baylis SA, Shah N, Jenkins A, Berry NJ, Minor PD (2002) Simian cytomegalovirus and contamination of oral poliovirus vaccines. Biologicals. 31: 63-73.

77. Smith KO, Thiel JF, Newman JT, Harvey E, Trousdale MD et al. (1969). Cytomegaloviruses as common adventitious contaminants in primary African green monkey kidney cell cultures. J Nat Cancer Inst. 42: 489-96.

78. Sabin AB. (1959). Present position on immunization against poliomyelitis with live virus vaccines. Brit Med J. 1 (5123): 663-80.

Table 1. SCMV related sequences in cloned DNA from cultures of stealth adapted virus-1

Genbank Accession Number: GI	Length (bp)	Coding Region (BlastX)	Matching to SCMV Strain GR2715				
			Nucleotide Identity*	%	Gaps	Score	Expect
121531683	28,199	UL28-UL48	26,523/28,245	94	210	43,062	0.0
121531687	8,407	UL48-UL54	8,027/8,414	95	17	13,402	0.0
121531684	6,328	UL72-UL76	5,112/6,367	80	168	5,615	0.0
121531682	25,023	UL84-UL102	23,941/25,092	95	104	40,010	0.0
121531685	12,375	UL115-UL132	6,222/6,654‡	94	67	10,013	0.0
121531686	16,011	US19-US28	11,583/12,242‡	95	54	19,111	0.0

*The nucleotide numbers shown include the gaps, which were inserted for optimal alignment using the BlastN program of NCBI.

‡ The entire length of the sequence did not align to SCMV GR2715 as a contiguous sequence, allowing for short gaps. The non-aligned regions could, however, generally be aligned to the Colburn strain of SCMV and/or to other SCMV strains for which only partial sequence data are currently available on GenBank.

Table 2. Mycoplasma fermentans species related sequences in culture of stealth adapted virus-1

CLONE	LENGTH	Matching to Mycoplasma fermentans			
		% Identity	Gaps	Score	Expect
3B35	2,142	99	2	3,833	0.0
3B512	2,345	99	1	4,213	0.0
3B520	2,797	99	0	4,985	0.0
3B528	2,043	99	1	3,577	0.0
3B622-T3	600	97	1	1,003	0.0
3B622-T7	600	97	2	1,003	0.0
3B627	328	99	0	587	2e-164
3B632	1,396	99	1	2,475	0.0

Note: All sequences aligned over their entire lengths to matching mycoplasma sp. sequences

Table 3. Ochrobactrum sp. Related DNA Sequences in Culture of Stealth Adapted Virus-1

CLONE NCBI Accession	Length	Matching to Ochrobactrum sp.*				
		Region	% Identity	Gaps	Score	Expect
3B23 U27612	8916	1-5016	84	16	5,380	0.0
		6777-8242	79	34	1,256	0.0
		8329-8915	78	17	450	1e-121
3B313 U27616	7985	1-6263	76	79	4493/4607**	0.0
		6497-7972	72	56	800/493	0.0
3B41 AF191072	2869	2575-2868	78	1	230	6e-56
3B43 AF191073	3620	1-1785	97	6	2,969	0.0
		1796-2303	91	7	690	0.0
		2378-3620	93	18	1,855	0.0
3B534 U27900	612	1-555	83	3	572	9e-160
		573-603	90	0	42.8	2.8
3B614 U27645	5062	1-378	83	0	385	2e-102
		935-5061	77	103	3,140	0.0
3B629 AF191078	165	1-165	93	0	248	1e-62
C1616 AF065660	4626	2-1346	77	16	1,058	0.0
		1460-1628	88	0	215	2e-51
		1719-2668	78	2	771	0.0
		2896-3702	75	3	547	3e-151
C16116-T3 AF065678	587	25-175	85	0	168/134	3e-38
		292-573	81	5	260/289	7e-66

C16116-T7 AF065679	548	11-112	85	1	114	5e-22
		218-539	79	8	255	3e-64
C16118-T3 AF065682	780	20-713	82	16	675	0.0
C16118-T7 AF065683	675	1-629	85	3	639	0.0
C16134 AF065710	4142	196-1099	82	1	890/911	0.0
		1246-1761	82	3	495/491	1e-135
		1864-3301	79	50	1,211/1,321	0.0
		3582-4142	86	0	655/645	0.0

** When two scores are listed, the clone contained at least one sequence, which showed a higher matching score to Brucella sp., than to Ochrobactrum sp. The first of the two scores is that to Ochrobactrum sp. (generally Ochrobacttrum anthropi) and the second score is to Brucella sp.

Table 4. DNA Sequences in Culture of Stealth Adapted Virus-1Which Match Better to Rhizobiaceae Than to Brucellaceae-Related Sequences Available on Genbank

Clone Accession #	Length	Region of Greatest Alignment to Sequences of Rhizobiaceae or Their Plasmids					
		Region	Best Match	% identity	Gaps	Score	Expect
3B23 U27612	8,916	5210-6411	Agrobacterium radiobacter K84	69	24	444	5e-120
3B41 AF191072	2,869	627-970	Rhizobium leguminosarum Plasmid pR132502	69	8	116	1e-21
3B47 AF191074	2,024	3-366	Sinorhizobium meliloti AK83	69	6	145	1e-30
		555-1917	"	66	54	324	3e-84
3B513 U27894	8,106	2688-2890	Rhizobium sp. Str. NT-26 plasmid	81	0	214	1e-50
		6101-6697	Methylobacterium extorquens AM1 (plasmid)	75	20	396	2e-105
C1616 AF065660	4,626	3741-4625	Agrobacterium rhizogenes, plasmid pRi1724	90	1	1,175	0.0
C16125-T3 AF065692	655	9-650	Chelativorans sp. BNC1 plasmid	81	3	607	4e-170
C16125-T7 AF065693	634	1-633	"	85	9	673	0.0

The Alternative Cellular Energy (ACE) Pathway in the Repair of the Cytopathic Effect (CPE) Caused by Stealth Adapted Viruses: In Vitro and In Vivo Evidence Supporting a New Therapeutic Paradigm

W. John Martin, MD, PhD.
Institute of Progressive Medicine South Pasadena CA 91030

Abstract

The alternative cellular energy (ACE) pathway is proposed as a major non-immunological defense mechanism against virus infections and other cellular energy depleting illnesses. It is mediated by energy transducing (converting) materials called ACE pigments, which can take the form of self-assembling particles and fibers. ACE pigments were initially identified in cultures of cell damaging (cytopathic) viruses obtained from patients with non-inflammatory illnesses, including the chronic fatigue syndrome (CFS). The formation of ACE pigments reverses the cytopathic effect (CPE) caused by these viruses when cultured in vitro. The viruses are designated stealth adapted because they do not trigger a cellular immune inflammatory response in patients. Complex intracellular and extracellular ACE pigments can, however, be identified in the tissues of stealth adapted virus infected patients. These findings provide a basis for endeavoring to enhance the ACE pathway in the therapy of infections and, especially those caused by stealth adapted viruses.

Key Words: Chronic fatigue syndrome, CFS, Stealth adapted viruses, Alternative cellular energy, ACE pathway, ACE pigments, Neutral red, Fluorescence, Ultraviolet light

Running Title: ACE Pathway

Author: W. John Martin, MD, PhD. Institute of Progressive Medicine, 1634 Spruce Street

South Pasadena CA 91030. Telephone 626-616-2868. E-mail wjohnmartin@hotmail.com

Introduction

This paper summarizes a series of experiments performed over the last 25 years on the detection and characterization of atypical viruses from patients with a range of neurological and psychiatric illnesses. The viruses are termed stealth adapted because they do not typically evoke an inflammatory response (1-15). As presented elsewhere, stealth adaptation is a generic immune evasion process involving the deletion or mutation of genes coding for the relatively few viral antigens normally targeted by the cellular immune system. Stealth adaptation can potentially occur with all types of viruses and has been definitively proven to have occurred with African green monkey simian cytomegalovirus (SCMV).

Unlike most clinical virus cultures, the cytopathic effect (CPE) developing in cultures of stealth adapted viruses is commonly transient, unless the cultures are frequently fed with fresh tissue culture medium. Indeed, the lack of progression of the CPE can easily mislead virologists to attributing early signs of cellular damage to non-viral causes of toxicity. Similarly, the lack of accompanying tissue inflammation can also divert diagnostic considerations away from an infectious disease process. Yet closer examination of the infected tissues indicates a striking similarity between the observed cellular changes and the in vitro CPE. Energy transducing particulate materials, referred to as alternative cellular energy (ACE) pigments, can also be identified in both the infected tissues and in the in vitro cultures. Consistent with the marked disruption of mitochondria seen in many of the infected cells, it is proposed that cells are increasing their utilization of the ACE pathway to help meet their added energy needs. These findings provide a basis for therapeutic endeavors to treat stealth adapted and conventional virus infected patients by enhancing the ACE pathway. The present paper is essentially intended to illustrate the basis for many of the above stated observations.

In Vitro Cultures Demonstrating Reversible CPE Caused by Stealth Adapted Viruses

Fig. 1-2 illustrate the striking capacity of cultures of stealth adapted viruses to undergo self-repair, even in cultures with very marked initial virus CPE. (These and many additional previously published figures are reproduced with permission from Elsevier.) Removal of the extracelluar materials by re-feeding "repaired" cultures can result in a rapid reactivation of the virus CPE (Fig. 3). The repair process coincides with the formation of structured intracellular materials that can coalesce and be extruded as extracellular particles and as material that can self-assemble into fibers, including long threads

and ribbons (Fig. 4-10). Intracellular and extruded extracellular particles are typically black, while the fibers can display a range of colors, including white, blue, green, yellow and red. The reactivation can be prevented if some of the medium from other repaired cultures is included with the re-feeding medium. Moreover, isolated particulate material from repaired cultures, added to the re-feeding medium can also inhibit reactivation (11). Because of the rapidity of action and the small amounts of particulate material needed, it is concluded that actual entry of the added material into cells is not necessary and that the material is mediating its protection via an indirect energy-based mechanism.

In Vivo Histological and Electron Microscopic Changes

A photomicrograph of a brain biopsy obtained from a patient in 1990 is shown in Fig. 11. In spite of being positive for virus sequences using the polymerase chain reaction (PCR), the biopsy showed no evidence of an inflammatory reaction. Yet damaged cells can be readily identified both histologically (Fig. 12) and by electron microscopy (Fig. 13). The electron microscopic features of the damaged cells include foamy vacuolization, marked disruption of mitochondria and the presence of variously structured complex intracellular inclusions. These electron microscopic features are further illustrated in Fig. 14-21, which show some of the striking cellular changes seen in a non-inflamed brain biopsy from a 4 y/o stealth adapted virus infected boy (10). Similar histological changes, including the presence of both intracellular and extracellular inclusions, were also identified in various tissues, including the brain, of stealth adapted virus inoculated animals. Marked clinical recovery was observed in cats following a severe acute illness and occurred in spite of the absence of any discernable inflammation (4).

Mitochondria Do Not Fully Explain Cellular Energy Production

The precise cause of the mitochondria damage is presently uncertain, but clearly attributable to the underlying virus infection since it is also readily observed in cultures of stealth adapted viruses. The apparent survival of cells without functional mitochondria led to the suggestion that the structured intracellular particles were providing an alternative (non-mitochondria) source of cellular energy. The following reasoning provides strong support for suggesting that the body utilizes an auxiliary source(s) of energy, beyond normal catabolism.

Thus, it is commonly stated that the calorie content of food is sufficient to sustain the energy needs of the body. In reality, humans and animals expend far more energy than that supplied by the metabolism of ingested food. For example, for a 50 Kg person to maintain a body temperature of 20°C above the environment temperature for 24 hrs (longer than the time after death for a body to return to room temperature) requires 1,000 calories. This is half of the energy in a typical basal diet of ~2,000 calories. A significant proportion of ingested nutrients is used for biosynthesis of bodily components. Particularly for proteins, subsequent catabolism of the synthesized components yields fewer calories than invested in biosynthesis. Normal auxiliary energy generating sources beyond food metabolism are, therefore, to be anticipated and the designation of the particulate materials observable in stealth adapted virus infected patients and their cultures as ACE pigments is reasonable. As a possible analogy, there have been numerous suggestions that

melanin may also function as a source of cellular energy (16-21). Actual data supporting a role for ACE pigments in energy conversion and supply are as follows:

Energy Transducing Properties of ACE Pigments

Stealth adapted virus culture-derived ACE pigments have been shown to have prominent energy transducing activities. The particles themselves and clusters of cells containing such particles are typically fluorescent when illuminated with a UV light and luminescent after exposure to white light (Fig. 22-26). They can also alter the normal fluorescing colors of various fluorescent dyes, such as acridine orange. Non-fluorescing or poorly fluorescing particles can commonly be rendered brightly fluorescent in the presence of certain non-fluorescent dyes, including neutral red and Stains-All. The particles display non-Brownian movements, especially when illuminated with a UV light. They respond to electrostatic forces and are occasionally ferromagnetic (Fig. 27). They can also evoke the formation of vapor bubbles when placed in water (not shown) and act as both electron donors (Fig. 28) and acceptors. Chemical analysis shows the presence of various functional groups, including both aliphatic and aromatic chemicals, most of which are undetectable in normal control cultures. Different minerals can also be identified within individual ACE pigment particles using energy dispersive X-ray analysis (EDX). For many particles, only a relatively few minerals are identified, but in aggregate a wide range of minerals are present. The actual results with a particle containing an unusually complex array of different minerals are shown in Fig 29.

Another very striking aspect of long maintained stealth adapted virus cultures is the abundant production of lipids, which can take the form of long needle-like linear trough-like formations, actual membranes, solid crystals, pyramids and more complex collections of golden-yellowish material (Fig. 25, 30-35). The lipid synthesis continues well after all of the living cells have disappeared. Extensive testing has excluded any contaminating microorganisms in these cultures, many of which have been studied over several years. The abiotic production of lipids is, therefore, attributable to the persisting ACE pigments within the culture tubes.

Summary

The basic in vitro and in vivo findings relating to the discovery of stealth adapted viruses are reiterated in this paper. This is being done, in part, to better explain the basis for a series of successful therapeutic endeavors undertaken in patients with various infectious diseases. Understanding the ability of the ACE pathway to suppress the CPE in virus cultures can also potentially facilitate the sensitivity of culture methods for virus detection. Similarly, the lack of inflammation in tissue biopsies should no longer be viewed as excluding an infectious process. Culture and patient derived ACE pigments provide valuable materials to further explore fundamental issues of cellular energetics, including the potential of abiotic synthesis.

References

1. Martin WJ, Zeng LC, Ahmed K, Roy M. (1994) Cytomegalovirus-related sequence in an atypical cytopathic virus repeatedly isolated from a patient with chronic fatigue syndrome. Am J Pathol. 145: 440-51.

2. Martin WJ. (1994) Stealth viruses as neuropathogens. CAP Today. 8: 67-70.

3. Martin WJ, Ahmed KN, Zeng LC, Olsen JC, Seward JG, Seehrai JS. (1995) African green monkey origin of the atypical cytopathic 'stealth virus' isolated from a patient with chronic fatigue syndrome. Clin Diagn Virol. 4: 93-103.

4. Martin WJ, Glass RT. (1995) Acute encephalopathy induced in cats with a stealth virus isolated from a patient with chronic fatigue syndrome. Pathobiology. 63:115-8

5. Martin WJ. (1996) Simian cytomegalovirus-related stealth virus isolated from the cerebrospinal fluid of a patient with bipolar psychosis and acute encephalopathy. Pathobiology. 64: 64-6.

6. Martin WJ. (1996) Severe stealth virus encephalopathy following chronic-fatigue-syndrome-like illness: clinical and histopathological features. Pathobiology. 64:1-8.

7. Martin WJ, Anderson D. (1999) Stealth virus epidemic in the Mohave Valley: severe vacuolating encephalopathy in a child presenting with a behavioral disorder. Exp Mol Pathol. 66:19-30.

8. Martin WJ. (1999) Bacteria-related sequences in a simian cytomegalovirus-derived stealth virus culture. Exp Mol Pathol. 66: 8-14.

9. Martin WJ. (1999) Stealth adaptation of an African green monkey simian cytomegalovirus. Exp Mol Pathol. 66: 3-7.

10. Martin WJ. (2003) Complex intracellular inclusions in the brain of a child with a stealth virus encephalopathy. Exp Mol Pathol. 74: 197-209.

11. Martin WJ. (2003) Stealth virus culture pigments: a potential source of cellular energy. Exp Mol Pathol. 74: 210-23.

12. Martin WJ. (2005) Alternative cellular energy pigments from bacteria of stealth virus infected individuals. Exp Mol Pathol. 78: 215-7.

13. Martin WJ. (2005) Alternative cellular energy pigments mistaken for parasitic skin infestations. Exp Mol Pathol. 78: 212-4.

14. Martin WJ. (2005) Progressive medicine. Exp Mol Pathol. 78: 218-20.

15. Martin WJ. (2005) Etheric biology. Exp Mol Pathol. 78: 221-7.

16. Zagal'skaia EO. (1995) The magnetic susceptibility of the melanin in the eyes of representatives of different vertebrate classes. Zh Evol Biokhim Fiziol. 31:416-22.

17. Menter JM, Willis I. (1997) Electron transfer and photoprotective properties of melanins in solution. Pigment Cell Res. 10: 214-7.

18. Zhernovoĭ MV, Lebedev SV, Grigoriuk AA, Krasnikov IuA, Kharitonskiĭ PV. (1998) Reaction of melanocytes in vertebrate retina to the exposure to constant magnetic field. Tsitologiia. 40: 676-81.

19. Nicolaus BJ. (2005) A critical review of the function of neuromelanin and an attempt to provide a unified theory. Med Hypotheses. 65:791-6.

20. Dadachova E, Bryan RA, Huang X, Moadel T, Schweitzer AD, Aisen P, Nosanchuk JD, Casadevall A. (2007) Ionizing radiation changes the electronic properties of melanin and enhances the growth of melanized fungi. PLoS One. 5: e457.

21. Goodman G, Bercovich D. (2008) Melanin directly converts light for vertebrate metabolic use: heuristic thoughts on birds, Icarus and dark human skin. Med Hypotheses. 71:190-202.

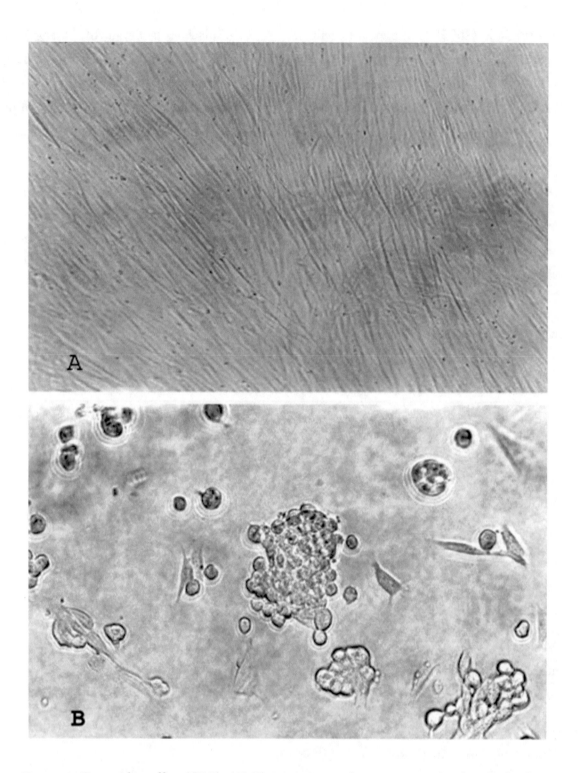

Figure 1. Cytopathic effect (CPE). (A) Phase-contrast photomicrograph of normal MRC-5 fibroblasts cells viewed through a 4× objective. The cells form a tightly coherent, nearly transparent monolayer of generally well-oriented, thin spindle-shaped cells (A). Within days of exposing such a monolayer to an extract of mononuclear cells from a stealth virus-infected patients, foci of rounded cells begin to develop that can be accompanied by considerable cell destruction. The rounded cells will typically form into small and larger clusters (B) 10× objective.

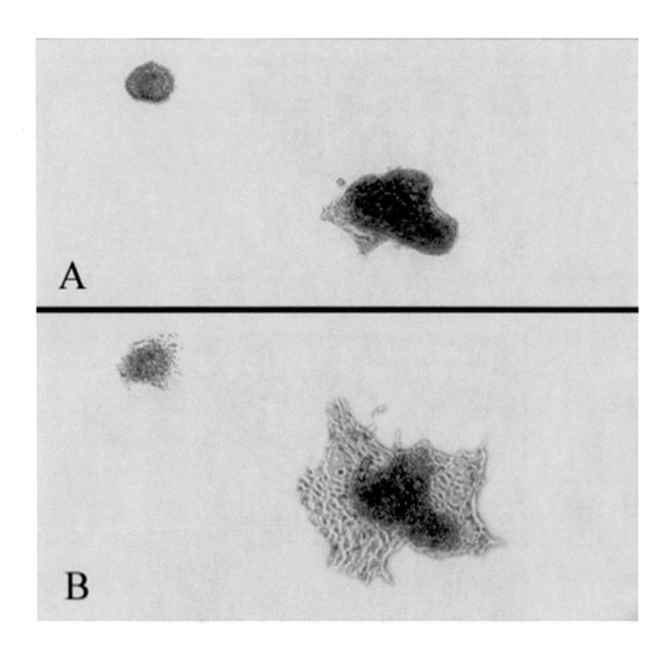

Figure 2. Repair process. The upper panel (A) shows low-power photomicrograph of two cell clusters from a positive culture of MRC-5 cells. The tissue culture medium was not changed over a 3-day period. During this time, relatively normal appearing cells grew out from the edges of both clusters (B). Note the prominent pigmentation present, especially in the larger cell cluster.

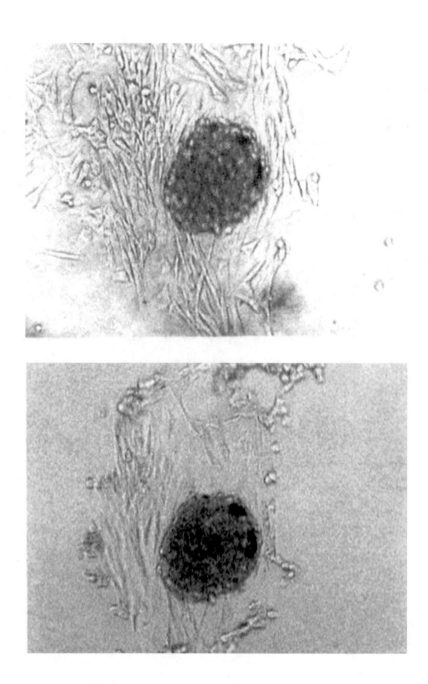

Figure 3. Reactivation. The upper panel shows relatively normal appearing viable cells in close vicinity to a small cell cluster. Within 10 min of replacing the culture medium with fresh medium, many of the cells surrounding the cluster had become rounded or detached or showed other signs of a loss of vitality (lower panel). This reactivation process does not occur if medium from another culture that had shown "repair" is mixed in near equal quantity with the fresh medium used for re-feeding. Note the overall pigmentation in the clusters with 3 small localized deposits of more intense pigment.

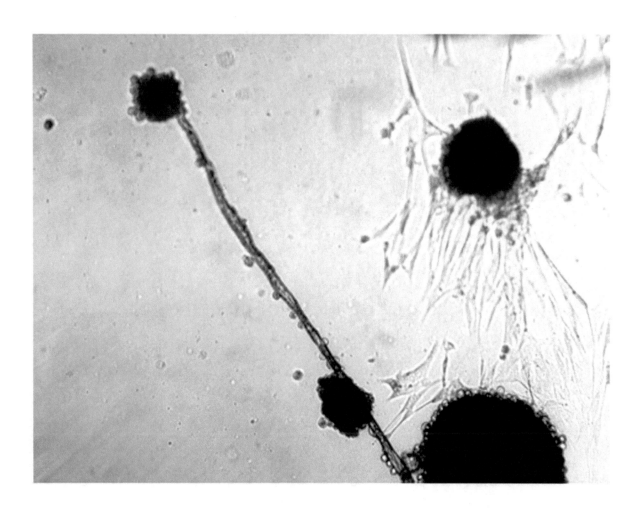

Figure 4. Ribbon. A long ribbon extending from a relatively small heavily pigmented cluster in the upper left corner. The ribbon runs adjacent to the small and the large pigmented cluster at the lower right side and terminated beyond the large cluster. Signs of cell repair can be seen around the remaining pigmented cluster in this photomicrograph.

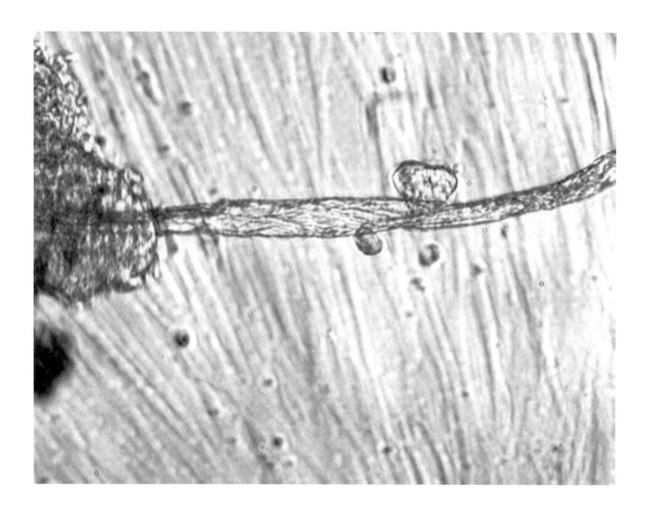

Figure 5. Ribbon in a repaired culture. This higher power photomicrograph shows the intricate, interwoven material that composes the ribbon emerging from a residual small cluster. Note the large numbers of normal appearing cells, indicating the extensive repair that had occurred in this culture.

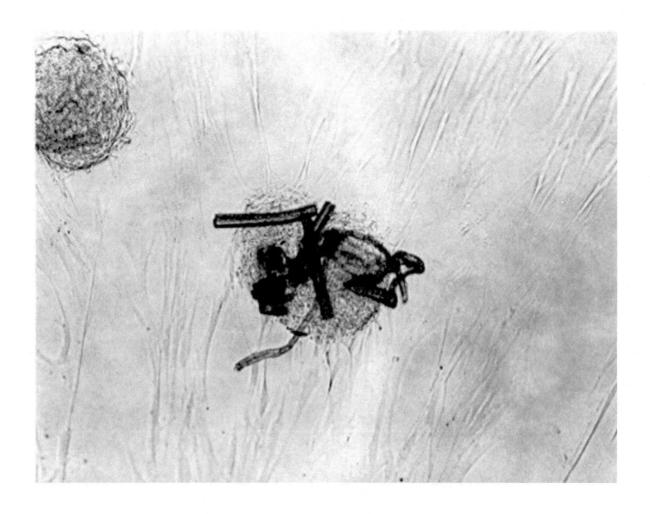

Figure 6. Multiple ribbons. This photomicrograph is presented to show the complex patterns of multiple ribbon components that can develop from some of the clusters. Another cluster in the same culture lacks any ribbons or thread-like structures. Note the signs of repair in the culture.

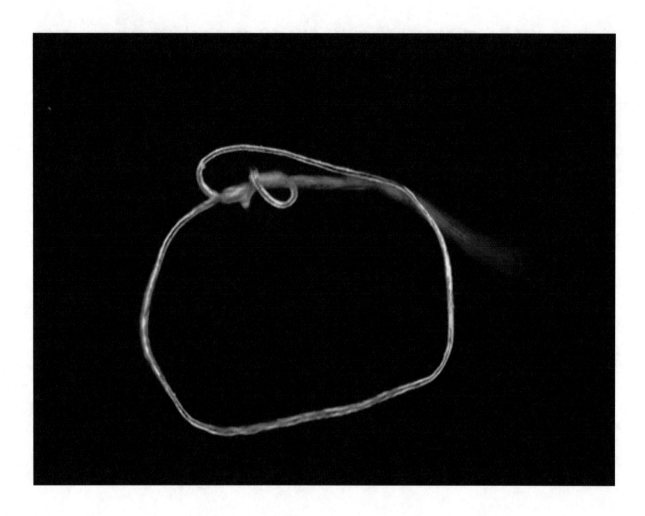

Figure 7. Thread. A freely floating knotted thread that formed in a long-term culture as seen under dark-field microscopy. Individual threads are usually slightly brownish in color but other colors, including blue, red and green, have been observed. The threads continue to grow even when no longer attached to cell clusters.

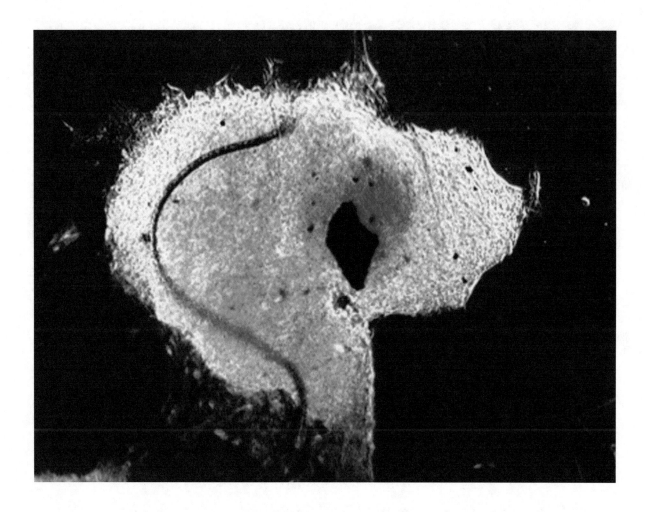

Figure 8. Colored cluster. A dark-field photomicrograph of an attached cluster containing a large black particle, several smaller pigmented deposits, and also a thread that curved through the cluster. The cluster showed a pronounced red hue under dark-field microscopy and the thread displayed a blue coloration.

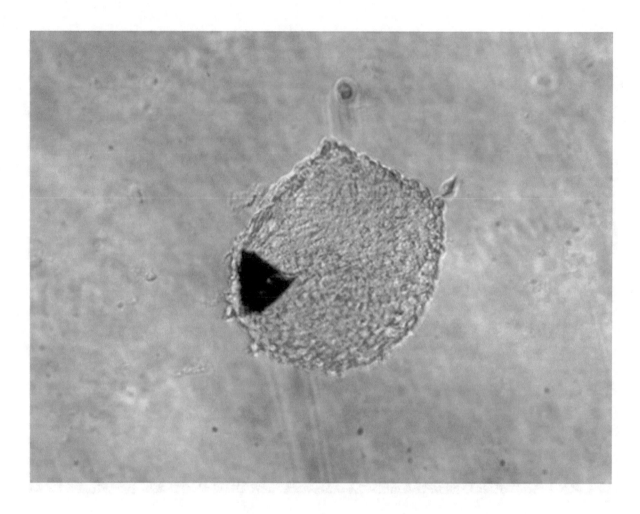

Figure 9. A triangular-shaped ACE pigment particle that formed from the coalescence of pigmented materials that were initially diffusely distributed within the cluster of MRC-5 cells in a stealth-adapted virus culture. The particle was subsequently extruded from the cell cluster. Similar types of particles can be seen in many cultures of stealth-adapted viruses. Magnification ×100.

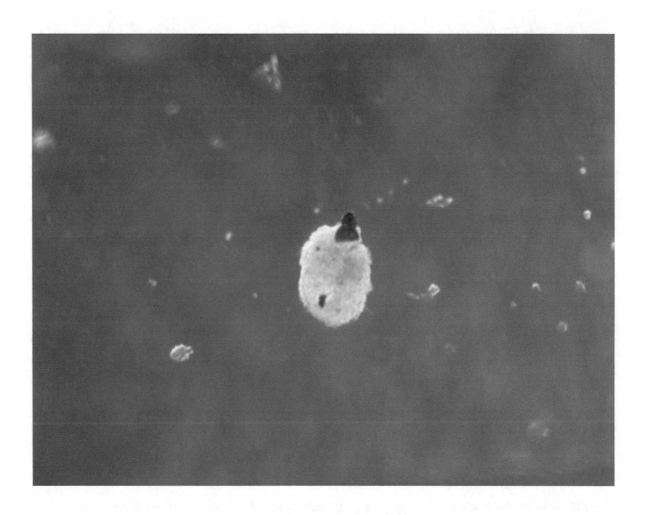

Figure 10. Extrusion of pigment. A small floating cell cluster viewed under dark-field microscopy. The cluster contained 3 localized deposits of dark pigment, the largest of which was being extruded from the cluster into the culture medium.

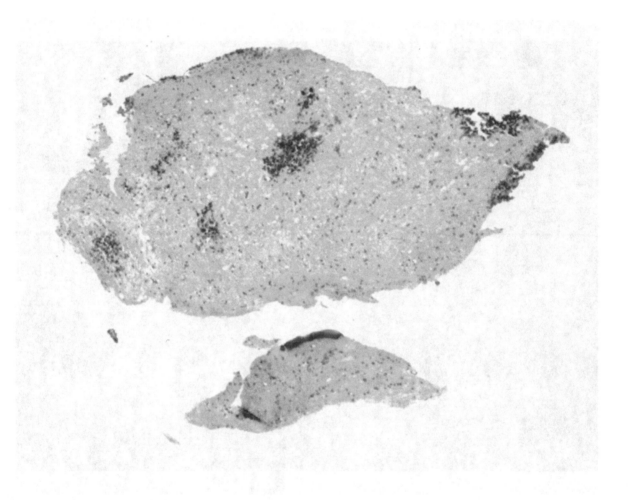

Figure 11. Low power photograph of a hematoxylin and eosin (H&E) stained needle biopsy obtained from the brain of stealth adapted virus infected female patient .Her illness began as CFS but she experienced progressively deteriorating cognitive function. The biopsy was obtained from an area of confluent periventricular density as seen on MRI. Note the absence of any inflammation.

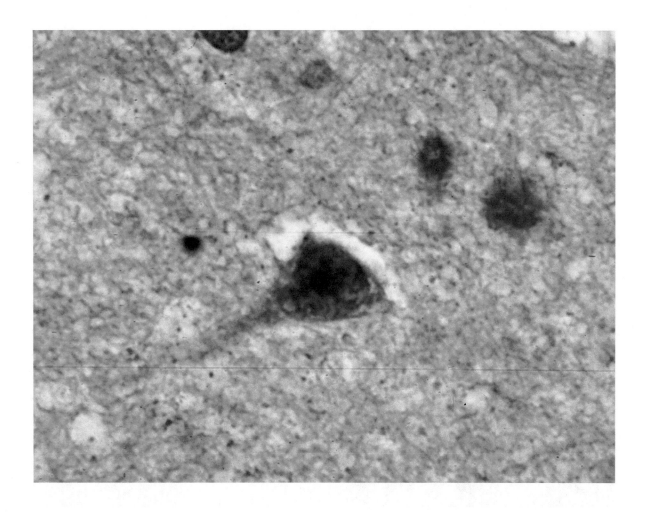

Figure 12. Abnormal cell in an H&E stained section of a brain biopsy obtained from a stealth adapted virus infected child. The cell is characterized by both cytoplasmic and nuclear changes; the former comprising vacuolization and stainable inclusions. Similar cellular changes occur in stealth adapted virus inoculated animals (x 400 magnification).

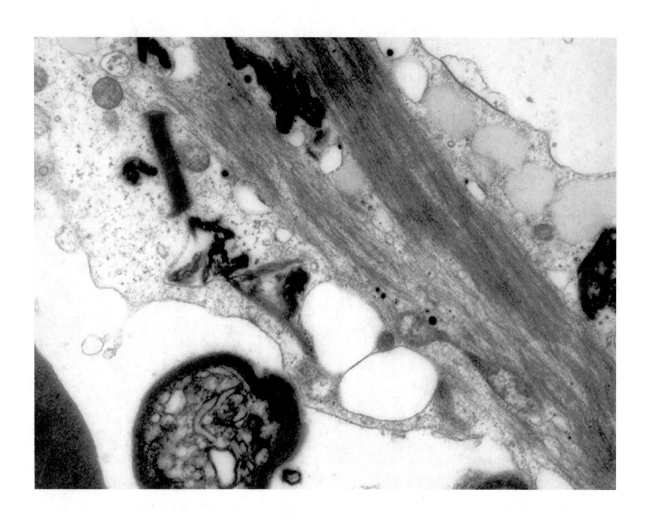

Figure 13. Electron micrograph of a glial cell in the brain biopsy that was shown in Fig. 11. Note the lipid filled vacuoles and the many irregularly shaped inclusions.

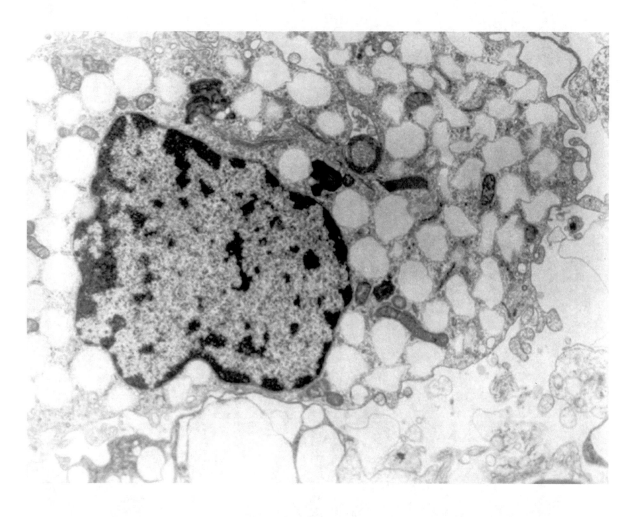

Figure 14. Electron micrograph of a markedly vacuolated cell from a brain biopsy of the stealth adapted virus infected child. A few distorted mitochondria can be seen between the vacuoles.

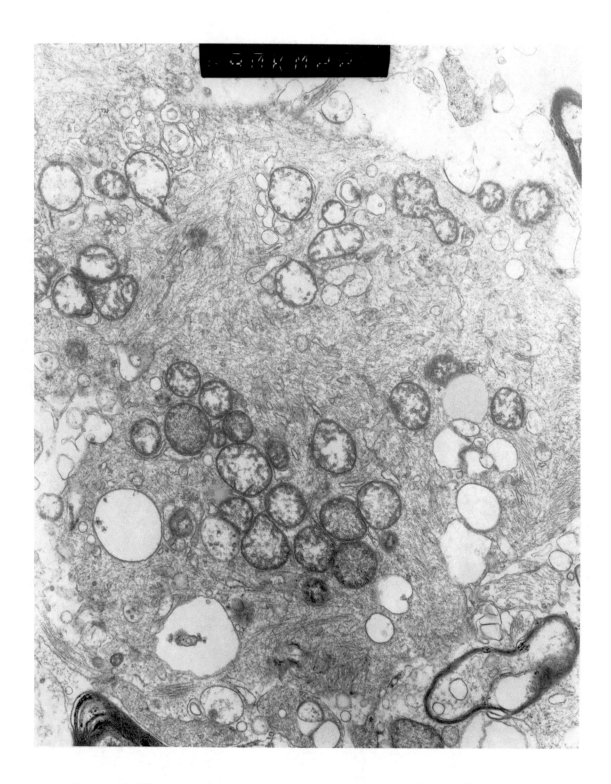

Figure 15. Electron micrograph showing a large number of markedly disrupted mitochondria within a brain cell from the stealth adapted virus infected child. Distended mitochondria with scant lamellar inner structures are seen in many of the affected cells. Nevertheless, the cells still show signs of having remained viable at the time of the biopsy.

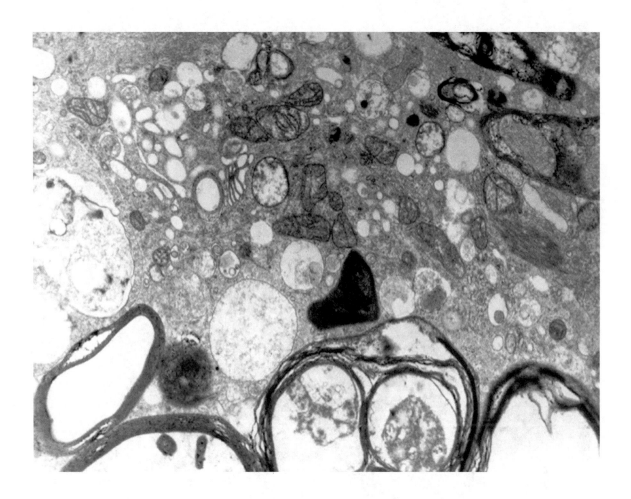

Figure 16. Electron micrograph of a brain cell, which contained a prominent inclusion showing internal fibrils. Vacuolization and distended mitochondria can also be seen. The empty myelin sheaths are reflective of neuron loss that had occurred in the brain of the stealth adapted virus infected child.

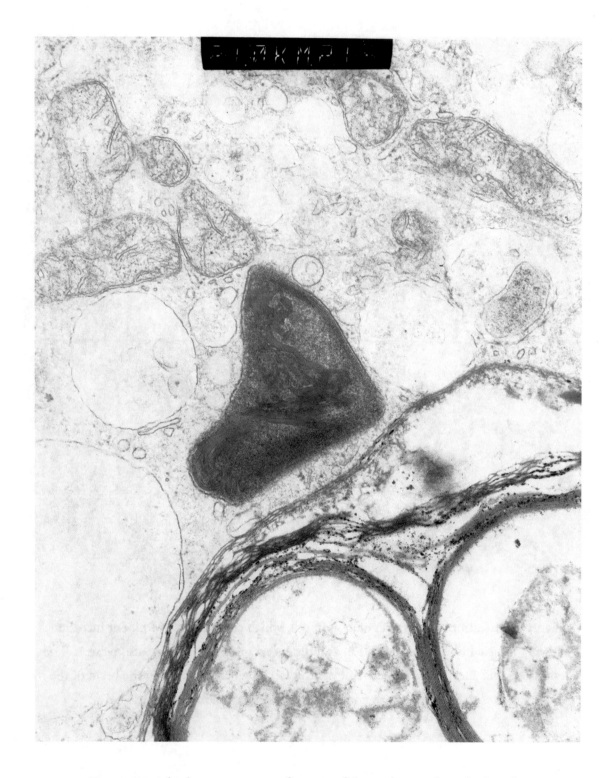

Figure 17. A higher power magnification of the inclusion described in the legend to Fig. 16.

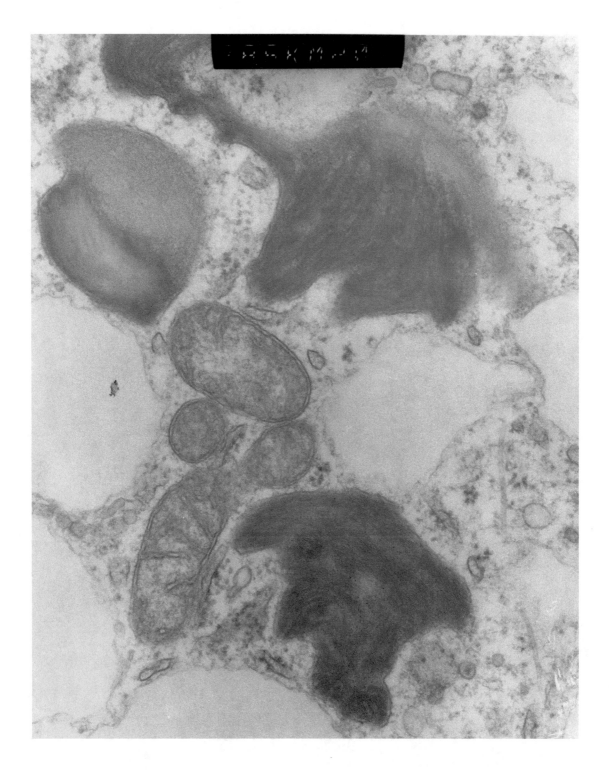

Figure 18. 18. Additional inclusions in another brain cell from the stealth adapted virus infected child. Several disrupted mitochondria are present.

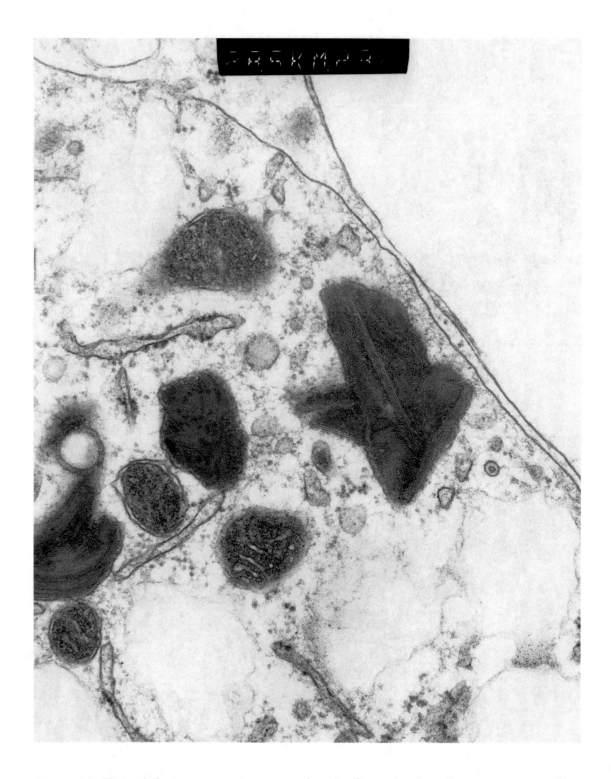

Figure 19. Brain inclusion containing narrow bands of very fine lamellar structuring within a more amorphous background of the remaining areas of the inclusion.

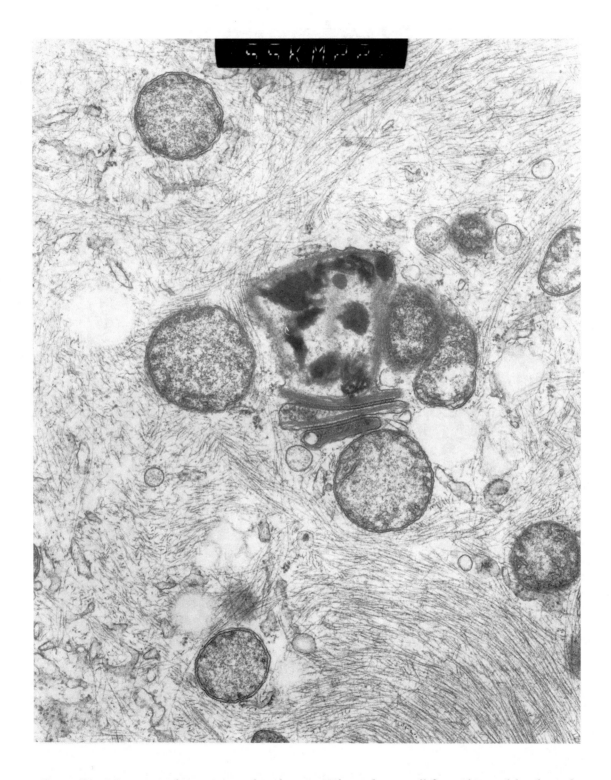

Figure 20. More complex structured inclusion within a brain cell from the stealth adapted virus infected child.

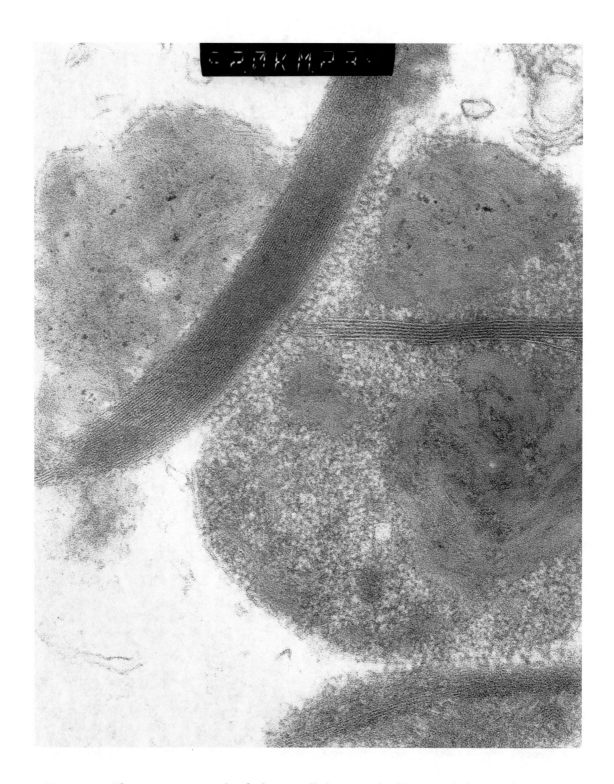

Figure 21. Electron micrograph of a brain cell showing double stranded spiral formation within a complex structured inclusion. Other areas of the inclusion show bands of very fine lamellar formations and fine particulate deposits within the more amorphous areas.

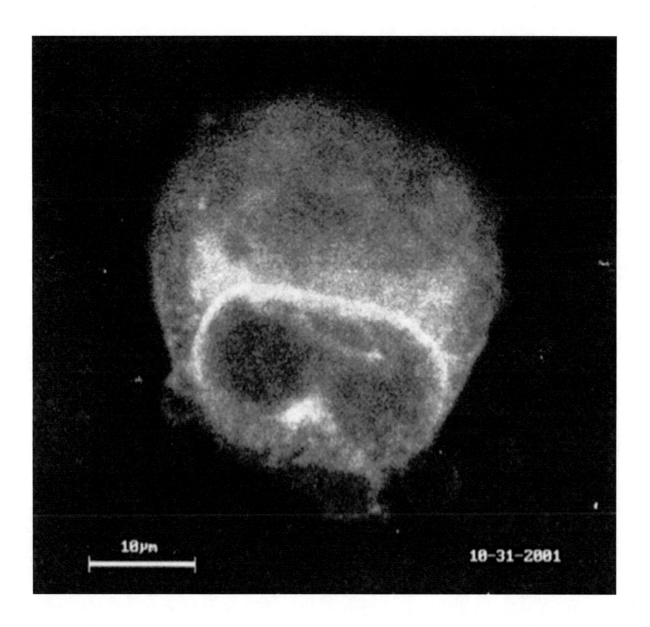

Figure 22. Cellular autofluorescence. A ghostly pattern of autofluorescence seen when examining a binucleate cell on Halloween, 2001. Both green and red autofluorescence was observed and is combined in this white-on-black representation. While there was considerable overlapping of the green and red autofluorescence, discrete areas of solely red and, to a lesser extent, solely green fluorescence were clearly seen. Maximum fluorescence occurred in the region surrounding the enlarged nuclei.

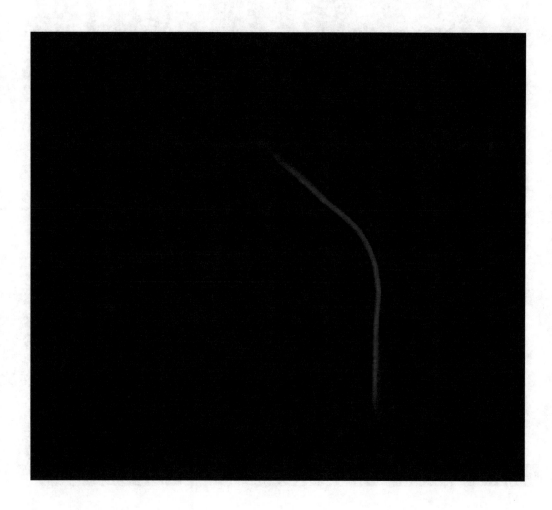

Figure 23. Autofluorescent thread. A strongly positive red autofluorescence seen when examining a thread from a positive culture using both blue and green laser illumination.

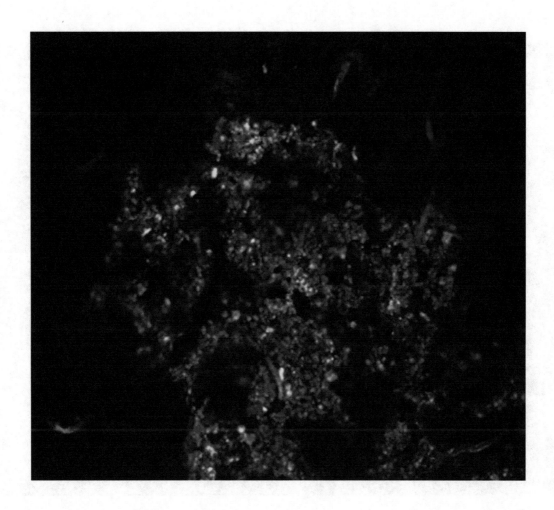

Figure 24. Autofluorescent particle. A speckled pattern of generally red admixed with some green autofluorescence seen on the illuminated surface of a particle retrieved from a stealth virus culture. Discrete dots of exclusively red and fewer dots with predominately green autofluoresnce were observed using helium/neon and argon laser excitations, respectively.

Figure 25. A floating cluster of cells from a positive stealth adapted virus culture viewed with white light under a microscope. The cluster is amidst some lipid needles, which are discussed later in regards to Figure 27.

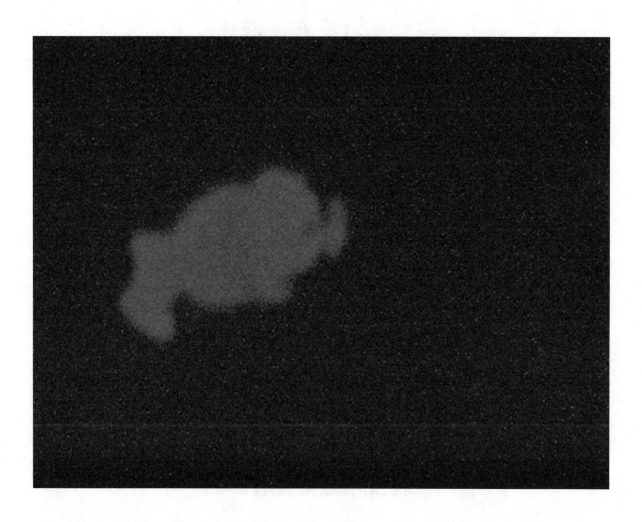

Figure 26. The same floating cluster as in Figure 24, showing luminescence following switching off the microscope. The luminescence in some clusters has persisted for over a minute.

Figure 27. Magnetism. A small minority of the particles floating in cultures of stealth adapted viruses are clearly magnetic. This figure shows the magnetic property of the particle, which was extruded from the cell cluster shown in Figure 9. The particle was placed into a droplet of phosphate buffered saline for microscopic viewing and photography. The particle could be rotated by bringing a small handheld magnet from differing directions, towards the microscope stage. The particle could be attracted or repulsed (flipped) by using different poles of the magnet, or by approaching from different sides with the same pole of the magnet. Cell clusters in stealth adapted virus cultures can, uncommonly, be drawn toward a magnet and easily rotated by reversing the polarity of the magnet.

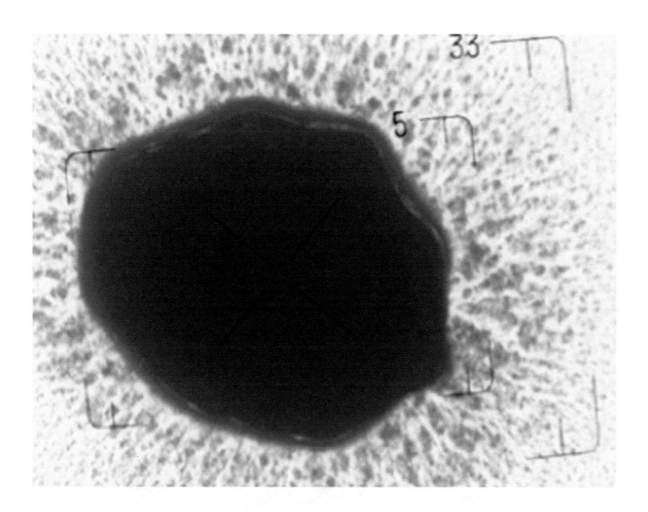

Figure 28. MTT reduction. Blue formazan precipitation occurred around a particle placed in a clear yellow solution of MTT (3-(4,5-dimethylthiazole-2-yl)-2,5-diphenyltetrazolium bromide) and observed over a 30-min period. No color change or precipitation occurred in control droplets containing MTT even over 24 h. This confirms the electron donating activity of the particle.

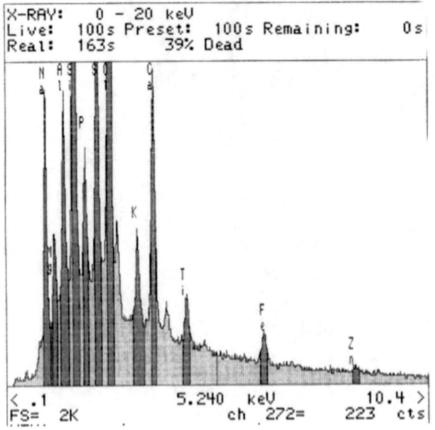

MEM1: minor deposit on membrane

WINDOW LABEL	START keV	END keV	WIDTH CHANS	GROSS INTEGRAL	NET INTEGRAL	EFF. FACTOR	%AGE TOTAL
na	1.02	1.16	8	9189	4045	1.00	10.92
mg	1.24	1.32	5	4409	574	1.00	1.55
al	1.44	1.60	9	11294	4571	1.00	12.34
SI	1.74	1.86	7	24608	5050	1.00	13.63
p	2.00	2.08	5	6597	625	1.00	1.69
s	2.28	2.40	7	15160	5024	1.00	13.56
cl	2.60	2.72	7	34680	12672	1.00	34.21
k	3.24	3.46	12	8054	2570	1.00	6.94
ca	3.66	3.74	5	8763	1121	1.00	3.02
ti	4.44	4.56	7	3587	451	1.00	1.22
fe	6.32	6.44	7	2146	358	1.00	.97
zn	8.54	8.66	7	878	-18	1.00	-.05

```
X-RAY:    0 - 20 keV
Live:  100s Preset:  100s Remaining:    0s
Real:  163s      39% Dead
```

```
< .1               5.240  keV         10.4 >
FS=  2K                ch 272=      223  cts
```

Figure 29. EDX analysis. Complex mineral composition of a small particle retrieved from a stealth virus culture. The upper table shows the windows used for detection and the relative percentages of the following minerals: sodium (na), magnesium (mg), aluminum (al), silicon (SI), phosphorus (p), sulfur (s), chloride (cl), potassium (k), calcium (ca), titanium (ti), iron (fe), and zinc (zn). The lower portion of the figure shows a graphical representation of the detected minerals. The two lightly shaded peaks were not analyzed since they reflect emissions from electrons entering L shell rather than K shell orbits. Different particles, even from the same culture, would commonly show highly diverse patterns of mineral composition, commonly with far fewer minerals than present in this particular example.

Figure 30. Lipid synthesis. This figure shows the characteristic double-edged, needle-shaped structure that can also commonly develop in cultures of stealth-adapted viruses. A film of more transparent lipid-like sheeting interconnects the parallel edges. The needle shaped structures can develop early in many cultures of stealth adapted cultures and continue to form well after the cells have disappeared. They are most easily seen using dark-field illumination. Magnification ×100.

Figure 31. A more detailed photomicroscope showing the linear trough-shaped structures in which lipid synthesis appears to be occurring. Fine particles can often be seen moving across the film connecting the edges. Similarly, the edges can occasionally be seen splitting and even disappearing.

Figure 32. Actual lipid membrane formation within a long term culture in which there were no remaining visible cells.

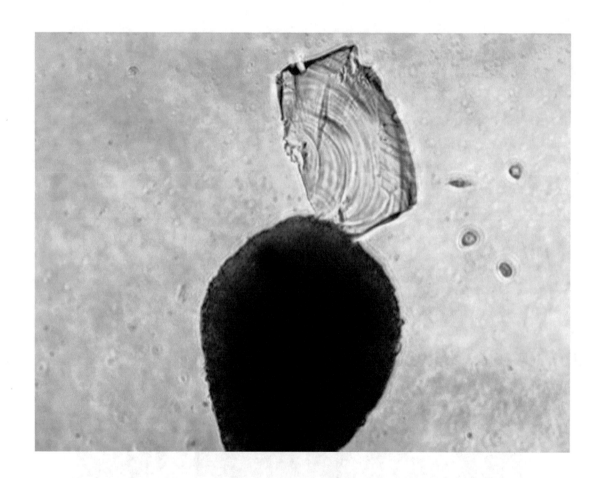

Figure 33. Solid crystal. A transparent lipid-like crystal developing in the vicinity of a heavily pigmented cell cluster. Similar appearing crystals can be seen in approximately a third of long-term cultures. They partition into chloroform and have other characteristics consistent with being mainly composed of cholesterol-related esters.

Figure 34. Pyramidal-shaped transparent crystals developing in long-term culture tubes that previously contained stealth-adapted virus-infected MRC-5 cells. Pyramid shaped crystals with 2 sets of 4 vertical planes would progressively increase in size by developing new planes perpendicular to their edges. They also typically showed a polygonal sloping vertex. Relatively large crystals are shown in this photograph. Magnification ×200.

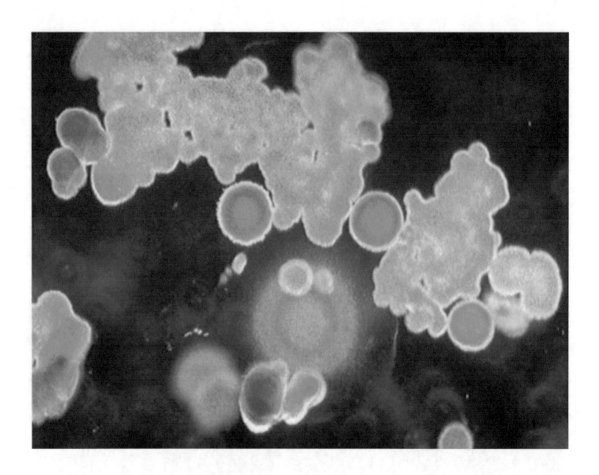

Figure 35. Collections of bright-yellow colored lipid-like material developing in the center of more transparent material giving the distinct impression of fried eggs. Over a year period, the amount of colored material increased considerably to yield complex multi-lobed collections that could readily be seen without magnification. No viable cells were remaining in the culture at the time of this photograph. A small transparent crystal overlies one of the yellow deposits. Close examination of the more transparent lipid-like material revealed shimmering due to rapid vibrations. Dark-field microscopy. Magnification ×200.

Neutral Red Dye/Ultraviolet Light Activation of the Alternative Cellular Energy (ACE) Pathway: A Historical Overview and Current Use of Neutral Red Dye Phototherapy of Herpes Simplex Virus Infections

W. John Martin, MD, PhD.
Institute of Progressive Medicine
South Pasadena CA 91030

Abstract

The potential benefits of light activated chemicals in the therapy of localized skin infections are generally ascribed to the formation of toxic oxygen free radicals, in a process termed photodynamic therapy. Studies on stealth adapted viruses, which are not effectively recognized by the cellular immune system, have identified an auxiliary (non-immunological) cellular defense mechanism. It is mediated by a form of cellular energy that is different from the metabolic energy obtained from mitochondria-based metabolism. The successful use of neutral red dye in conjunction with ultraviolet (UV) light in the therapy of herpes simplex virus (HSV), herpes zoster virus (HZV) and human papillomvirus (HPV) skin infections can be explained by the activation of the alternative cellular energy (ACE) pathway. Dye/light activation of the ACE pathway is being extended to other illnesses, including those caused by stealth adapted viruses.

Key Words: Photodynamic therapy, Stealth adapted viruses, Neutral red, Proflavine, Herpes, HSV, Shingles, HZV, Human papillomavirus, HPV, Alternative cellular energy, ACE, ACE pathway, Autism, Ultraviolet light, Enerceuticals™

Running Title: Dye/Light Activation of the ACE Pathway

Author: W. John Martin, MD, PhD. Institute of Progressive Medicine, 1634 Spruce Street South Pasadena CA 91030. Telephone 626-616-2868. E-mail wjohnmartin@hotmail.com

Historical Origins of Photodynamic Therapy

The therapeutic use of sunlight, in conjunction with chemicals applied to the skin, was reportedly practiced by the Arabian physician, Ebn Baithar, in the 13th century (1). Studies on the biological influences of light on particular chemicals were greatly extended in the mid to late 19th century with the development of coal tar derived aniline-based dyes (2). Microorganisms and tissues were soon shown to stain distinctively with various synthetic and naturally occurring dyes. Dr. Christian Gram used crystal violet dye to differentiate bacteria based on the ease of washing the dye from the stained bacteria (3). Hematoxylin and eosin dyes are still widely used to differentially stain the nucleus and cytoplasm, respectively, of cells within animal and human formalin-fixed tissues (4). Certain non-toxic stains did not appreciably affect the viability of cells and could, therefore, be used to study living cells. A medical student, Oscar Rabb noted, however, that acridine stained paramecia would lose their viability if the experiments were performed in brightly lighted areas (5).

A somewhat arbitrary distinction was made between the essentially direct photographic reaction of light on silver compounds; and the more intriguing and extensive biological effects of certain chemical compounds on living organisms in the presence, but not in the absence of external lighting. Some of the biologically effective chemical compounds, including acridine and eosin, fluoresce when illuminated with UV light. The demonstration of fluorescence along with the 1902 observation of Dr. Ledoux-Lebards that the cell killing effect of several light sensitive compounds was oxygen dependent (6), suggested a complex ionic photo-energetic process. The term photodynamic therapy was, accordingly, coined in 1904; distinguishing it from phototherapy, which was reserved for the general benefits of light on human health (7).

The anti-microbial and tissue damaging effects of photodynamic therapy have subsequently been explained by the generation of free radicals within the light illuminated compounds and/or within nearby air molecules. Specifically, many light activated photosensitizing agents become chemically reactive and generate singlet oxygen ($^1O_2$), superoxide anion ($O_2^{.-}$) and .OH radicals from atmospheric oxygen (8-10) and may also generate related nitrogen radicals. Several of the body's naturally occurring molecules were also identified as being photosensitive and capable of releasing oxygen free radicals. Examples include porphyrins in heme and cytochrome enzymes; 5-aminolevulic (ALA), the precursor of heme porphyrin; riboflavin (vitamin B2); etc. (9) While excessive free radicals are generally regarded as

toxic for many biological processes, both positively and negatively charged radicals clearly participate in the intracellular transfer of cellular energies and may also mediate specific signal messaging (11).

Antibacterial and Antiviral Therapies

Photodynamic therapy for superficial bacterial infections was extensively pursued through to the mid 20th century. Compounds were synthesized that could preferentially absorb visible, rather than UV light. Proflavine (diamino-acridine) is one such dye and was widely used to treat wounds incurred in both World Wars (12-14). Other tricyclic dyes in widespread prior use were methylene blue, toluidine blue and gentian violet. Sulfa drugs and antibiotics came to replace most of the photodynamic therapies being employed for superficial bacterial infections. Yet, even today, a triple dye compound comprising gentian violet, brilliant green and proflavine hemisulfate, is routinely used to prevent umbilical cord sepsis in approximately half of newborn babies in the United States (15).

Since antibiotics are not indicated for viral infections, a movement arose to see whether photodynamic therapy might be useful for superficial virus infections. Support for this possibility was provided in 1930 with the demonstration that dye plus light could inhibit the infectivity of bacteriophages, followed by a report in 1934 on photodynamic inactivation of a plant virus (16, 17). Neutral red, along with other dyes, were considered as potential anti-virus agents if sufficiently illuminated with light. Neutral red stains living cells and was commonly being used in monolayer virus cultures to identify plaques of non-viable and, therefore, non-stainable cells from the surrounding viable and, therefore stainable cells (18). Interestingly, for cytomegalovirus (CMV), the infected cells tended at first to show enhance staining by neutral red prior to being killed (19). Numerous experiments confirmed that neutral red, proflavine, toluidine blue, methylene blue and other dyes could significantly reduce the infectivity of various viruses when used with a concentrated source of lighting (20-26).

Therapy of Herpes Simplex Virus (HSV) Infections

A major impetus to anti-viral studies came at a dermatology conference in 1971. Dr. Troy Felber, working with Dr. Joseph Melnick of Baylor University, reported that neutral red dye applied to HSV lesions and illuminated with a cool fluorescent light provided symptomatic relief to the patients and expedited the healing of their skin lesions (27). Dr. Friedrich and Dr. Melnikoff separately published on the efficacy of Dr. Felber's protocol in the treatment of genital herpes (28, 29). A brief report also confirmed benefit of neutral red in oral herpes (30) and in a long-standing case of recurrent stomatitis (31). It is noteworthy that Dr. Melnick and his colleagues soon, thereafter, reported using proflavine, as an alternative to neutral red, in HSV infected humans (32) and in a rabbit model of HSV keratitis (33).

A formal report by Dr. Felber and his colleagues contrasted the benefits of neutral red as an experimental dye with that of phenol sulfonphthalein as a control (placebo) dye (34). "Neither the patient nor the physician who administered the dye-light treatment and conducted the follow-up examinations knew whether the experimental dye or the placebo dye was used." Within several hours of treatment, 18 of 20 neutral red treated patients and 6 of 12 control patients experienced a reduction in pain, tingling

and burning "superior to any mode of therapy previously used" and were symptom-free by 24 hrs. More striking was the 50% or greater reduction in recurrences noted in 17 of the neutral red treated patients compared to a similar reduction occurring in only 2 of the control patients. Moreover, only 11% of recurrences in the neutral red dye treated patients occurred at the previously involved skin site. In contrast 83% of the recurrences in the control patients involved the previously affected area of skin. The follow-up period of the study was 24-30 months during which time many other HSV infected patients were being treated with neutral red dye plus light at Baylor University and at other dermatology clinics throughout the country (35, 36)

Dr. Fred Rapp, a former graduate student of Dr. Melnick, pursued pioneering research in the possible virus origins of human cancers (37-39). He argued that the cell killing effect of HSV normally masked its potential cancer causing (oncogenic) ability. The oncogenic activity could potentially be unmasked if the virus was only partially inactivated. His group as well as others supported this concern by reporting on the enhanced growth (transformation) of human, hamster and rat fibroblasts exposed to neutral red plus light treated HSV (40-43). It was further shown that transformed hamster fibroblasts could form tumors in animals (40). Dr. Rapp also reported that UV light, by itself, could effectively unmask the oncogenic activity of HSV (44).

A bigger setback to dye/light therapy of HSV occurred in 1975 with the failure of controlled studies to confirm efficacy of applying neutral red to HSV followed by light stimulation on HSV lesions. Dr. Roome and his colleagues published the first negative study in the British Journal of Venereal Diseases (45). Sixteen women and 4 men were enrolled in the study with no appreciable benefits to any of the patients. The second larger negative study appeared in November in the New England Journal of Medicine by Dr. Martin Meyers and his colleagues (46). These reports led to the closure of many of the dye/light therapy clinics and a return to the use of topically applied idoxuridine for more severe HSV infections (47). Additional negative, or only minimally effective photodynamic studies for HSV using neutral red, proflavine and/or methylene blue as photosensitzers, were published in the late 1970's (48-53). In a status report publication in 1977, Dr. Melnick and his colleagues continued to advocate the use of neutral red and proflavine for HSV photodynamic therapy (54). In this report, he noted that proflavine was proving superior to neutral red, yet he subsequently acknowledged in the following year that the proflavine dye/light procedure failed to suppress HSV when evaluated in a double blind randomized study (55). There have been some efforts in recent years to try other potential photosensitive chemicals, including ALA, in the tissue culture destruction of HSV (56, 57). One group of investigators has reported benefits in a total of 6 patients with recurrent oral herpes using laser light activated methylene blue (58, 59).

Therapy of Human Papillomavirus (HPV) Genital Warts

In a related early study, neutral red and proflavine were separately tested on patients with recalcitrant, symmetrical genital warts (verrucae vulgares). Warts on only one side of the body were treated with the experimental dye and those on the other side treated with a control dye. A 300 Watt Osram full spectrum incandescent light was used to illuminate the neutral red and its control dye, while a sunlamp plus a UV light was used to illuminate the proflavine and its control dye. Ten of the 27 patients treated with proflavine and 10 of the 23 patients treated with neutral red were cured by the end of an 8 week period, with the warts disappearing simultaneously from the actively as well as the placebo-treated side (60). A minor consideration was given to the possibility of the therapy activating a systemic immune response, but essentially dismissed because anti-HPV antibodies developed in only one patient. A recent study has reported regression of hand and foot warts with the repeated application of ALA followed by intense red light illumination. The warts were also pared and treated with a keratolytic agent during treatment period. Fifty six percent of ALA treated warts, compared to 42% of the warts not treated with ALA (some on the same patients) were no longer present at 18 weeks after starting the study (61). Pain in the treated lesions was a major side effect of the therapy and is attributed to the toxicity of oxygen free radicals generated during the procedure (62).

Procedural Differences Between the Early Positive and Later Negative HSV Studies

There were several erroneous suppositions in the design and interpretations of the later neutral red-light therapy studies on HSV and HPV skin lesions. The first was the assumption that viruses were being locally destroyed by free radicals generated by the light activated dye. For example, to explain the lower frequency of recurrences at the site of therapy, Dr. Melnick argued that reduced numbers of HSV particles were possibly returning to the specific region of the ganglia innervating the treated site, but that virus from other parts of the ganglia would still be able to subsequently travel to the skin to induce disease (36). In reporting his negative results, Dr. Roome and his colleagues also commented that since HSV was commonly detectable in swabs obtained from the cervix that any local control of HSV lesions on the vulva would still prove to be inadequate treatment (45). Based on the principle of photodynamic therapy it was further assumed that the most effective wavelength of the light would be that which was best absorbed by the dye itself (24). Depending on pH, the maxima absorption of aqueous solutions of neutral red ranges from 480-570 nm (green light) (66). Incandescent tungsten lights provide these wavelengths. It is not clear why Dr. Felber fortuitously chose cool fluorescent lighting in his initial study. Interestingly, he advised his patients that they could follow up on his treatment at home using either a fluorescent or an incandescent light (34). Dr. Meyers and his colleagues clearly stated that they had used an incandescent light bulb in their non-confirmatory study. The source of lighting is relevant since cool fluorescent lighting, which emphasized the blue end of the light spectrum is known to also have much more of an UV component at 365 nm than does incandescent lights (67).

The switch from fluorescent lighting to incandescent light bulbs was probably not the entire explanation for the negative findings. For example, Dr. Roone and his colleagues used a 15 Watt fluorescent light in their negative study (44). The actual light used was not specified as being a "cool" (blue light spectrum enriched) bulb similar to that used by Dr. Felber (34). Of possible greater relevance is that Dr. Felber referred to using fresh solutions of neutral red while Dr. Roone and his colleagues sterilized their neutral red solution (45). Dr. Meyers and his colleagues also stated that they used their solutions within three months of being

prepared (46). Unless maintained anaerobically, neutral red solutions oxidize and long-term storage of the solution may have been an additional factor influencing the results (68).

Use of UV Light and Freshly Prepared Neutral Red

Working outside of mainstream academic medicine, an optometry student, Jon Stoneburner, was conducting tissue culture experiments on the possible value of various anti-herpes agents. He learned of unpublished English studies suggesting a benefit of using neutral red plus UV light to suppress disease in tadpoles caused by a recently isolated frog Lucké herpes virus. Extending this experiment, he observed the apparent destruction of HSV infected cells when neutral red containing cultures were examined using a microscope equipped with a UV light. He, therefore, decided to follow the neutral red protocol being used by some dermatologists, but using a UV light rather than using a white fluorescent light. This procedure was quite successful in achieving symptomatic relief in many of his tested patients. He further noted the importance of making up a fresh solution of neutral red and also found benefit in adding a small quantity of sodium hypochlorite (bleach) to the solution. He reasoned that the bleach might provide better skin penetration of the dye. For the same reason, he also followed Dr. Felber procedure of gently scrubbing the skin to remove any overlying covering of the ulcerated skin lesions.

Dr. Stoneburner extended his protocol to the treatment of patients with shingles, caused by herpes zoster virus (HZV) and to patients with HPV induced genital warts. The warts would be repeatedly pierced with a fine needle prior to applying the neutral red solution. Apart from this minor trauma, none of the patients reported pain in the treated herpes or HPV skin lesions. With each of these diseases, Dr. Stoneburner was presuming that the UV light illuminated neutral red dye was directly interacting with and destroying the virus particles.

Reinterpreting the Mode of Action of Neutral Red Dye – Light Therapy

I was interested in Dr. Stoneburner's protocol since it could potentially give support to my research regarding a non-immunological anti-viral defense mechanism (69). I had proposed that this defense mechanism is critically involved in the recovery from illnesses caused by viruses that evade recognition by the cellular immune system through a process termed stealth adaptation. Arguably, the defense mechanism could be as effective against conventional viruses for which there was also a potentially effective immune response. The non-immunological defense mechanism is mediated through an alternative cellular energy (ACE) pathway and involves energy transfer reactions among mineral containing ACE pigments. I have shown that these pigments will commonly fluoresce when exposed to UV light and that the fluorescence can be enhanced using various dyes, including neutral red and acridine.

Dr. Stoneburner had also incidentally noted fluorescence in the herpes and HPV skin lesions during his treatment protocols. We jointly conducted an Institutional Review Board (IRB) approved collaborative study. It confirmed the fluorescence and the expedited healing of UV light illuminated HSV, HZV and HPV induced skin lesions when neutral red dye was applied, but not when a red food dye was used as a control (70). All of the food dye control patients were subsequently successfully treated using neutral red dye. The fluorescence of the HSV lesions is typically bright yellow, while that of HZV and HPV skin lesions is more yellow-orange to light purple, respectively. An example of a HSV lip lesion showing UV induced yellow fluorescence after treatment with neutral red dye and the subsequent expedited healing of the lesion is shown in Fig. 1. In subsequent studies, we observed that fluorescence was not always restricted to the lesion on which the neutral red had been applied. In several patients, UV inducible fluorescence could also be readily seen in the skin adjacent to the actual lesion and on which no dye had been applied. Moreover, in those patients in whom multiple herpetic or papillomavirus skin lesions were present, UV illuminating a single lesion on which neutral red dye was applied, slowly led to discernable UV inducible fluorescence occurring in the other skin lesions. As I had expected, expedited healing occurred in both the neutral red treated and neutral red untreated skin lesions. Of particular interest, direct UV illumination of the secondary lesions on which no dye has been administered was not required for these lesions to heal. An additional incidental finding was the marked swelling of the lesions during therapy from an apparent ingress of fluid, which could sometimes be seen oozing from larger lesions.

Overcoming Regulatory Objections to the Skin Application of Neutral Red Dye

The US Food and Drug Administration (FDA) determined that the neutral red was being used as a drug and that extensive animal and other toxicity studies would be required in order to even continue with clinical trials. Rare instances of allergic reactions to neutral red had been reported in the medical literature (71, 72). FDA had also published a Drug Bulletin in 1975 referring to the oncogenic potential of using neutral red in treating HSV (73). Based on my understanding of ACE pigments, I reasoned that these pigments could still be effective if collected from a patient and activated in close proximity to, but not necessarily in direct contact with an active herpetic skin lesion. Moreover, I knew from previous studies on stealth adapted virus infected patients that either directly fluorescing, or more commonly dye inducible fluorescing material, was present in various bodily fluids, including saliva, perspiration and/or urine. Deficiency in the ACE pathway can be readily monitored in these and other patients based upon saliva and/or perspiration fluorescence in the presence or absence of neutral red dye. The testing is not diagnostic for an underlying illness; however, since vigorously exercising athletes can show occasional positive reactions for direct or dye inducible body fluid fluorescence.

In a series of IRB approved studies, I was personally able to confirm that active herpes skin lesions would fluoresce and undergo expedited healing if material collected from the lesion was mixed with freshly prepared neutral red solution and UV illuminated in close proximity to the lesion (74). Neutral red reactive perspiration, saliva and/or urine were similarly tried and shown to be effective in indirectly inducing UV fluorescence in an underlying herpetic skin lesion. Once the lesion begins to fluorescence, the UV illumination can be directly applied to the lesion till the fluorescence either fades or at least 30 minutes illumination has been provided. Using a Halco 13 watt condensed spiral UV light bulb is more convenient than using BLB UV tubes as the source of lighting. A 365 nm LED penlight has also been successfully used.

Therapy for Latent HSV Infection and for Post Herpetic Neuralgia

The same basic procedure using perspiration mixed with neutral red was extended to treat herpes infected patients who were not presently experiencing an active outbreak. These included patients with a high expectation of one or more HSV recurrences within the next few months. It also included patients with post-herpetic neuralgia resulting from prior shingles caused by HZV. A wet or dried surgical towel was used to wipe the area of skin in which the herpes outbreaks had occurred. Using a plastic sheet as a waterproof barrier, the towel on which perspiration had been collected was placed onto the area of previously affected skin, sprayed with a neutral red solution and UV illuminated. The procedure was deemed successful if, upon temporarily removing the towel, the underlying skin area would show UV fluorescence. More importantly, on direct inquiry over several months, none of the HSV infected patients had experienced any further outbreaks, even through periods of significant stress. A similar approach was also used in patients with post herpetic neuralgia following shingles. Perspiration from the affected area was collected onto a surgical towel, stained with neutral red to yield UV fluorescence and placed onto the patient skin. With continued UV illumination, a photo-biological reaction occurs in the underlying skin, as shown by clearly discernable direct UV fluorescence. The surgical towel can then be dispensed with and direct UV illumination of the skin continued till the fluorescence fades or time becomes limiting. As confirmed by Dr. Stoneburner, a single procedure has provided rapid and long lasting relief to several patients. Occasionally, the procedure was repeated on a second or even a third day, if marked direct fluorescence was still observable in the treated area after the 30 minutes therapy session.

Replacing the Need for Collecting Patient's ACE Pigments

A basic modification of the protocol has since been made by removing the need for collecting ACE pigments from the patients. Various natural products have been identified with ACE pigment-like activities. They are being grouped under the term enerceuticals.™ Some require the addition of neutral red while others appear to work well in evoking direct fluorescence of the body's ACE pathway. A Lidocaine containing homeopathic herbal formulation was quite effective in a small series of patients in suppressing both acute and latent herpes virus infections. It was either sprayed onto neutral red soaked surgical towels or placed into a Ziploc plastic bag with the addition of freshly dissolved neutral red. Various other formulations are currently being evaluated for their capacity to indirectly activate the body's ACE pathway.

Application of Protocol to Stealth Adapted Virus Infected Patients

The new protocols allow for testing in presumptively stealth adapted virus infected patients (75-87). Patients with CFS, including two patients with Morgellons disease (88) showed considerable improvements when treated in a medical clinic. More interestingly, parents of several children with autism were provided details of the IRB approved procedure of using UV illuminated neutral red added to an ACE pathway activating solutions. The preliminary reports from these parents are highly encouraging with definite skin and/or mouth fluorescence being achieved during the therapy, along with marked behavioral improvements following the therapy.

Other Applications of ACE Pathway Activating Protocols

Several incidental observations made during the therapy of adult patients have suggested possible beneficial effects of activation of the ACE pathway in patients with presumed metabolic energy deficiency illnesses, including emphysema. Indeed, the ACE pathway is likely to be a natural adjunct, or alternative, to food metabolism in supplying the energy needs of the body with additional specific contributions in the healing process.

Monitoring of the ACE Pathway

As noted above, sampling of dried perspiration from skin areas of prior HSV outbreaks, can be used as a potential source of ACE pigments. Either the direct spraying of neutral red dye onto these areas or preferably the neutral red staining of materials collected from these areas can, when viewed under UV illumination, reveal periods of heightened production of ACE pigments. In preliminary studies with Dr. Stoneburner, data were obtained correlating the ease of detecting UV fluorescence with HSV shedding, as determined using the polymerase chain reaction (PCR). For more systemic illnesses, dried perspiration, hair samples, saliva and/or urine can be tested for both direct and neutral red dye triggered UV fluorescence. Other possible causes of fluorescence need to be considered, as does the possibility of ACE pathway activation in response to strenuous exercise and possibly to dieting. For more detailed clinical studies, dried blood smears can be repeatedly collected over time and studied before and after neutral red staining for intracellular UV fluorescence, typically seen in polymorphonuclear cells. Live cell microscopic analysis can similarly be performed to show motility of fluorescing extracellular materials. These methods can also help in the evaluation of various therapeutic endeavors aimed at enhancing the ACE pathway.

Summary

An understanding of the ACE pathway is providing fresh insight into earlier and generally inconsistent reports of dye-light therapy of localized virus induced skin lesions. These earlier studies incorrectly assumed that success could be achieved through the photodynamic generation of free radicals leading to virus inactivation. There is evidence that ACE pigments accumulate within virus skin lesions and can be triggered to respond to UV light by the addition of freshly prepared neutral red dye. ACE pigments can also be present in perspiration, saliva and urine of ACE deficient individuals and can similarly be triggered by neutral red dye to respond to UV illumination. Activated ACE pigments and certain other formulations, placed in close proximity to the skin can enhance the energy levels of cells throughout the body sufficient to suppress virus induced cellular damage, including that caused by stealth adapted viruses. This approach may potentially have far wider benefits for many illnesses in which there is an ongoing deficiency in the patient's ACE pathway. The efficacy of such interventions can be monitored by sequentially assessing the levels of direct and neutral red-triggered UV fluorescence of ACE pigments sampled from patients undergoing the interventions.

Acknowledgement. I am grateful to those who expressed their confidence in my early reports on the existence of stealth adapted viruses. Dr. Jon Stoneburner kindly described many of his prior observations and encouraged our collaboration. It has also been a pleasure working with parents of children with autism.

References

1. Szeimies R-M, Drager J, Abels C, Landthaler M. (2001) History of photodynamic therapy in dermatology. In "Photodynamic Therapy and Fluorescence Diagnosis in Dermatology" Eds. EG Calzavara-Pinton, RM. Szeimies and B. Ortel. 3-16. Elsevier. ISBN 0-444-50828-7

2. Johnston WT. (2008) The discovery of aniline and the origin of the term "aniline dye". Biotech Histochem. 83: 83-7.

3. Gram C. (1884) Ueber die isolirte Färbung der Schizomycetin in Schnitt- und Trockenpräparaten. Fortschritte der Medicin. 2:185-189

4. King DF, King LA. (1986) A brief historical note on staining by hematoxylin and eosin. Am J Dermatopathol. 8:168.

5. Raab O. Z (1900) Biol. 39: 524.

6. Ledoux-Lebards C. (1902) Action de la lumiSre sur la toxicitg de l'eosin et de quelques autres substances pour les paramecies. Annal Institut Pasteur. 16, 593.

7. Von Tappeiner H, Jodlbauer A. (1904) Uber die Wirkung der photodynamischen

(flureszier-enden) Stoffe auf Infusorien. Dtsch. Arch. Klin. Med. 80: 427-487.

8. Dalle Carbonare M, Pathak MA. (1992) Skin photosensitizing agents and the role of reactive oxygen species in photoaging. J Photochem Photobiol B. 14: 105-24. Müller-Breitkreutz K, Mohr H, Briviba K, Sies H. (1995) Inactivation of viruses by chemically and photochemically generated singlet molecular oxygen. J Photochem Photobiol B. 30: 63-70.

9. Wainwright M. (2004) Photoinactivation of viruses. Photochem. Photobiol. Sci., 3, 406-411.

10. Gutierrez J, Ballinger SW, Darley-Usmar VM, Landar A. (2006) Free radicals, mitochondria, and oxidized lipids: the emerging role in signal transduction in vascular cells. Circ Res. 99: 924-32.

11. Browning CH, Gulbransen R, Thornton LH. (1917) The antiseptic properties of acriflavine and proflavine, and brilliant green: With special reference to suitability for wound therapy. Br Med J. 21:70-75.

12. Berkeley C, Bonney V. (1919) Proflavine oleate in the treatment of open wounds. Br Med J. 1:152-153.

13. Heggie JF, Warnock GB, Nevin RW. (1945) "Neutral Proflavine Sulphate" in Infected Wounds involving Bone. Br Med J. 1: 437-440.

14. Suliman AK, Watts H, Beiler J, King TS, Khan S, Carnuccio M, Paul IM. (2010) Triple dye plus rubbing alcohol versus triple dye alone for umbilical cord care. Clin Pediatr (Phila). 49: 45-8.

15. Brendel M.(1930) Different photodynamic action of proflavine and methylene blue on bacteriophage. Mol Gen Genet. 120: 171-180.

16. Birkeland JM. (1934) Photodynamic action of methylene blue on plant viruses. Science 80: 357-358.

17. Dulbecco R, Vogt M. (1954) Plaque formation and isolation of pure lines with poliomyelitis viruses. J Exp Med. 99:167-82.

18. Schmidt NJ, Dennis J, Lennette EH. (1976) Plaque reduction neutralization test for human cytomegalovirus based upon enhanced uptake of neutral red by virus-infected cells. J Clin Microbiol. 4: 61-6.

19. Hiatt, C. W., E. Kaufiman, J. J. Helprin, and S. Baron. (1960) Inactivation of viruses bv the photodynamic action of toluidine blue. J. Immunol. 84: 480-484.

20. Crowther D, Melnick JL. (1961) The incorporation of neutral red and acridine orange into developing poliovirus particles making them photosensitive. Virology. 14: 11-21.

21. Tomita Y, Prince AM. (1963) Photodynamic inactivation of arborviruses by neutral red and visible light. Proc. Soc. Exptl. Biol. Med. 112: 887-890.

22. Wallis C, Melnick JL. (1964) Irreversible photosensitization of viruses. Virology. 23: 520-527.

23. Wallis, C, Melnick JL. (1965) Photodynamic inactivation of animal viruses: a review. Photochem. Photobiol. 4: 159-170.

24. Wallis, C., C. Scheiris, and J. L. Melnick. (1967) Photodynamically inactivated vaccines prepared by growing viruses in cells containing neutral red. J.Immunol. 99: 1134-1139.

25. Wallis C, Trulock S, Melnick JL. (1969) Inherent photosensitivity of herpes virus and other enveloped viruses. J.Gen.Virol. 5: 53-61.

26. Felber TD. (1971) Presentation to Scientific Assembly of AMA Section on Dermatology, June 21,1971; Reported as "Photoinactivation may find use against herpesvirus" JAMA. 217: 270.

27. Melnikoff RM. (1972) Photodynamic inactivation of herpes simplex infections. Calif Med. 116: 51–52.

28. Friedrich EG. (1973) Relief for herpes vulvitis. Obstet Gynecol. 41:74-77, 1973.

29. Lefebvre EB, McNellis EE. (1973) JAMA. 224: 1039.

30. Padval DG, Nat R. (1972) Cure of stomatitis. Arch Dermatol. 106: 421-422.

31. Kaufman RH, Gardner HL, Brown D, Wallis C, Rawls WE, Melnick JL. (1973) Herpes genitalis treated by photodynamic inactivation of virus. Am J Obstet Gynecol. 117:1144-6.

32. Moore C, Wallis C, Melnick JL, Kuns MD. (1972) Photodynamic treatment of herpes keratitis. Infect Immun. 5:169-71.

33. Felber TD, Smith EB, Knox JM, Wallis C, Melnick JL: (1973) Photodynamic inactivation of herpes simplex: report of a clinical trial. JAMA. 223: 289-292.

34. Jarratt MT, Hubler WR, Knox JM, Melnick JL. (1974) Dye-photoinactivation and herpes simplex. Arch Dermatol. 109: 570.

35. Wallis, C, Melnick JL. (1975) Photodynamic inactivation of herpes virus. Perspect Virol. 9: 287-314.

36. Rapp F, Duff R. (1973) Transformation of hamster embryo fibroblasts by herpes simplex viruses type 1 and type 2. Cancer Res. 33: 1527-34.

37. Rapp F. Buss ER. (1974) Are Viruses Important in Carcinogenesis? Am J Pathol. 77:85-102.

38. Rapp F, Li JH. (1974) Demonstration of the oncogenic potential of herpes simplex viruses and human cytomegalovirus. Cold Spring Harb Symp Quant Biol 39: 747-763.

39. Rapp F, Li JH, Jerkofsky M. (1973) Transformation of mammalian cells by DNA-containing viruses following photodvnamic inactivation. Virology 55:339-346,

40. Li JL, Jerkofsky MA, Rapp F. (1975) Demonstration of oncogenic potential of mammalian cells transformed by DNA-containing viruses following photodynamic inactivation. Int J Cancer. 15: 190-202.

41. Kucera LS, Gusdon JP, Edwards I, Herbst G. (1977) Oncogenic transformation of rat

embryo fibroblasts with photoinactivated herpes simplex virus: rapid in vitro cloning of transformed cells, J Gen Virol. 35, 473–485.

42. Bockstahler LE, Lytle CD, Hellman KB. (1974) A review of photodynamic therapy for herpes simplex: benefits and potential risks. DHEW Publication (FDA) 1974; No. 75-8013. Reprinted in N Y J Dent. 45: 148–157.

43. Duff R, Rapp F. (1971). Properties of hamster embryo fibroblast transformed in vitro after exposure to ultraviolet-irradiated herpes simplex virus type 2. J Virol. 8: 469-477.

44. Roome PCH Tinkler AE, Hilton AL, Montefiore DG, Waller D. (1975) Neutral red with photoinactivation in the treatment of herpes genitalis. Br J Vener Dis. 51: 130-133.

45. Myers MG, Oxman MN, Clark JE, Arndt KA. (1975) Failure of neutral-red photodynamic inactivation in recurrent herpes simplex virus infections. N Engl J Med. 293: 945-9.

46. Hall-Smith SP, Corrigan MJ, Gilkes MJ. (1962) Brit Med J. 2: 1515-1516. : 13952144

47. Taylor PK, Doherty NR. (1975) Comparison of the treatment of herpes genitalis in men with proflavine photoinactivation, idoxuridine ointment, and normal saline. Br J Vener Dis. 51: 125-9.

48. Myers MG, Oxman MN, Clark JE, Arndt KA. (1976) Photodynamic inactivation in recurrent infections with herpes simplex virus. J Infect Dis. 133 Suppl: A145-50.

49. Chang TW, Fiumara N, Weinstein L. (1975) Genital herpes: Treatment with methylene blue and light exposure. Int J Dermatol. 14: 69-71.

50. Yen GS, Simon EH. (1978) Photosensitization of herpes simplex virus Type 1 with neutral red. J Gen Virol 41, 273–281.

51. Thomas JV, Dunlap WA, Rich AM. (1973) Photodynamic inactivation in the treatment of experimental herpes simplex keratitis. Brit J Ophthal. 57: 336-339. Bartholomew RS, Clarke M, Phillips CI. (1977) "Dye/light" Dye-induced photosensitization of herpes virus. A clinical trial on humans. Trans Ophthalmol. Soc UK. 97: 508-9.

52. Melnick JL, Wallis C. (1977) Photodynamic inactivation of herpes simplex virus: A status report. An NY Acad Sci. 284: 171-181.

53. Kaufman RH, Adam E, Mirkovic RR, Melnick JL, Young RL. (1978) Treatment of genital herpes simplex virus infection with photodynamic inactivation. Am J Obstet Gynecol. 132: 861-9.

54. Tao JN, Duan SM, Li J. (2007) Experimental studies on treatment of HSV infections with photodynamic therapy using 5-aminolevulinic acid. Zhonghua Shi Yan He Lin Chuang Bing Du Xue Za Zhi. 21: 79-82.

55. Smetana Z, Mendelson E, Manor J, van Lier JE, Ben-Hur E, Salzberg S, Malik Z. (1994) Photodynamic inactivation of herpes

viruses with phthalocyanine derivatives. J Photochem Photobiol B. 22: 37-43.

56. Marotti J, Aranha AC, Eduardo Cde P, Ribeiro MS. (2009) Photodynamic therapy can be effective as a treatment for herpes simplex labialis. Photomed Laser Surg. 27: 357-63.

57. Marotti J, Sperandio FF, Fregnani ER, Aranha AC, de Freitas PM, Eduardo Cde P. (2010) High-intensity laser and photodynamic therapy as a treatment for recurrent herpes labialis. Photomed Laser Surg. 28: 439-44.

58. Veien NK, Genner J, Brodthagen H, Wettermark G. (1977) Photodynamic inactivation of verrucae vulgares. II. Acta Derm Venereol. 57: 445-7.

59. Stender IM, Na R, Fogh H, Gluud C, Wulf HC. (2000) Photodynamic therapy with 5-aminolaevulinic acid or placebo for recalcitrant foot and hand warts: randomised double-blind trial. Lancet. 355: 963–966.

60. Stender IM, Borgbjerg FM, Villumsen J, Lock-Andersen J, Wulf HC. (2006) Pain induced by photodynamic therapy of warts. Photodermatol Photoimmunol Photomed. 22: 304–309.

61. Appleton N. (2002) "Rethinking Pasteur's Germ Theory: How to Maintain Your Optimal Health." Frog Ltd, Berkeley CA ISBN 1-58394-051-0

62. Murphy KM, Travers P, Walport M. (2007) "Janeway's Immunobiology," 7th Ed. Garland Sci. New York. ISBN 978-0815341239

63. Cooper MD, Alder MN. (2006) The evolution of adaptive immune systems. Cell. 124: 815-22.

64. Durda F. The Fluorescent Lighting System. http://nemesis.lonestar.org/reference/electricity/fluorescent/index.html

65. Chignell CF, Sik RH, Bilski PJ. (2008) The photosensitizing potential of compact fluorescent vs incandescent light bulbs. Photochem Photobiol. 84: 1291-3.

66. McKinlay JB, Zeikus JG. (2004) Extracellular iron reduction is mediated in part by neutral red and hydrogenase in Escherichia coli. Appl Environ Microbiol. 70: 3467-74.

67. Martin WJ. (2003) Stealth virus culture pigments: a potential source of cellular energy. Exp Mol Pathol. 74: 210-23.

68. Martin WJ, Stoneburner J. (2005) Symptomatic relief of herpetic skin lesions utilizing an energy-based approach to healing. Exp Mol Pathol. 78: 131-4.

69. Mitchell JC, Stewart WD. (1973) Allergic contact dermatitis from neutral red applied for herpes simplex. Arch Dermatol. 108: 689.

70. Goldenberg RL, Nelson K. (1975) Dermatitis from neutral red therapy of herpes genitalis. Obstet Gynecol. 46: 359-60.

71. Oncogenic potential of new herpes simplex therapy. FDA Drug Bull 1975; Jan-March; 3.

72. Martin WJ. (2010) ACE Pathway Activation as Natural Therapy for Herpes infections. BIT's Ist World Congress of Viral infections July 31-Aug. 3 Busan, South Korea (Abstract).

73. Martin WJ, Zeng LC, Ahmed K, Roy M. (1994). Cytomegalovirusrelated sequences in an atypical cytopathic virus repeatedly isolated from a patient with the chronic fatigue syndrome. Am. J. Pathol. 145: 441452.

74. Martin WJ. (1995) Stealth virus isolated from an autistic child. J Aut Dev Dis. 25: 223224.

75. Martin WJ. (1996) Severe stealth virus encephalopathy following chronic fatigue syndromelike illness: Clinical and histopathological features. Pathobiology. 64:18.

76. Martin WJ. (1996) Stealth viral encephalopathy: Report of a fatal case complicated by cerebral vasculitis. Pathobiology. 64:5963.

77. Martin WJ. (1996) Simian cytomegalovirusrelated stealth virus isolated from the cerebrospinal fluid of a patient with bipolar psychosis and acute encephalopathy. Pathobiology. 64: 6466.

78. Martin WJ, Anderson D: (1997) Stealth virus epidemic in the Mohave Valley. Initial report of viral isolation. Pathobiology. 65:51-56.

79. Martin WJ, Anderson D. (1999) Stealth virus epidemic in the Mohave Valley: Severe vacuolating encephalopathy in a child presenting with a behavioral disorder. Exp Mol Pathol. 66:19-30.

80. Martin WJ. (1999) Stealth adaptation of an African green monkey simian cytomegalovirus. Exp Mol Pathol. 66: 3-7.

81. Martin WJ. (2003) Complex intracellular inclusions in the brain of a child with a stealth virus encephalopathy. Exp Mol Pathol. 74: 179-209.

82. Martin WJ. (2005) Alternative cellular energy pigments mistaken for parasitic skin infestations. Exp Mol Pathol. 78: 212-4.

83. Martin WJ. (2005) Progressive Medicine. Exp Mol Pathol. 78: 218-220.

84. Martin WJ. (2005) Etheric Biology. Exp Mol Pathol. 78: 221-227.

85. Martin WJ. (2008) Stealth adapted viruses and discovery of the alternative cellular energy (ACE) pathway. Best Syndication 8-7-2008. http://www.bestsyndication.com/?q=20080807_stealth_virus.htm

86. Martin WJ. (2013) ACE pigment dermatitis: Possibly a healing process gone awry? (Submitted).

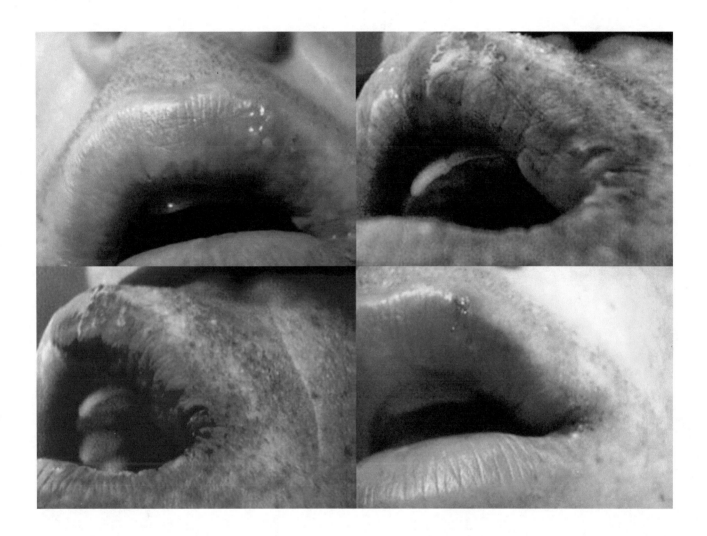

Legend to Figure 1. HSV lip lesions prior to dye-light therapy; UV illumination of the lesions at 1 min. after applying neutral red dye to opened vesicles, note the marked yellow fluorescence; fading of the UV fluorescence by 15 min. after the application of the dye; and near complete resolution of the lesions at 8 hrs. after therapy. Photographs kindly provided by Dr. J. Stoneburner.

CHAPTER 4

Alternative Cellular Energy (ACE) Pathway Activation as Natural Therapy for Autism

W. John Martin, MD, PhD.
Institute of Progressive Medicine

Abstract

Symptomatic and quality of life improvements occur in autistic children and adults following activation of the alternative cellular energy (ACE) pathway. This can be achieved by spraying an energized water based product onto neutral red dye moistened paper towels placed onto the skin of the patients and illuminating the towels with ultraviolet-A (UV-A) light. Activation of the ACE pathway is shown by the appearance of UV-A evoked fluorescence within the underlying and even distant areas of the patients' skin and also commonly within the oral cavity. For several weeks following activation of the ACE pathway, autistic individuals seemingly express a greater self-awareness and mindfulness, with a striking reduction in many of the clinical manifestations of autism, including seizures. ACE based therapy is suitable for home use and can be easily administered by parents of autistic children.

Key words: autism, dye-light therapy, phototherapy, ACE Water, Fluorescence, ultraviolet light, stealth virus, enerceutical, alternative cellular energy, ACE therapy, energy medicine

Running Title: ACE Pathway Activation

Author: W. John Martin, MD, PhD. Institute of Progressive Medicine, 1634 Spruce Street, South Pasadena CA 91030 wjohnmartin@hotmail.com

Abbreviations: ACE; Alternative cellular energy

CFS: Chronic fatigue syndrome

UV-A, Ultraviolet-A

The increasing incidence of autism is consistent with a relatively silent virus epidemic that in pregnant women can cause brain damage of the fetus predisposing to the subsequent development of autism (1, 2). Deletion or mutation in viruses of the relatively few components that are targeted by the cellular immune system can lead to a persistent virus infection in the absence of an accompanying inflammatory reaction (3, 4). This immune system evading process has been termed "stealth adaptation" and has been shown to occur with several viruses, including the simian cytomegalovirus of African green monkeys (SCMV) (5-7). Evidence for the presence of stealth adapted viruses has been consistently obtained in blood samples from the majority of autistic children tested using a cell culture method (2).

Studies on stealth adapted viruses led to the identification of an auxiliary (non-immunological) cellular repair mechanism mediated by particulate materials produced by virus infected cells (8). These mineral containing organic materials are postulated to provide a non-mitochondria source of cellular energy, since the surviving cells can show markedly disrupted mitochondria (9). The particulate materials display a range of energy transducing properties; responding to electromagnetic, magnetic, sound and other energy inputs, along with the capacity of donating electrons and of inducing vapor bubbles in water (9, 10). The materials are fluorescent when illuminated by ultraviolet-A (UV-A) light (9) and this fluorescence can be enhanced using various dyes, including neutral red. The culture derived materials are referred to as alternative cellular energy pigments (ACE-pigments). Similar materials are produced by the body in response to a variety of illnesses, including both conventional and stealth adapted virus infections (11-13). ACE pigments, identified as microscopically particulate materials with fluorescing and other energy transducing properties, can also be recovered from perspiration, saliva, urine and blood of many patients, including those presumptively infected with stealth adapted viruses.

The term enerceutical™ has been introduced to describe therapeutic products, which similar to ACE pigments are capable of promoting the ACE pathway. An early test system was to explore whether various natural products could assist in reversing the cytopathic effect (CPE) in stealth adapted virus cultures. These studies led to interest in a supposedly homeopathic preparation termed HANSI (homeopathic activator of the system immune). While originally produced in Argentina, the product formulation was sold to a United States corporation called World Health Advanced Technology. I reported my positive findings on testing HANSI in stealth adapted virus cultures to the US manufacturer and on how I believed the product was working. The company accordingly changed the name of HANSI to Enercel™. I was impressed by the company's broadly based reports of clinical efficacy of their seemingly "energized water" product in various clinical conditions. Furthermore, I directly participated in a convincingly successful controlled study using the product, as a form of "ACE Water" in the therapy of tropical childhood diarrhea.

Although, not disclosed in its labeling, GC-MS analysis of Enercel™ revealed the presence of trace quantities of Lidocaine. Additional indications arose suggesting that the various herbal tinctures listed as the starting ingredients of the homeopathic formulation may have been somewhat misleading. Still based on the listed and other suggested tinctures and with the addition of measurable quantities of Lidocaine, various solutions were formulated and tested for their capacity to brightly fluoresce when mixed with neutral red dye. A two-gallon batch of solution containing the diluted homeopathic tinctures listed in the Materials and Method in 10% ethanol and <0.1% Lidocaine and mineral salts was subsequently prepared for clinical testing. This energized water formulation proved effective when added to neutral red dye and UV illuminated in the close proximity of herpes simplex virus (HSV) and herpes zoster virus (HZV) skin infections. Moreover, within minutes, the underlying herpetic lesions and surrounding skin areas would also fluoresce when directly exposed to the UV light. The indirectly induced fluorescence was comparable to that which occurs by directly applying neutral red to herpetic skin lesions (11). A similar approach using the solution plus neutral red dye was also able to evoke localized skin fluorescence in patients without active disease but in whom there was a history of prior recurrent lesions. Treating the areas of prior recurrences seemingly led to a reduction in the frequency of future herpetic outbreaks. This energy approach was, therefore, tried in children with autism.

Material and Methods

The energized formulation used in the reported studies comprised a 10% ethanol in water blend of homeopathic dilutions (>6X, i.e. six-10 fold) of the following tinctures: Aconite napellus; Amica montana; Arsenicum album; Byronia alba; Chelidonium major; Cinnamomum; Lachesis mutus; Pulsatilla; and Thuja occident. Diluted sea salts and Lidocaine were added to the initial 10-fold dilution such that their final concentration was each estimated to be approximately 0.1% (minus some of the Lidocaine, which precipitated out as long needle shaped crystals upon successive sucussions). The homeopathic tinctures were purchased from Boiron, Newtown Square, PA. The ethanol and Lidocaine were purchased from Sigma, St. Louis, MO. The solution was non-irritating even when applied directly to pain sensitive HSV and herpes zoster virus (HZV) skin lesions. The solution did not fluoresce when exposed directly to UV-A light but does so in the presence of small quantities of added neutral red dye (Dudley Corporation, Lakewood, New Jersey). The energized water-based solution was stored in a 5-gallon carboy and succussed (repeatedly jolted) vigorously each time before aliquots were taken for both clinical and laboratory testing.

In the clinical studies, neutral red powder was added to tap water in a 12" bowl to an estimated concentration of approximately 0.5 mg/ml. Absorbent surgical towels, measuring 17" x 20" with either a bonded or non-bonded internal polyethylene water impermeable layer, were obtained from Little Rapids Corporation, Green Bay, WI and Henry Schein Inc. NY., respectively. Either a whole towel or several four to six inch squared portions of either whole thickness towel or the detached paper layer from a non-bonded towel were moistened in the neutral red solution. Several sections of neutral red moistened towels were then placed onto the patient, but separated from direct contact with the skin by either an intervening dry towel or the sheet of polyethylene removed from a non-bonded towel. The body areas chosen for placing the moistened varied between patients and included the upper back, abdomen, popliteal fossa, soles of the feet and back of the head. A UV-A light was placed approximately 3" above the moistened paper towels and the area illuminated. The UV lights used included fixtures containing either a single or double 2 foot long, 20 watt bulbs, (F20T12/BLB), single 4 foot long 40 watt 48" bulb (F40T12/BLB) or in most of the home based studies, a single compact 13 watt spiral UV-A light bulb that fits into a regular power socket (Halco Lighting, GA). UV protective goggles were provided and the children were positioned so as to not directly view the UV light.

Once the UV illumination was arranged, a freshly succussed spray bottle containing the energized water-based solution, was used to spray approximately 5 ml onto the still moistened neutral red stained towel. The addition of the energized solution immediately resulted in a bright yellow fluorescence occurring on the neutral red stained towel. Additional 5 ml sprayings of the energized solution onto the illuminated towel was repeated at 15-minute intervals during the one-hour treatments.

In the Medical Clinic phase of the study, a nurse experienced in the care of children and adults with autism clinically assessed each patient prior to each of the two treatments, which she administered on consecutive days. The nurse further examined the patients on several occasions during the ensuring months. Her basic assessments included observations, responses to questions and reading ability. Input was also obtained from parents and occasionally from schoolteachers and social workers. Several parents agreed to a video recoded interview in the company of their treated child. The author, along with an interested Canadian parent of a 4-year-old autistic boy, conducted the interviews.

Having been quite impressed by what he learned during the interviews, the parent was instructed in the procedure and subsequently applied the therapy at home to his son. This being successful, he further volunteered to coordinate home-based studies by other parents of autistic children. The parents were required to provide a baseline assessment of their child that comprised answers to a series of 194 behavioral questions; the majority of which were obtained from a list compiled by Dr. Sidney Baker and published on the Autism Research Institute internet site www.autism.org The range of questions comprehensively addresses the key areas of social interaction, communication skills and aberrant behavioral patterns, which impact on the quality of children with autism.

The parents were instructed to give hour-long therapies for 5 consecutive nights and to re-answer the questionnaire a week after beginning the therapy and at various times thereafter. They were specifically asked to record each item as either showing no significant change or a clearly noticeable improvement, including whether the issue of concern had been essentially corrected by the treatment. The parents were also invited to openly post their comments regarding the therapy to a publicly accessible web site.

All phases of the study were approved and overseen by an Institutional Review Board (IRB). The Food and Drug Administration was also consulted on the issue of allowing parents to act as clinical investigators in using what were classified by the IRB as a Class I medical devices.

Results

The medical clinic based study comprised 13 functionally compromised patients who were receiving various forms of governmental support for their autism. In addition, two infants born to autistic parents were included in the study. The patients received two treatments given one day apart. Clinical improvements were clearly noted immediately after the first and the second treatments, by both the nurse who administered the therapy and the attending family member(s). Further improvements were generally also noted over the ensuing 2-3 weeks.

The following specific changes were attributed to the therapy: Two girls (7 and 9 years of age) and a 22 year old man, who had been involuntarily urinating both during the day and at night, did not do so for an entire month after treatment. The 7 year old girl's grades in school switched from F's to B's. Another child was moved from special education classes into the regular school system.

The two infants born to autistic parents improved neurologically, enabling the 5 month old infant to hold her head up without support and the 6 month old boy to sit up for the first time. Again, motor difficulties gradually returned during the second month of observation. Their treated autistic mothers, as well as the other participants reported much improved sleep, far less forgetfulness and a greater awareness of themselves as individuals.

Reading skills in the older children markedly improved both because of the ability to concentrate and because of the apparent joy in being able to read aloud to others. The younger children were now able to calmly follow television programs appropriate for their age and education. Several began to speak in whole sentences and to actually focus their listening to what was being said by others. Another notable comment provided by a parent is how her child was now able to respond to inquiries without the earlier need for long pauses and hesitations. Yet, another mother noted how wonderful it was that her daughter was now willing and able to help with household chores. Two teenagers began part time employment. All of the patients appeared to be in better physical and emotional control of their body. Overall, if normal behaviors are given a value of 100 and the prior behaviors considered as the zero point background, then every treated patient had achieved somewhere between an estimated 20-40% recovery to the level expected for their age and received education.

The markedly improved behaviors persisted for at least a month with significant gains in learning and social advancement. While there was clearly a gradual reversion back towards the pre-therapy levels of intellectual and social impairments, all of the patients continued to perform above their earlier functioning levels. The patients and/or their parents uniformly attributed the observed and unmistakable benefits to the energized water based therapy.

This conclusion was confirmed when an additional six patients were treated in a similar manner with UV-A illumination of neutral red stained surgical towels but without the addition of the energized water. None of the patients achieved any appreciable benefit. This failure to respond was in marked contrast to the striking improvements that occurred in the autistic children on whom the energized water was used. Moreover, when the energized water was used, not only was there bright UV-A induced fluorescence on the neutral red dye moistened surgical towel, but the recipients also developed patchy areas of deep skin fluorescence both in the area on which the towel had been placed and elsewhere on their body. (Fig. 1 and 2). Distant direct fluorescence was particularly noticeable on the palms and on the soles of the feet. The oral cavity also commonly showed UV-A inducible fluorescence during the procedure and for periods of time thereafter (Figure 3).

The next phase of the study was to adapt the therapy for home use. The first patient so treated was the 4 year old boy diagnosed autistic at age 3, at which time he began intensive Applied Behavioral Analysis (ABA) therapy. In spite of the therapy, he remained essentially non-verbal except for a couple of words that required much prompting and reinforcement. He did not relate to others, nor did he respond to his name or other vocal cues. His medical condition was further complicated by generalized (Grand Mal) seizures that were occurring 3-4 times a day and not controlled with Trileptal (oxcarbazepine) plus Valproic acid, necessitating the use of Diazepam (Valium) sedation. Moreover, the more severe seizures left him paralyzed from the waist down for several hours. He had nine emergency room hospital visits from February through May of that year. Six of the hospital visits were for uncontrolled epileptic seizures, one for post-seizure paralysis and two for vomiting/dehydration. He began 5 days of one-hour nightly therapy in June. He has also received subsequent single treatments at the discretion of his father, at approximately monthly intervals for 6 months. His response to the therapy was quite remarkable. For example, he has not experienced any seizures since beginning ACE therapy and, on medical advice following EEG screening in August 2008, was taken off all seizure medications. Many of his earlier symptoms also disappeared. He would come when called, make sustained eye contact, play happily with his sister and spontaneously use socially appropriate 3-4 word phrases and occasional whole sentences. Even from his initial treatment, he was found to have surpassed his ABA pre-set goals and this therapy was soon discontinued.

The second home based therapy was administered by the mother of a 14 year old child who had been practically non-verbal till she was nearly 6 years of age. She was diagnosed as having an autism spectrum disorder with marked expressive and receptive language impairments and an attention deficit hyperactivity disorder. She was also epileptic with both petit mal (absence) seizures, and complex partial seizures, for which she has been on medication since age 3. She initially responded well to the therapy administered in the medical clinic. To her mother's amazement, her child returned to the clinic's waiting room after her first treatment and immediately attempted to read from a magazine. The next day, she received a second treatment at the medical clinic. The improvements noted by her mother far exceeded any benefit she had seen with many previous attempts at therapy. Also striking was the immediate cessation of absence seizures that were previously occurring several times each day.

Approximately a month after the initial therapy, her mother began noticing increasingly longer periods of social disengagement with others, along with a reduction of her recently acquired alertness and interest in everyday activities. Instead, she was reverting to repetitive stereotypic behaviors (stimming) and to non-purposeful activities. She was also again experiencing repeated absence seizures and occasional more uncontrolled twitching accompanied by an altered state of consciousness.

Within days of the home administration of the protocol in July 2008, her mother once more observed the same marked improvements as seen with the initial therapy. Her daughter again showed willingness to engage in and to demonstrate success with activities that her parents or therapists had been trying to have her learn. For example, she would dress herself at home, including tying her shoe laces; brushing her teeth and preparing her own breakfast. She was able to try on new clothes in a Department Store, using the dressing room without assistance. Her balance and bike riding skills were much improved. More pointedly, she became verbally interactive and progressively acquired a sense of humor and empathy. Her reading and writing skills improved considerably and the frequency of the focal twitching and absence seizures was markedly reduced, especially within the first few weeks following therapy. To quote her mother "She is (now) able to focus, learn, communicate, enjoy and be aware of her own life and others." Upon completion of the questionnaire at the end of the first week of the home based therapy, the mother recorded 56 of the previously recorded 104 impairments to have improved and 22 to have been completely corrected.

Parents of 10 of the 12 additional children who were home treated during July and August, 2008 also reported on remarkable behavioral improvements. These improvements were acknowledged even by those parents who had openly expressed a high level of skepticism prior to beginning the study. The remaining 2 parents also reported many minor improvements in their children, although some bothersome aberrant behaviors were still persisting at the end of the first week after beginning the therapy. Interestingly, one of these parents began to notice marked improvements 2-3 weeks after the therapy. Two other parents received kits but did not follow the basic protocol, allowing the neutral red dye moistened towels to completely dry before applying the energized water. In neither case was there any reported improvement.

All of the parents provided narrative descriptions of the major changes seen in their child. In addition, 9 of the 14 parents, including the parents of the two initially home treated children, completed the questionnaire at the end of the first week following the ACE therapy. The answers were compared with the baseline data provided prior to beginning the study to determine how many of the preexisting issues had improved and those that were considered to be fully corrected by the ACE therapy. These numbers are recorded in Table 1. Some of the parents' specific comments about their children are recorded in Table 2.

The most consistent observation by the participating parents was that their children had rapidly gained a greater sense and awareness of themselves as individuals. This was sometimes preceded by what appeared to be excitable hyperactivity for 1-2 days following the beginning of treatment. As the hyperactivity subsided, the children were noticeably better able to communicate their feelings to their parents and to other people. There was also realization that the treated children were beginning to observe aspects of their surroundings in a manner that was different from before the treatment. Specifically, they appeared to be much more observant and interested in items that attracted little attention in the past. The improved social and other interactions brought considerable joy to the parents and seemingly greatly enhanced the child's quality of life.

Particularly striking were the reports of children now making direct eye contact, smiling and laughing without having to be tickled, singing with and generally enjoying the company of others, using new words and in the proper context, sitting calmly without the need to run around or to self stimulate, watching television shows with far greater interest and attention than ever previously expressed, etc. Four children became verbal within 2-3 weeks of therapy with another younger child beginning to babble for the first time in his life. In addition to improved behavioral and cognitive skills, several parents commented that their child was showing improved motor skills. These included both fine movements as required for drawing and coordinated major muscle activities such as in running and catching a ball.

The parents of the treated children began to observe increasing periods when their child would forego many of the achievements and other skills acquired following the earlier therapy. This was gradually occurring from 2-4 weeks after the treatment and was being effectively addressed by the parents of the two initially home treated children who had a sufficient supply of energized water and neutral red dye for repeated treatments.

To continue with the study and to enroll additional patients, a late summer shipment of presumably energized water-based solution was sent to 5 parents who had previously treated their children using the established protocol as well as to 7 parents who had more recently enrolled their children in the study. Based on what they had either seen previously in their own child, or on what they had read from some of the other parents, there was a high expectation of continued success. Instead of this, the 5 parents that had previously noted considerable benefits from a previous treatment with the earlier shipment of energized water-based solution adamantly reported that they were seeing no changes with the newly arrived material. Similarly, of the 7 parents who were trying the protocol for the first time, only 3 parents make mention of possible minor improvements in their child's performance; none of which was considered to be clinically significant nor lasting beyond a few days after beginning the protocol.

Although not well understood at the time, I have subsequently realized that the energized water-based solution was left in a hot environment throughout the summer months. On numerous occasions, I would remove small aliquots of fluid for research purposes. Each time before doing so, I would vigorously succuss the solution within the carboy. I now realize that the high temperature of summer and the repeated succussion allowed for the dissipation of much of the energy from the water. This loss occurs through the evaporation of the more volatile (less hydrogen-bonded) water molecules. The activity in the previously energized water-based solution had apparently expired and the water was no longer clinically effective. A comparison of the results using the expired product with the earlier results is provided in Table 3.

Discussion

The cellular immune system does not defend against stealth adapted viruses. Nevertheless, both in vitro cultures and the results from animal inoculation studies pointed to an effective recovery mechanism operating against these viruses. As shown elsewhere, the recovery is mediated through the ACE pathway.

Guided by studies showing that activation of the ACE pathway could lead to the suppression of both active and latent HSV infections; a similar approach to activating the ACE pathway has now been shown to be beneficial in patients with autism. The approach involves the direct transmission of energy to the patient's ACE pathway, from UV illuminated energized water plus neutral red dye positioned in close proximity to the patient's skin. The procedure proved to be quite robust with positive findings reported from both a medical clinic and from several parents independently performing the procedure on their own children.

The use of energized water-based solution was an extension of an early preliminary study in which a moistened paper towel was used to collect dried perspiration from the skin of an autistic child. The towel brightly fluoresced under UV illumination, especially with the addition of a few drops of neutral red dye. When placed onto the skin, but avoiding direct contact, the illuminated paper towel clearly evoked direct fluorescence within the underlying skin of the child. In a similar manner, both saliva and urine from an autistic child were shown to potentially provide usable sources of ACE pigments in the dye-light ACE activation protocol.

The suggestion that the clinic nurse and parents falsely believed that they were witnessing marked improvements in the treated family member is highly unlikely. Most of the early participant volunteers in the home based studies, and especially their spouses, were initially highly skeptical that ACE therapy held any real promise; having seen only marginal changes with many prior biomedical interventions. Nevertheless, they uniformly became convinced of the efficacy of the procedure, many even after the first of the five daily treatments.

The most consistent description of the beneficial changes seen with the successful ACE-enhancing protocol is that treated autistic patients appeared to rapidly gain a heightened level of consciousness, becoming more aware of themselves as individuals. An intriguing possibility, therefore, is that the ACE pathway is more specifically involved in sustaining the capacity for high levels of consciousness than is cellular energy derived from the metabolism of foods.

ACE therapy also seemingly allowed for the improved expression of latent capacities apparently already potentially present in the children, such as speaking, reading and social interactions. Both fine and major motor skills, that are not infrequently impaired in children with autism (14-16), were also improved in the patients receiving effective ACE therapy.

Of additional interest is the complete elimination of grand mal seizure activity in a treated child with epilepsy. There was also a marked reduction in the occurrence of petit mal and complex partial seizures in another epileptic child. Epilepsy is relatively common among autistic children and may be an additional contributing factor to ongoing brain damage (17-19). Epilepsy can have a direct effect on consciousness and it is possible that both uncontrolled neuronal activity and impaired consciousness are the results from an inadequacy of the ACE pathway. Certainly it is important to determine whether enhancement of the ACE pathway can also suppress epilepsy in patients beyond those with autism.

While it is reasonably proposed that the primary cause of the energy insufficiency in autistic patients is a congenitally acquired stealth adapted virus illness (1,2), the rationale for utilizing ACE based therapies in autistic patients is not dependent upon this proposition. In performing the present study, the author has encountered many mothers of autistic children who freely admit to having a prior or even a present cognitive and/or emotional disorder, not uncommonly labeled as chronic fatigue syndrome (CFS). Based on earlier studies, it can be predicted that many if not the vast majority of these individuals are infected with a stealth adapted virus that has been transmitted to their child. Even presumptive evidence for virus transmission to a child could be an argument for proactive ACE therapy prior to the socializing challenges faced by the child during the first 2-3 years of life. In this regard, the adequacy level of the ACE pathway can potentially be routinely assessed based on fluorescence reactions observed with UV-A illumination of the skin and body fluids, with and without the addition of small quantities of neutral red dye. The preferred "fully charged" situation shows no fluorescence either with or without added dye. "Partially charged" ACE pigments are seemingly able to fluoresce upon direct UV illumination, while "uncharged" ACE particles require the "triggering" presence of neutral red dye. A less than a "fully charged" ACE pathway may also provide justification for delaying many of the routine vaccinations, some of which can seemingly lead to brain damage from a persisting stealth adapted virus infection.

The failure of the expired shipment of water-based solution to provide any discernable benefit was clearly disappointing. Nevertheless, the loss of clinical activity in the expired product in all 12 later-treated children helped validate the extremely promising earlier results from initially highly skeptical parents. This was especially so, since as information concerning the clinical trial became available to other parents, an expectation of success was generated among those who received the expired shipment of energized water. Similarly, the inability of neutral red dye in regular water to provide any benefit when used by itself served as an additional control.

Since the time of this study, much simpler formulations of ACE Water have been produced and kindly assayed with the assistance of a mother of an autistic teenager. In addition to the direct reporting of mother and daughter, the appearance of strong intra-oral fluorescence after therapy sessions have provided useful markers of efficacy. The use of paper towels has been replaced by using neutral red solutions contained within a sealed plastic bag. The soles of the feet have also been chosen as the preferred location for applying the dye-light therapy. The therapy works well even with an intervening black plastic barrier is placed between the solution and the skin. Similarly, the whole paper towels used in the earlier patient studies did not allow for direct light transmission between the illuminated energized water-based solution and the patient's ACE pathway. In a further development, consumable liquids have also been energized using a similar approach as applied to the patients' skin. These developments are opening up new pathways of therapy applicable well beyond the treatment of autism.

Acknowledgement: I am pleased to acknowledge the cooperation of the parents who participated in the study. I am especially thankful to the parent who helped coordinate the home-based clinical study and who provided useful training videos for other parents. I am also particularly thankful to the mother of the teenager who insightfully reported on her child's continuing improvement in her mindfulness.

References

1. Martin WJ (1994) Stealth viruses as neuropathogens. *CAP Today* 8: 67-70.

2. Martin WJ (1995) Stealth virus isolated from an autistic child. *J Autism Dev Disord* 25:223-4.

3. Martin, WJ, Glass RT (1995) Acute encephalopathy induced in cats with a stealth virus isolated from a patient with chronic fatigue syndrome. *Pathobiology* 63:115-8.

4. Martin, WJ (1996) Severe stealth virus encephalopathy following chronic-fatigue-syndrome-like illness: clinical and histopathological features. *Pathobiology* 64:1-8.

5. Martin WJ, Ahmed KN, Zeng LC, Olsen J-C, Seward JG Seehrai JS (1995) African green monkey origin of the atypical cytopathic 'stealth virus' isolated from a patient with chronic fatigue syndrome. *Clin Diag Virol* 4: 93-101.

6. Martin, WJ (1999) Stealth adaptation of an African green monkey simian cytomegalovirus. *Exp Mol Pathol* **66**: 3–7.

7. Martin, WJ (1996) Simian cytomegalovirus-related stealth virus isolated from the cerebrospinal fluid of a patient with bipolar psychosis and acute encephalopathy. *Pathobiology* 64: 64-6.

8. Martin, WJ (2003) Complex intracellular inclusions in the brain of a child with a stealth virus encephalopathy. *Exp Mol Pathol* 74: 197-209.

9. Martin, WJ (2003) Stealth virus culture pigments: a potential source of cellular energy. *Exp Mol Pathol* 74: 210-23.

10. Martin, WJ (2005) Etheric biology. *Exp Mol Pathol* 78: 221-7.

11. Martin WJ, Stoneburner J (2005) Symptomatic relief of herpetic skin lesions utilizing an energy-based approach to healing. *Exp Mol Pathol* 78:131-4.

12. Martin WJ (2005) Alternative cellular energy pigments mistaken for parasitic skin infestations. *Exp Mol Pathol* 78:212-4.

13. Martin WJ (2005) Alternative cellular energy pigments from bacteria of stealth virus infected individuals. *Exp Mol Pathol* 78: 215-7.

14. Rinehart NJ, Tonge BJ, Iansek R, McGinley J, Brereton AV, Enticott PG, Bradshaw JL (2006) Gait function in newly diagnosed children with autism: Cerebellar and basal ganglia related motor disorder. *Dev Med Child Neurol* 48: 819-24.

15. Dziuk MA, Gidley Larson JC, Apostu A, Mahone EM, Denckla MB, Mostofsky SH (2007) Dyspraxia in autism: association with motor, social, and communicative deficits. *Dev Med Child Neurol* 49: 734-9.

16. Goldman S, Wang C, Salgado MW, Greene PE, Kim M, Rapin I (2009) Motor stereotypies in children with autism and other developmental disorders. *Dev Med Child Neurol* 51: 30-8.

17. Tuchman R, Moshé SL, Rapin I (2009) Convulsing toward the pathophysiology of autism. Brain Dev 31: 95-103.

18. Deonna T, Roulet E (2006) Autistic spectrum disorder: evaluating a possible contributing or causal role of epilepsy. *Epilepsia* 47 Suppl 2: 79-82.

19. Peake D, Notghi LM, Philip S. (2006) Management of epilepsy in children with autism. *Cur Paediatr* 16: 489-494.

20. Dawson G, Murias M. *Autism* (2009) *Encyclopedia of Neuroscience* Editor *Larry R. Squire Academic Press 779-784.*

21. Rapin I, Tuchman RF. Autism: definition, neurobiology, screening, diagnosis (2008). *Pediatr Clin North Am* 55: 1129-46.

22. Geschwind DH (2008) Autism: Many genes, common pathways? *Cell* 135: 391-395.

23. Walsh CA, Morrow EM, Rubenstein JLR (2008) Autism and brain development. *Cell* 135: 396-400.

24. Raymond F. Palmer RF, Blanchard S, Wood R (2009) *Proximity to point sources of environmental mercury release as a predictor of autism prevalence. Health & Place* 15: 18-2.

25. Waldman M, Nicholson S, Adilov N, Williams J (2008) Autism prevalence and precipitation rates in California, Oregon, and Washington counties. *Arch Pediatr Adolesc Med* 162: 1026-34.

26. Rubin DH, Ruley HE (2006) Cellular genetics of host susceptibility and resistance

to virus infection. *Crit Rev Eukaryot Gene Expr* 16: 155-70.

27. Arkwright PD, Abinun M (2008) Recently identified factors predisposing children to infectious diseases. *Curr Opin Infect Dis* 21: 217-22.

28. Stancek D, Kosecká G,. Oltman M, Keleová A, Jahnová E (1995) Links between prolonged exposure to xenobiotics, increased incidence of hepatopathies, immunological disturbances and exacerbation of latent epstein-barr virus infections. <u>International Journal of</u> *Immunopharmacology* 17: 321-8.

29. Head JL, Lawrence BP (2009) The aryl hydrocarbon receptor is a modulator of anti-viral immunity. *Biochem Pharmacol* 77: 642-53.

30. Schaeffer C, Robert EI, van Breugel PC, Leupin O, Hantz O, Strubin M (2008) Hepatitis B virus X protein affects S phase progression leading to chromosome segregation defects by binding to damaged DNA binding protein 1. *Hepatology* 48: 1467-76.

Table 1

Improvements in Autistic Children One Week After ACE Therapy

Pt. No. Sex and Age	Pre-therapy Issues Identified as Problems	No. Issues that Therapy Improved	No. of Issues that Therapy Completely Corrected
1 Male 2	192	123	84
2 Male 4	80	58	20
3 Male 4	135	13	6
4 Male 4	120	60	8
5 Male 4	82	61	1
6 Male 5	124	111	32
7 Male 6	104	76	23
8 Male 8	138	124	18
9 Female 14	104	56	22

Table 2

**Examples of Recorded Benefits Seen in Autistic
Children One Week After ACE Therapy**

Pt. No.	Prominent Examples of Improvements
1	More purposeful behavior, far less aloof; much better eye contact
2	Now engages in purposeful play, responds to pain, sleeps in own bed
3	Now makes eye contact, speech much improved, more self-awareness
4	Now makes good eye contact; responds to name; less fastidious
5	Grand mal seizures stopped; improved speech and feelings for others
6	Now has feelings for others; not distressed by sounds, better fine work
7	Responsive to his name, speech no longer repetitive, names items
8	Hand flapping and tantrums ceased. More aware of himself as a person
9	Focal seizures and fingertip squeezing have ceased; now enjoying life

Table 3

Number of Patients Showing a Marked Clinical Improvement Within One Week of Therapy

Product Used on UV Illuminated Towels

Energized Water-Based Solution with Neutral Red Dye	27/31*
Neutral Red Dye Used Alone	0/6
Expired Water-Based Solution with Neutral Red Dye	0/12**

* Comprising 15 patients treated with energized water at medical clinic, including 2 infants; and 16 home treated patients. A delayed beneficial response was apparent in one of the patients who failed to show a response within the first week after treatment. In 2 other patients the probable cause of the failure to respond was that the neutral red dye containing water was mistakenly allowed to dry before the energized water was applied.

** Includes 7 patients who were treated for the first time and 5 patients who had previously reported marked benefits from the original energized water.

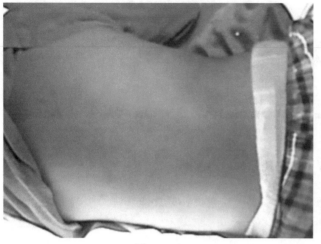

Figure 1

Figure 2

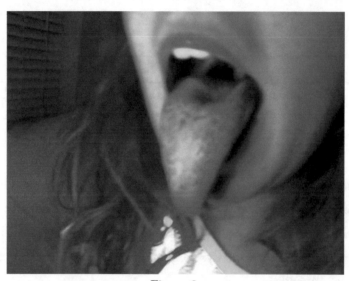

Figure 3

Fig. 1. An example of UV fluorescence seen in the skin of the back of a 4 year old child following therapy applied to this area. Note, that there was no direct contact of the UV illuminated soaked paper towels and the underlying skin because of an intervening covering of the skin with Saran™ wrap. The induced capacity of the skin to show direct fluorescence persists for several hours following the therapy.

Fig. 2. The same patient as shown in Figure 1, also showed striking UV induced fluorescence of the skin at the side of a foot, easily discernable at the end of the therapy.

Fig. 3. Tongue fluorescence seen using UV illumination during and following the procedure performed on a 14-year-old girl with autism. The directly treated area was the soles of the feet. In this example a Ziploc bag containing the activated solution plus neutral red dye was used instead of paper towels laid onto Saran™ wrap covered skin.

CHAPTER 5

Alternative Cellular Energy Based Therapy of Childhood Diarrhea

Rafael Ruiz Izaguirre,[1] Miguel Reyes Guzman,[1] Rafael Chacon Fuentes,[1] Carlos E. Mena,[1] Emillio Penate[1] and W. John Martin[2] Benjamin Bloom National Children's Hospital, El Salvador[1] and Institute of Progressive Medicine, South Pasadena CA USA 91030[2]

RUNNING TITLE: ACE Therapy of Diarrhea
Corresponding author

W. John Martin, MD, Ph.D.
1634 Spruce Street
South Pasadena CA 91030
Telephone 626-616-2868
e-mail wjohnmartin@hotmail.com

Additional Clinical Information
Rafael Ruiz-Izaguirre, MD, FAAP,
e-mail ruizmd65@yahoo.com

Key Words: Enercel™, Diarrhea, Alternative Cellular Energy, Rotavirus, Homeopathy, ACE Water™, Tropical medicine, Yale Observation Scale

Abstract

Acute diarrhea is a major killer of children under 5 years of age in developing countries. Enercel is a licensed energized water product that can provide a source of alternative cellular energy (ACE). Two intramuscular injections of Enercel were administered as an adjunct to routine care to a randomized group of children (n=58) with acute diarrhea. Clinical progress was assessed during a 24 hour hospitalization period and again at 48 hours after hospital discharge. Compared to an initially well matched randomized control group (n=53), at 48 hours post discharge there were significantly fewer Enercel treated children with persisting increased peristalsis (p<0.001), dehydration (p=0.0224), fever (p=0.0126) or continued multiple bowel movements (p=0.0035). Benefits occurred in both rotavirus antigen positive and negative children. Enercel represents a class of ACE Water™ products that can provide an effective adjunct to the routine care of young children in developing countries with acute diarrhea.

INTRODUCTION

Acute infectious diarrhea is a major cause of morbidity and mortality in the developing world; especially among children less than 5 years of age (1). While there are multiple causes, including both bacterial and viral infections, the diarrhea essentially reflects an inability of the small and/or large intestinal epithelial cells to absorb the excessive fluid secreted into the intestinal lumen (2). Restoring normal cellular function is an energy-requiring process. Therefore, patients with diarrhea could potentially benefit from receiving an alternative (non-mitochondria) source of cellular energy (3-7). Alternative cellular energy (ACE) generating activity is contained in the energized water product Enercel (8). This paper documents that the simple addition of two intramuscular injections of Enercel as an adjunct to routine patient care of children with diarrhea significantly improves clinical outcome.

MATERIALS AND METHODS

Enercel is a water product that has acquired the capacity to convert physical energies into a biological energy that can be utilized by cells to help maintain normal function (4,7). It is produced by successively diluting a mixed solution of various minerals and diluted products from the Homeopathic Pharmacopoeia into water containing 4.0% v/v alcohol (see footnote*). At least six ten-fold dilutions of the already diluted starting material are made into pharmaceutical grade water for injection containing alcohol. The total dissolved solids of the final product is <0.005%, meaning that the product is essentially free of the starting homeopathic material. The product used in the present study was bottled using Good Manufacturing Practice (GMP) procedures and quality controlled by Vijosa Laboratory, at San Salvador, El Salvador.

Children who presented with acute diarrhea to the Emergency Unit of the Benjamin Bloom National Children's Hospital in El Salvador during the hours of 7 am to 4 pm, Monday to Friday, from August 20, 2003 to April 1, 2004, were admitted for 24 hours and instructed to return for a follow-up evaluation 48 hours after discharge. Each patient was assigned a patient number. Children eligible for inclusion in the study with an even patient number were assigned to receive Enercel, while those with an odd patient number provided the controls. Enercel was administered intramuscularly at a dose of 3.0 ml given at the time of admission and 12 hours later. Except for this difference, all patients were provided the same routine diagnostic and clinical care appropriate to their illness. All clinical evaluations and follow ups were done by independent clinical staff on duty in the Emergency Unit during the time of the study and who accepted to follow the protocol guidelines. A member of the research team was present at all times to maintain quality control.

The eligibility requirements were as follows:

- An age from 6 months to 5 years at the time of admission

- Two or more watery bowel movements during the preceding 24 hours

- Clinical evidence of mild or moderate dehydration, with or without electrolyte imbalance

- Written consent from parents or guardian for the child to participate in the study.

The exclusion criteria were:

- Unwillingness to participate in study

- Duration of illness greater than 24 hours

- Severe dehydration, continuous peristalsis or ileus

- One or more blood-stained bowel movements

- Systemic sepsis, urinary tract infection or pathogenic amoeba in feces

- Chronic underlying illness including malnutrition.

- Any medication given during the previous 24 hours before admission to the study that may alter the natural evolution of the illness, including antibiotics, anti-parasitic, antipyretics, antispasmodics and laxatives.

Observed and recorded parameters: In addition to identification number; demographic data and date and time of admission were recorded. Medications taken during the 24 hours of admission time that could possibly alter the natural course of the illness were subsequently noted; these included antipyretic and antibiotics that the hospital house staff considered necessary. The overall severity of the illness was assessed at admission using the Yale Observation Scale, YOS (8), and by separately gauging the following parameters: Refusal of oral intake; recent vomiting; abdominal cramps and tenderness; peristalsis; dehydration; fever; and number of bowel movements over the preceding period of time. A distinction was made between mild and moderate dehydration, as evidenced by the extent of dryness of mucus membranes, loss of skin turgidity and fontanel and/or eye socket depression. No child with an estimated >10% total body dehydration was included in the study. Peristalsis was categorized as abnormal if there were >5 peristaltic waves over a 15 second interval, with a further distinction of mild (5-15 waves) and moderate (>15 waves) peristalsis waves over the 15 second time period. Children with continuous peristalsis or ileus were excluded from the study. A temperature within the range of 37.5°C to 38.5°C was considered a mild fever while >38.5°C but <40.5°C was considered as moderate fever. Children with a high fever indicative of sepsis were excluded from the study. The clinical history was used to assess the illness prior to admission while direct examinations were made by the attending hospital staff clinicians on admission and at 6, 12,

18 and 24 hours post admission. Fecal output was measured between these examination times. A final clinical assessment was made on the return visit at 48 hours after hospital discharge. Thirteen of 124 patients failed to make the scheduled appointment and were examined 72 hours after their hospital discharge.

All participants had one or more complete blood counts, urine analysis and stool examination. Included in the latter were Wright and acid fast staining for parasites, rotavirus antigen detection (using VIROTECT-ROTA from Omega Diagnostics), and salmonella, shigella and E. Coli 0157 detection using a VITEK system.

The data were entered into the EPI-INFO computer program for subsequent analysis.

RESULTS.

Of 177 children initially identified as being potentially eligible for the study, 35 were excluded because of a positive urine culture suggestive of a coexisting urinary tract infection. Twenty-five suspected urinary infections were seen in female patients and 10 in males. Another 18 patients were ruled ineligible primarily because the parent decided not to proceed after having signed the Informed Consent form. These exclusions left 124 eligible patients of whom 59 were randomly assigned to the Control Group and 65 assigned to the Enercel Treated Group. One hundred and eleven (111) patients returned at 48 hours after hospital discharge for their final clinical assessment. An additional 13 patients (7 Enercel treated and 6 controls) did not return at 48 hours but were examined 24 hours later.

Retrospective analysis of the illness prior to admission and clinical and laboratory examination on admission of the 124 patients revealed no significant differences between the two groups. Table 1 summarizes demographic and preadmission data on the entire group of 124 patients. Table 2 indicates the clinical findings on admission of the 111 patients who returned for their final evaluation 48 hours after discharge. While the Enercel treated group showed a tendency towards being more symptomatic with a somewhat higher percentage with mild leukocytosis, the two groups were well matched at the onset of the study. Antipyretics were clinically indicated and, therefore, administered during hospitalization to 53% of the children in the Control Group and 47% of the children in the Enercel Treated Group. The corresponding percentages of Control and Enercel treated children administered antibiotics were 8.5 and 4.7, respectively.

Significant overall clinical differences were not seen between the Enercel treated and control groups of children at 6, 12 and 18 hours post admission. By 24 hours, however, a quantitative measure reflecting the combined severity of multiple symptoms within the Enercel treated children was statistically significantly less than that of the control children. Analyses of individual symptoms and signs at this time point confirmed the trend towards improvement but not at statistically significant levels. For example, 28 of 65 Enercel treated children had passed less than 100 cc of feces during the preceding 6 hours compared to only 18 of the 59 control children (p = 0.19).

Major improvements in multiple clinical parameters were, however, readily detected when the two groups of children were reassessed 48 hours after discharge. In addition to an overall clinical evaluation, each child at this time point was evaluated for residual symptom of dehydration, peristalsis, abdominal tenderness and fever. In addition, the number of bowel movements (defecations) and episodes of vomiting since hospital discharge were noted. Nine of the control group of children had a YOS of 8 or greater compared to only 2 of the Enercel treated group (Chi Square Fisher exact p value 0.0519). Statistically significant reductions were noted in the frequency of peristaltic waves (p = 0.0001), persisting moderate dehydration (p = 0.0224), fever (p = 0.0126) and multiple bowel movements since leaving hospital (p = 0.0035). These data are provided in Tables 3 and 4.

No significant differences were seen between the groups of remaining children seen at 72 rather than at 48 hours post hospital discharge. All were afebrile with mild dehydration. Three of the Enercel treated and two of the control children had moderate peristalsis while the remaining children had mild peristalsis.

Among the 111 children evaluated 48 hours post hospital discharge, 12 control patients and 13 Enercel treated patients had a positive fecal rotavirus antigen test on admission (Table 2). Although the numbers are small, there was no apparent distinction between the rotavirus antigen positive and negative children within each of the two study groups. The data are, therefore, consistent with a benefit of Enercel therapy for childhood diarrhea regardless of the rotavirus status (9).

No detectable adverse side effects of Enercel therapy were found in any of the children in the treated group.

DISCUSSION.

This paper establishes a therapeutic benefit from administering an energized water product, termed Enercel, to children with acute infectious diarrhea.

While the mode of action of Enercel requires further investigation, it does provide a non-nutrition based source of cellular energy through the alternative cellular energy (ACE) pathway (3-7). Activation of this pathway has been shown to expedite healing of skin lesions due to conventional herpes and papillomaviruses (5) and to play a major role in the recovery from stealth-adapted virus infections that are not effectively recognized by the cellular immune system (3,4).

The most notable benefits were seen on the return visit of the children 48 hours after they had been discharged from the hospital. Compared to the control group, some of whom had actually regressed since discharge; fewer children treated with Enercel had persisting increased peristalsis, moderate dehydration or an elevated temperature. They had also experienced less frequent bowel movements since discharge from the hospital. The data underscore the lack of complete resolution of symptoms by those receiving standardized care and to a lesser extent even in those receiving Enercel as an adjunct to standard care. Persisting symptoms are not uncommon with episodes of childhood diarrhea and can contribute to the faltering growth and malnutrition seen in many of these children (10,11).

Enercel and other cellular energy boosting products differ from typical pharmaceutical agents in that they are not directed against a particular biochemical pathway or disease process. Rather, they are intended to assist in the body's own recovery process (6). Enercel can also be distinguished from many products within the field of alternative and complementary medicine in being proven to have efficacy in a well controlled clinical trial. Non-energized water was not used in the controls because, unlike Enercel, intramuscular injection of non-energized water with alcohol is quite painful. In more recent observations, children with both moderate and severe diarrhea are showing marked clinical improvement when the initial dose of Enercel is slowly administered intravenously and subsequent 12 hour doses are given sublingually. This revised protocol has also been found effective in alleviating the symptoms of acute respiratory illnesses in children (unpublished data).

Enercel should not be classified as a homeopathic preparation. While it is routinely succussed before administration and succussion is used at each step of the preparative diluting process, it is not intended to act according to the homeopathic "Laws of Similars" (12). Thus, even with high dosages (50 – 150 ml) administered intravenously, it causes no diarrhea, or indeed any effect not attributable to the low percentage of alcohol. It is interesting, however, that individualized homeopathic preparations (13), a plant extract (14) and zinc (15) have been reported as having modest, yet significant, therapeutic benefits in children with diarrhea. Possibly these products are also acting via the ACE pathway.

A surprisingly large number (19.8% total of whom 70% were girls) of children presenting with diarrhea had a positive urine culture. While fecal contamination of urine samples cannot always be prevented, there may be some association between urinary and gastrointestinal infections. Although they were excluded from the present study, prior nearly uniformly favorable experience with Enercel suggests that these children with positive urinary cultures would have also benefited from the therapy. Indeed, products such as Enercel may have a role in the therapy of many diseases of both infectious and non-infectious origin. As with the present study, progress is to be expected by following strict guidelines for controlled clinical trials. This concept is embodied in the growing field of Progressive Medicine (6).

References

1. O'Ryan M, Prado V, Pickering LK. (2005) A millennium update on pediatric diarrheal illness in the developing world. Semin Pediatr Infect Dis 16: 125-136.

2. Thapar N, Sanderson I R. (2004) Diarrhoea in children: an interface between developing and developed countries. Lancet 363: 641-653.

3. Martin WJ. (2003) Complex intracellular inclusions in the brain of a child with a stealth virus encephalopathy. Exp Mol Pathol 74: 179-209.

4. Martin W J. (2003) Stealth virus culture pigments: A potential source of cellular energy. Exp Mol Pathol 74: 210-223.

5. Martin WJ, Stoneburner J. (2005) Symptomatic relief of herpetic skin lesions utilizing an energy based approach to healing. Exp Mol Pathol 78: 131-4.

6. Martin WJ. (2005) Progressive Medicine. Exp Mol Pathol 78: 218-220.

7. Martin WJ. (2005) Etheric Biology. Exp Mol Pathol 78: 221-228.

8. Bonadio WA. (1998) The history and physical assessments of the febrile infant. Pediatr Clin North Am 45: 65-77.

9. Katyal R, Rana SV, Singh K. (2000) Rotavirus infections. Acta Virol 44: 283-8.

10. Checkley W, Epstein LD, Gilman RH, Cabrera L, Black RE. (2003) Effects of acute diarrhea on linear growth in Peruvian children. Amer J Epidemiol 157: 166-175.

11. Pelletier DL, Frongillo Jr EA, Schroeder DG, Habicht JP. (1995) The effects of malnutrition on child mortality in developing countries Bull World Health Organ 73: 443-8.

12. Hahnemann S. (1842) "Organon" 6th Edition (W. B. O'Reilly, Ed.) Bird Cage Press, Palo Alto.

13. Jacobs J, Jonas WB, Jimenez-Perez M, Crothers D. (2003) Homeopathy for childhood diarrhea: combined results and metaanalysis from three randomized, controlled clinical trials. Pediatr Infect Dis J 22: 229-34.

14. Subbotina MD, Timchenko VN., Vorobyov MM, Konunova YS, Aleksandrovih YS, Shushunov S. (2003) Effect of oral administration of tormentil root extract (Potentilla tormentilla) on rotavirus diarrhea in children: a randomized, double blind, controlled trial. Pediatr Infect Dis J 22: 706-11.

15. Bhatnagar S, Bahl R, Sharma P K, Kumar GT, Saxena SK, Bhan MK. (2004) Zinc with oral rehydration therapy reduces stool output and duration of diarrhea in hospitalized children: a randomized controlled trial. J Pediatr Gastroenterol Nutr 38: 34-40.

Footnote to text on page 3:

Enercel is a licensed therapeutic product in El Salvador. The registered components used in its manufacture include: Cactus grandiflora; Aloe socotrina; Abies nigra; Amica; Lachesis; and Lycopodium.

Acknowledgement: Funding for the clinical aspects of the study and for the supply of Enercel was received from World Health Advanced Technologies, Inc. Sarasota, FL.

Table 1. Retrospective Clinical Analysis of the Children Prior to Hospital Admission*

Parameter	Control Group	Enercel Treated Group
Total number of patients	59	65
Percentage of male patients	76	74
Median age (months)	13	12
Duration of illness prior to admission (median hours)	13	15
> 3 bowel movements (%)	78	83
Fever (Temp. > 38.5°C)	76	83
Frequent vomiting (%)	51	48
Refusal of oral feeding (%)	53	55
Frequent abdominal cramps (%)	7	5

* Information obtained on admission and recorded in patients' medical charts

Table 2. Clinical Data on Admission of Patients Subsequently Evaluated 48 Hours After Hospital Discharge*

Clinical Parameter	Control Group (n=53)	Enercel Treated Group (n=58)
YOS (points)		
6	27	27
8	9	14
10	8	8
12	4	1
14	1	5
16	3	2
18	1	1
Refusal of oral feeding	49	47
Temperature > 38.5°C	12	17
Dehydration**		
Mild	34	34
Moderate	19	24
Peristalsis (15 seconds)		
<5	23	19
5 - 15	26	33
>15	4	6
Leukocyte count / cu mm		
<10,000	22	13
10,000 – 15,000	20	30
>15.000	11	14
Rotavirus antigen positive	12	13

* Analysis restricted to the 111 patients who were evaluated 48 hours after being discharged from the hospital. Thirteen patients included in Table 1 did not return as scheduled but were examined 72 hours following hospital discharge.

** Note patients presenting with severe dehydration on admission (estimated at > 10% loss of total body fluid) were excluded from the present study

*** Severe indicates continuous or >15 peristalsis waves over a 15 second period.

Table 3. Comparison of Three Clinical Parameters between the Control and Enercel Treated Groups 48 Hours after Hospital Discharge.

Treatment Group	Peristalsis*			Dehydration**		Fever***	
	<5	5-15	>15	Mild	Mod	<38.5°C	>38.5°C
Control	28	22	3	48	5	9	1
Enercel	51	6	1	58	0	2	0
Chi Square Analysis	p <0.0001			p = 0.0224		p = 0.0126	

*Peristaltic waves recorded over a 15 second interval.

** Mild and moderate dehydration reflect a clinically estimated reduction in total body fluid of up to 5% and from 5-10% reduction in total body fluid, respectively. None of the children had regained normal hydration status.

***The remaining children were afebrile (<38.5°C).

Mild abdominal tenderness was recorded in 10 of the controls and in 8 of the Enercel treated group (not significantly different). To facilitate using the robust 2 x 2 Chi Square analytic method, the following comparisons were made between the Control and the Enercel treated groups. For peristalsis, the number of patients with 5 – 15 and > 15 peristaltic waves were combined and compared with the number of children with < 5 peristaltic waves. The number of children in each group with mild dehydration was compared with those with signs of persisting moderate dehydration. Children with mild or moderate fever were compared to afebrile children within the same treatment group. The p values were obtained using Fisher's Exact test.

Table 4. Comparison of the Incidence of Persisting Diarrhea between the Control and Enercel Treated Groups 48 Hours after Hospital Discharge

Treatment Group	Number of Bowel Movements					
	0	1	2	3	4	5
Control	13	23	9	6	0	2
Enercel	33	20	3	1	0	1
Chi Square Analysis	p = 0.0035					

An episode of vomiting was reported in 15 of the controls and in 11 of the Enercel treated group (not significantly different). Combining the number of children with 2 or more bowel movements allowed for a Chi square analysis of this symptom.

Alternative Cellular Energy Based Therapy of Childhood Diarrhea

Rafael R. Izaguirre 1 , Miguel R. Guzman 1 , Rafael C. Fuentes 1 , Carlos E. Mena 1 , Emillio Penate 1 , W. John Martin 2
1 Benjamin Bloom National Children's Hospital, San Salvador, El Salvador, 2 Center for Complex Infectious Diseases, Rosemead, CA, United States

Abstract

Enercel is a licensed water product that is able to convert physical energies into a biological energy that can be utilized by cells to help maintain normal function. It is generated by successively diluting a mixed solution of various minerals and diluted products from the Homeopathic Pharmacopoeia into water containing 4.0% v/v alcohol. *Enercel* is bottled using Good Manufacturing Practice (GMP) procedures and quality controlled by Vijosa Laboratory, at San Salvador, El Salvador. In the present study, *Enercel* was evaluated as an adjunct to the routine care of children under 5 years of age presenting with acute diarrhea. An intramuscular injection of 3 ml of *Enercel* was provided to a randomized group of children (n = 58) upon hospital admission and again 12 hours later. The children were discharged at 24 hours and reexamined 48 hours later. Compared to the initially well matched randomized control group (n = 53), at 48 hours post hospital discharge the *Enercel* treated group had fewer children with persisting increased peristalsis (p<0.0001), dehydration (p=0.0224), fever (p=0.0126) and continued multiple bowel movements (p=0.0035). Benefit occurred in both rotavirus antigen positive and rotavirus negative children with acute diarrhea. *Enercel* represents a class of broadly acting non-toxic therapies that can seemingly enhance the body's capacity to regain normal cellular function through a recently defined alternative cellular energy (ACE) pathway. *Enercel* and related products have potential application in the prevention and therapy of many of the major illnesses in both developing and developed countries.

Introduction

Acute infectious diarrhea is a major cause of morbidity and mortality in the developing world; especially among children less than 5 years of age. While there are multiple causes, including both bacterial and viral infections, the diarrhea essentially reflects an inability of the intestinal epithelial cells to absorb the excessive fluid secreted into the intestinal lumen. Restoring normal cellular function is an energy-requiring process. Therefore, patients with diarrhea could potentially benefit from receiving an alternative (non-mitochondria) source of cellular energy. Alternative cellular energy (ACE) generating activity is contained in the energized water product *Enercel*. This presentation records that the addition of *Enercel* as an adjunct to routine patient care of children with diarrhea significantly improves clinical outcome.

Materials and Methods

Children presenting with acute diarrhea to the Emergency Unit of the Benjamin Bloom National Children's Hospital in El Salvador were admitted for 24 hours and instructed to return for a follow-up evaluation 48 hours after discharge. Children eligible for inclusion in the study were randomized to either receive or not receive Enercel, intramuscularly at a dose of 3.0 ml given at the time of admission and 12 hours later. Except for this difference, all patients were provided the same routine diagnostic and clinical care appropriate to their illness. Clinical evaluations were done by independent clinical staff.

Patients

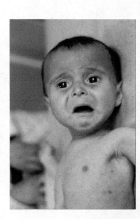

Summary of Randomized Stud

177 children initially enrolled
  35 excluded for urinary tract infection
  18 parents decided not to proceed

124 patients evaluated while in hospital
  65 received Enercel plus standard care
  59 received only standard care
    Overall improvement seen at 24 hrs.

111 patients evaluated 48 hr post discharge
  58 received Enercel plus standard care
  53 received only standard care

Marked improvement in multiple clinical parameters observed in the Enercel treated children

Results

Retrospective analysis of the illness prior to admission and clinical and laboratory examination on admission of the 124 patients revealed no significant differences between the two groups

Nor were significant overall clinical differences seen between the *Enercel* treated and control groups of children at 6, 12 and 18 hours post admission.

By 24 hours, however, a quantitative measure reflecting the combined severity of multiple symptoms within the *Enercel* treated children was statistically significantly less than that of the control children. Analyses of individual symptoms and signs at this time point confirmed the trend towards improvement but not at statistically significant levels. For example, 28 of 65 *Enercel* treated children had passed less than 100 cc of feces during the preceding 6 hours compared to only 18 of the 59 control children (p = 0.19).

Major improvements in multiple clinical parameters were, however, readily detected when the two groups of children were reassessed 48 hours after discharge. In addition to an overall clinical evaluation, each child at this time point was evaluated for residual symptom of dehydration, peristalsis, abdominal tenderness and fever. In addition, the number of bowel movements and episodes of vomiting since hospital discharge were noted. Nine of the control group of children had a YOS of 8 or greater compared to only 2 of the *Enercel* treated group (Chi Square Fisher exact p value 0.0519). Statistically significant reductions were noted in the frequency of peristaltic waves (p = 0.0001), persisting moderate dehydration (p = 0.0224), fever (p = 0.0126) and multiple bowel movements since leaving hospital (p = 0.0035 movements since

Enercel administration benefits children with acute infectious diarrhea. It provides a non-nutrition based source of cellular energy through an alternative cellular energy (ACE) pathway. Activation of this pathway has been shown to expedite healing of skin lesions due to conventional herpes and papillomaviruses5 and may play a major role in the recovery from stealth-adapted virus infections that are not effectively recognized by the cellular immune system

The most notable benefits were seen on the return visit of the children 48 hours after they had been discharged from the hospital. Compared to the control group, some of whom had actually regressed since discharge; fewer children treated with *Enercel* had persisting increased peristalsis, moderate dehydration or an elevated temperature. They had also experienced less frequent bowel movements since discharge from the hospital. The data underscore the lack of complete resolution of symptoms by those receiving standardized care and to a lesser extent even in those receiving *Enercel* as an adjunct to standard care. Persisting symptoms are not uncommon with episodes of childhood diarrhea and can contribute to the faltering growth and malnutrition seen in many of these children.

Enercel and other cellular energy boosting products (enerceuticals) differ from typical pharmaceutical agents in that they are not directed against a particular biochemical pathway or disease process. Rather, they are intended to assist in the body's own recovery process by providing an added biophysical source of cellular energy through the alternative cellular energy pathway.

Additional information is available at www.s3support.com

Conclusion

CHAPTER 6

KELEA Activated Water – Enhancing the Alternative Cellular Energy (ACE) Pathway

W. John Martin, MD, PhD.
Institute of Progressive Medicine South Pasadena CA 91030

Author Information:
Name: W. John Martin
Institution: Institute of Progressive Medicine
Address: 1634 Spruce Street, South Pasadena, CA 91030 USA
Telephone: 626-616-2868
e-Mail: wjohnmartin@hotmail.com

Abstract

This article defines a kinetic energy quality of water, which can enhance the body's alternative cellular energy (ACE) pathway. The article also reinterprets the practice of homeopathy by challenging its premise of specificity. The wide range of methods for energizing water is reviewed and a unifying hypothesis relating enhanced kinetic activity to a reduction in intermolecular hydrogen bonding is suggested. The term "KELEA," referring to "kinetic energy limiting electrostatic attraction," is proposed as a fundamental force, which may primarily function in the prevention of annihilation of opposing electrical charges through fusion.

Key Words: Homeopathy; Stealth adapted virus; mitochondria; metabolism; static electricity; etheric energy, Brown's gas, humic acid, ceramics, hydrogen, Tesla, radiant energy

Abbreviations: ACE – alternative cellular energy; CPE – cytopathic effect; HSV – herpes simplex virus; KELEA – kinetic energy limiting electrostatic attraction

Introduction

The majority of researchers automatically reject suggestions that water can be anything more than a random collection of H_2O molecules[1]. Yet, there is substantial evidence supporting those who ascribe functional differences between water obtained from different geographic locations; or who further claim that regular water can be modified so as to significantly enhance its life supporting capacity. Unfortunately many of the proponents of "energized" or "structured" water do not exercise the scientific rigor of clearly distinguishing proven facts from wishful suppositions. This uncertainty is further compounded by a near-uniform unwillingness among water entrepreneurs to be forthright in disclosing manufacturing details of possible commercial value in a competitive marketplace. Even acknowledging that competitors' products may have benefits comparable to one's own water product is vigorously avoided; adding to an overall under appreciation of the extensive evidence that the life supporting quality of water can indeed be improved.

Water is fundamental to many of life's processes. It deserves high quality objective analyses to at least identify the major points of contention between advocates[1] and skeptics[2]. This article is intended to highlight the issues, which if true, will lead to a rethinking of certain basic principles of physics, chemistry and biology. The review begins with a brief description of the major characteristics of purportedly energized or structured water. This is followed by a summary of many of the widely used methods of creating such water. Homeopathic formulations are discussed in the context of essentially being further examples of energized water. Evidence is then presented that energized water likely achieves many of its claimed clinical benefits through enhancement of the alternative cellular energy (ACE) pathway. Relatively simple assays are described to assess the energy component of water and of the body's ACE pathway. Finally, a testable hypothesis is offered regarding a proposed "kinetic energy limiting electrostatic attraction" (KELEA) force of Nature, which may explain the energizing of water, other liquids and various compounds of presumed therapeutic benefit. The delivery of KELEA force is also consistent with the mechanism of effective action of various devices used in energy medicine.

General Characteristics of Energized Water

Laboratory Parameters: To help justify claims of water being beneficially altered, it is frequently stated to differ from ordinary water in specific physicochemical characteristics. An aggregate summary of these reported changes includes the following: Diminished viscosity; reduced surface tension; lower boiling and freezing points; increased absorption and emission of infrared radiation (heats and cools faster); bluish luminosity; smaller and more mobile clusters of water molecules (as determined by a shorter range of realignment frequencies in nuclear magnetic resonance, NMR); more negative charge with a greater electron donating capacity (reflected in higher pH and negative redox measurement); increased electrical conductivity, observed even in comparisons using highly purified water; reduced solubility for most solutes, yet greater miscibility of oil in water (emulsions); increased oxygen carrying capacity; and selective toxicity for pathogenic bacteria and fungi.

Dr. Gerald Pollack[3] (born 1940) has described exceptions to some of these characteristics for the type of "structured" water, which forms as a layer adjacent to reverse osmosis hydrophilic membranes. He refers to this layer as exclusion zone (EZ) water since, as with energized water, it can exclude many solutes. Unlike energized water, however, the exclusion layer has a higher than normal viscosity and slower realignment time by NMR, when compared with regular water. Also, when compared to regular water, EZ water has a net negative charge due to absorption of light energy (mainly from the infrared region), which results in the removal (separation) of some of the hydrogen atoms from the water, as it apparently coalesces into a liquid crystal[4], which Dr. Pollack refers to as the fourth state of water. He proposes that membranes of living cells may have the same structuring effects on much of the body's water. Qualitative differences may still exist, however, between different water samples, even when in this seemingly more highly structured, hydrogen (proton) depleted, biological state.

Certain specified forms of energized water have been laboratory tested in rather unique ways and the information derived from these studies may not necessarily be generalized to all forms of energized water. Indeed, a limitation in the interpretation of many of the sophisticated assays is that each particular assay has typically been applied to only a single commercial brand of energized water. This has hindered meaningful, direct comparisons being made between competing brands. Still the suggestions are sufficiently intriguing to warrant broader consideration.

One topic is the apparent expansion of energized water[5]. This has been variously described as i) an overall increase in the comparative volume of the water, especially when frozen into ice; ii) an increase in the ratio of low density (high energy) domains compared with more compact high density (lower energy) domains, with individual domains comprising minute clusters of water molecules[6]; and iii) a possible wider separation up to approximately 114° from the normal 104° bond angle between the two hydrogen atoms in individual H_2O molecules[5], along with a lengthening of the oxygen to hydrogen (O-H) covalent bonds.

A related topic is the greater performance of certain enzymes when tested in energized water compared to regular water[7]. It has even been suggested that the conversion of low-density to high-density water domains can directly mediate the formation of adenosine triphosphate (ATP) from its precursors (ADP plus oxygen-bonded phosphate in the form of KH_2PO_4). Conversely, the biochemical energy provided by the conversion of ATP to ADP could be conveyed to other molecules via an initial localized change in water density[6].

High power dark field microscopy on water reveals random minute flashes of light, referred to as biophotons[8]. A conceivable source of the light is the slowing of released subatomic particles, such as can occur in the generation of neutrino-induced Cherenkov light[9] from water. More likely is the probable collapsing of vapor bubbles as can be seen, for example, when water is vaporized with sound (sonoluminescence[10]). The frequency of the flashes is increased approximately 100 fold when water is passed through an energizing turbulence process comprising rapidly rotating blades[8]. More revealing is the apparent reformation of structured liquid water within each burst of light observed in this particular system. The reforming water takes the form of concentric layers of hexagons, which may actually be dodecahedrons in three-dimensional views[8]. Another investigator has reported on mobile light emitting water particles (possibly as liquid crystals) moving in an apparent vortex within his "Vortical Energy Water"[11]. Both observations imply a continuing source of energy input into water in a non-homogeneous manner. The notion of structural heterogeneity within water is further suggested by the identification of gel-like particles in the post-evaporation residue from a brand of highly purified energized water called Double Helix Water[12]. These structures have not formally been identified as only comprising water molecules and, as will be mentioned later, energized water may possibly have biosynthetic capacities.

Another provocative observation is that water can be electrolytically vaporized into a high-energy state, which can be ignited. The flame from an ignitable form of vaporized energized water, referred to as Brown's gas[13], can be used to create particularly high temperatures in various metals. It is also stated to significantly reduce the radioactivity of certain isotopes, including [60]cobalt[13].

Biological Parameters: A major impetus for studying energized water is the potential of improved human and animal health, as well as greater agricultural and industrial productivity.

i) Human Studies: Consuming water is essential for human survival. Moreover, significantly increased incidences of various illnesses have been noted among those who consume only minimal quantities of water (< 1,000 ml per day) compared to those consuming larger quantities (~ 3,000 ml)[14]. The marketing of more specialized water is predicated on the further assertions that it is not only safer to consume, but that it can provide functional benefits well beyond those of ordinary drinking water. Mainly because of imposed restrictions by the United States Food and Drug Administration (FDA), claims relating to drinking water preventing or curing specific diseases are assiduously avoided. Still, it is commonly portrayed that energized water is effective in one or more of the following ways; Alleviating fatigue and renewing one's motivation for engaging in purposeful activities; allowing for more sustained mental concentration and clarity of thought; improving the capacity for social interactions; enabling better physical endurance along with accelerated recovery from strenuous exercise; diminishing emotionally-driven appetite for foods; providing deeper nighttime sleep, sometimes accompanied by vivid enjoyable dreams; developing a clearer skin complexion; and most of all, providing a greater sense of wellbeing and an increased zest for living.

Two clinical analyses are commonly cited as confirming systemic biological effects from drinking energized water. Microscopic dark field analysis of blood smears shows reduced red blood cell adhesion (rouleaux formation[15]). This is probably related to the increase in capillary blood flow, which can be observed using nailfold video capillaroscopy. The other commonly cited assay is Kirlian photography[16]. It uses a high voltage electric field to induce an ionization layer over the surface of an object, including parts of the body. The resulting "Gas Discharge Visualization" (GDV) shows that energized water itself and those who consume such water have more intense and expanded ionization "biofields" when compared with corresponding controls. Neither dark field analysis of blood nor Kirlian photography in humans (often performed on finger tips) is used in regular medicine and the authenticity of many reported findings using these procedures remains somewhat questionable. A similar uncertainly extends to reported body changes after ingesting energized water, using other body-energy sensing procedures, such as kinesiology, and using medical devices, such as Ondamed, Scenar, Larcher, etc.

A more quantitative, but less commonly performed biological analysis involves measuring electrical bioimpedance to assess the relative volumes of intracellular to extracellular body fluids[17]. Seemingly, rather small quantities of ingested water from some manufactures can lead to a substantial increase in total intracellular water in tested individuals. This is consistent with the ingested water overcoming an illness-related insufficiency in the passage of extracellular water molecules through the aquapores of cellular membranes[18]. Conversely, however, it has been reported that consumption of energized water can lead to urination well in excess of the amount of water consumed. This too is favorably interpreted as an enhanced urinary excretion of previously retained water, along with accumulated toxins. Rigorous clinical studies using defined patient populations are, however, rarely performed making it difficult to draw reliable conclusions.

Some individuals have speculated that there may be functionally unique types of biologically energized water, each comprising distinct changes within one or more of a complex array of unrelated qualitative characteristics. This would imply the capacity of water to acquire unique sets of functional characteristics, such as different vibrational frequencies or memories, precisely determined by the various energizing processes[19]. Even more exacting, some researchers contend without compelling data that water can acquire information or memory corresponding to the precise chemical composition of specific pharmaceutical products. In this way, it is argued that the modified water can mediate the same specific biological effect as the particular pharmaceutical used to energize the water[19]. It was also recently suggested that dissolved DNA molecules could impart a unique signature into nearby water, which corresponds to their precise DNA nucleotide sequence. In other words, the DNA sequence data was transferable via a water memory process to a distantly placed sample of water, such as to facilitate in this water, the assembly of random nucleotides into the exact sequence as the originating DNA[20]. Rigorous specificity studies, which include the exclusion of possible observer bias, have yet to be reported to warrant the conclusion that water can acquire differing arrays of multiple, independent energy-based parameters, comparable to the diversity of chemical structures. Similar uncertainties exist with regards to reports of energized water preparations being able to selectively and differentially alter the gene transcription patterns in cultured cells. The presence of bacterial endotoxins and/or other minor contamination in water samples can potentially explain many of the reported findings. The issue of qualitative differences between energized water products is discussed later in relation to the specificity claims for homeopathy.

ii) Animal Studies: More convincing studies have been performed using energized water in a variety of animal production facilities. Notable examples of ongoing quantitative studies have confirmed significantly enhanced growth and reduced pestilence when water exposed to Kiko Stones (pellets formed from finely ground and preheated volcanic material) has been consumed by both land based (e.g. chickens) and aquatic based (e.g. trout) animals[21]. Seemingly not all similarly designed experiments produce equally impressive results and this could reflect underlying variation in the intrinsic energy levels of the water sources available at different locations.

iii) Agricultural Studies: Beyond its purported benefits for humans and animals, energized water is seemingly superior to regular water in numerous agricultural applications. Although still to be rigorously confirmed, it seems that energized water is capable of achieving many of the following benefits: Increasing crop yields; shortening the time to harvest; improving taste and extending the shelf life of commercial produce; enhancing the coloring of flowers and of crops; allowing for more efficient germination of seeds; enhancing root formation and the sturdiness of stalks; achieving greater water penetration into tightly packed soils; and, most interestingly, reducing bacterial, fungal, insect and rodent infestations. These last two observations will be discussed later since they imply an innate capacity of insects and rodents to discern a "vitality" characteristic of plants cultivated in beneficial water. Agricultural benefits from using energized water are illustrated on numerous commercial internet sites[21-22]. As will be emphasized later, there is no fundamental distinction between efforts to energize water for human, animal or agricultural purposes.

iv) Industrial Studies: Indeed, there is further overlap of water energizing technologies into the industrial arena. Specifically, various technologies have been marketed to render water less corrosive to metal pipes and to be actually capable of reversing existing scale deposition[23]. These effects can greatly improve the longevity and efficiency of water-based heat exchanges. Energized water may be used to more efficiently wash clothes and other materials. It has also be used to help deodorize contaminated environments; allow for more even dispersion of cement; provide for smoother application of paints; diminish friction, thereby leading to more efficient drilling into rocks; and, possibly, it can significantly enhance the process of fracking.

Methods for Energizing Water

These methods can be divided into i) those in which certain chemical or other components are added to, or produced from within the water; ii) those in which an outside physical force or energy is applied to the water; iii) various combinations of these two approaches; and iv) other methods including the simple collection of water from certain renowned locations. These locations include Lourds (France); Nordenau (Germany): Hunza (Pakistan); Nadana (India), Tlacote (Mexico); Marcial (Russia) and the Great Salt Lake (USA). Undisclosed sites in the Atlantic Ocean provide a marketed seawater product called Quinton™ Marine Plasma[24].

i) Additives: Among the common components added to regular water to enhance its life-supporting properties are mineral-rich zeolites[25] and humic/

fulvic acids[26]. Both groupings of molecules can exist as clathrates[27], referring to the ability of silicon in zeolites and carbon in humic/fulvic acids, along with oxygen and other atoms, to assemble into honeycomb-like collections of adjoining molecular cages. Humic acids differ from fulvic acids in having a higher percentage of direct carbon-to-carbon bonds with fewer intervening oxygen atoms. Humic acids also form into larger complexes and precipitate in acidic solutions. Many different cations (positive charged atoms) can reside within the clathrate cages and are held in place by electrostatic forces. Depending upon its source, zeolites also commonly contain varying amounts of aluminum and other minerals as part of the cage structure. An extremely wide range of minerals is incorporated as cations within naturally occurring zeolites and their various proportions can be experimentally modified using cation exchange processes[28]. Electrostatic neutrality is favored by the external attachment of anions (negatively charged ions) including oxygen, chloride, phosphate and sulfate. Locally, however, the differing charges remain spatially separated because of inherent repulsion between like charges and because of residual spacing between opposing charges. The term dielectric is used to denote the spatial separation of positive and negative charges and can apply to atoms, individual molecules and multi-molecular complexes.

Humic and fulvic acids are largely derived from decomposition and subsequent reassembly of organic matter. The terms are also sometimes used for certain complex organic molecular aggregates in living plants; and even in animals and bacteria, for example worm castings and so called "Effective Microorganisms™,"[29] respectively. Bentonites are considered zeolite-humic/fulvic acids hybrids with their clathrate shells comprise both silica and carbon atoms, together with oxygen and aluminum atoms. As with zeolites and humic/fulvic acids, bentonites can also incorporate essentially all of the different types of minerals as cations, as well as attach to various anions. Rather than assembling into cage-like formations, these and related compounds can also form into multiply branched macromolecules termed dendrimers. Multiple carbon atoms can also cross-link into spherical cages termed fullerenes; a common example of which has sixty covalently bounded carbon atoms.

These multi-molecular compounds fit within a broader category of poorly electrical conductive mineral-rich complex materials termed ceramics[30]. Most ceramics contain a more restricted range of additional cations and anions than do zeolites. Common dominant cations include sodium, potassium, carbon and hydrogen along with a relatively higher proportion of oxygen atoms than in zeolites. The inclusion of certain other dominant cations, such as magnesium or vanadium, can yield various transparent ceramics, within a category called spinals. Either covalent or electrostatic bonds link the molecules, with the proportion of covalent bonding being reduced by heating to a high temperature ($\sim 1,000^{oC}$), followed by slow (~ 3 hours) cooling, treating with strong acids or other methods. Examples of common ceramics include clays, mica, tourmaline, charcoal, graphene, fullerenes (including shungite), feldspar and volcanic rocks[30]. Most of these complexes are thought to be either insoluble in water or to incompletely dissociate into colloids rather than fully dissolving into single molecules. All have been used to energize water, with comparable reports of success.

Vaporizing and subsequently condensing of water (distillation) can provide essentially mineral-free water. This approach can help address the issue of whether mineral-rich materials added to energize water are benefiting those consuming the water by simply correcting preexisting mineral deficiencies. Several studies, for example, have reported on improved health by drinking water with higher than usual calcium and/or magnesium levels. The carryover amounts of minerals in some examples of energized water are, however, rather miniscule and when tested, energy activity is typically retained after distillation. Furthermore, dense, insoluble complexes of magnesium oxide (termed prills) have water-enhancing effects, which overlap with those of more complex ceramics and which continue even with the repeated replacement of water covering the same insoluble prills over many months[31].

As noted above, ceramics have limited electrical conductivity, although they uniformly have the dielectric property of minor electrical charge separation, especially when re-formed within an electrostatic field. Indeed, many ceramics assemble into particles on the basis of dielectric hydrogen bonding, as can various organic polymers, including those composed of terpenes (essential oils) and phenolics. This is noted since a particular terpene-rich, water based plant extract developed in Japan and termed HB-101, has exceptional water enhancing properties, even when used at 1:20,000 dilution (5 drops per gallon)[32]. It is also useful to mention that certain ceramics and other materials, possibly including liquid crystallized water, display a piezoelectric and/or pyroelectric effect in that their internal charges can become further separated in response to pressure and heat, respectively. Photons of light of particular wavelengths can also induce charge separation involving the outward movement of external electrons of certain specific molecules. The orbitally-shifted electrons spontaneously return to their original energy level with the release of photons, typically of a somewhat longer wavelength than that of the inducing photons. If the emitted photons are induced by ultraviolet (UV) light and are at a visible wavelength, the process is termed fluorescence[33]. In some circumstances, isolated magnetic and/or electrical energies can be extracted from the emitted photons by various chemical compounds leading to photoelectric and photomagnetic properties of the compounds being illuminated.

High heat, for example as provided by electrical arcing, can, in the absence of oxygen, partially hydrolyze and dehydrate complex organic polymers leading to the formation of largely hydrogen bonded dielectric porous insulators. Edgar Cayce (1877-1945) used this approach in producing his water-enhancing "animated ash" from bamboo[34]. Residues from the electric arcing of coconut fiber and of lignite coal have similarly been used in combinations with surfactants to create "Miracle II Soap" and "Willard Catalyst Activated Water," respectively).

Electrical conductivity does not, however, exclude materials from enhancing the quality of water. Thus benefits have been ascribed to water into which particles of electrically conductive silver, gold, titanium, germanium and ferric/ferrous iron complexes have been added[35]. The term "monatomic" is rather loosely used to describe minerals, especially gold, silver and rare earths, when they exist in a chemically non-reactive form. It has been speculated that the nuclei in these minerals are expanded and that some of their electrons are orbitally rearranged. Such "ORMUS" elements[36] can be very finely dispersed in water and seemingly convey some colloid-like dielectric properties. Unfortunately, reports of detailed science are lacking on monatomic water-enhancing minerals.

A far better defined additive, which is able to enhance the wellness quality of water is hydrogen[37-38]; basically a dielectric (one proton, one electron) atom. The diatomic H_2 form is easily generated as a gas from the cathode (negative electrode) during the electrical hydrolysis of water; with oxygen being generated from the anode (positive electrode). Most of the hydrogen gas readily escapes from the water, with relatively small amounts of soluble hydrogen remaining after the end of the electrolysis. A more steady longer-term production of hydrogen is achievable using water permeable containers of magnesium particles, especially in the presence of small amounts of magnesium hydroxide[39]. A third approach is to use hydrides (hydrogen complexes) of various atoms, including silica. The hydrogen content of zeolites and humic/fulvic acids can also be enriched by cation exchange[28].

Although, not pursued at the time and by now largely forgotten, a compelling series of observations were reported in the 1930's on the therapeutic use of diluted hydrochloric acid (HCl)[40]. Polymorphonuclear cells from treated patients are increased in numbers and show markedly enhanced motility. Of further interest, the amount of oxygen being metabolically consumed by HCl treated patients appeared to be reduced as shown by greater residual oxygenation of venous blood.

Hydrogen is regarded as a potent reducing (electron donating) agent in redox reactions. This probably does not explain its action in activating water, however, since equally impressive claims have been made for water treated with powerful oxidizing (electron absorbing) agents, such as ozone gas, chlorine dioxide and other oxides of chlorine. Indeed, some additives to water are said to provide both reducing and oxidizing agents in co-existing mixtures, e.g. ASEA™. It is also unlikely that the pH of the energized water is relevant since added benefits are reported with acidic, alkaline and neutral pH water.

The most intriguing of water additives is the use of small quantities of previously energized water. For example, water treated with a low concentration of a not fully disclosed ferric/ferrous iron compound is termed "Pi Water." Only a few drops of Pi Water are supposedly needed to quickly activate several ounces of drinking water[41]. Similar claims have been made for additional water products, such as "M-water," "Double Helix Water", "Cell Food," "Vortical Energy Water" and others.

Another example of water energizing itself involves a form of water vapor, included in Brown's gas[13]. The gas mixture comprises hydrogen and oxygen gases obtained by electrolysis together with water in a vaporized state (discussed later). Small amounts of Brown's gas are seemingly capable of energizing large additional quantities of water by simply being bubbled into the water.

Homeopathy: It is worthwhile to focus attention on those water additives with the best evidence for clinical efficacy. These are undoubtedly in the field of homeopathy[42]. Dr. Samuel Hahnemann (1755-1843) formulated remedies on the basis of the components, when used in larger amounts in healthy individuals, would induce symptoms similar to those being experienced by his patients. Consequently, he employed many different formulations using multiple compounds selected from the Materia Medica and other sources, including known toxins. Dr. Hahnemann concluded that agitation occurring with the horse and buggy traversing of cobbled stoned streets was enhancing the therapeutic efficacy of his homeopathic formulations. This observation

led to the practice of succussion (abrupt jolting) of the fluid during the preparation of homeopathic formulations and just prior to their administration.

The actual preparation of homeopathic formulations involves the initial addition of relatively small quantities of tinctures (ethanol extracts) of different herbs plus various minerals, into water, often with a low percent (~5%) of ethanol. The solution is succussed 20 to 40 times and further ten or more fold further dilutions are progressively made into water (with or without a small quantity of added ethanol). The amounts of residual original tincture are considered miniscule. From 4 to 6, ten to one hundred fold dilutions, are said to be more potent than either earlier dilutions (<4) or later dilutions (>6). Factors, such as the aggregate duration of succussion and the possible value of increasing the duration of time between later dilutions, may be relevant but are not commonly considered.

Proof of efficacy of homeopathics is now available from the peer-reviewed medical literature[43-45]. Rigorous comparisons between homeopathic and untreated ordinary solutions have justified the assertion that the observed benefits are not psychogenic (placebo effect). Still, homeopathy generally entails a reassuring relationship between practitioner and patient, such that it is commonly stated that little or no benefit comes to the disgruntled, highly skeptical patient.

It is important to note that there are no published crossover data affirming that homeopathics differ in their selectivity for treating particular diseases. In other words, formal proof of disease specificity for different homeopathic formulations does not exist. It is quite possible that Dr. Hahnemann and his many followers have mistakenly assumed specificity of their remedies or have otherwise misled others to believing their professional input in the formulation of products was an essential element in patient care. In other words, effective homeopathics may simply be further examples of non-specifically energized water.

The wide diversity of tinctures used to produce homeopathic formulations is a further statement that many additives can be used to initiate the water activation process and that, once energized, the activated solutions can be used to further energize additional water solutions. In terms of actual costs of production, the many fold dilutions essentially obviate any costs of the initiating formulations and it essentially only leaves the cost of water and its packaging.

Claims of beneficial activities of homeopathics in cancer patients have typically evoked unfair retaliation by regulatory agencies protecting the "cut, burn or poison" triad of conventional cancer control. Legitimate manufacturers of homeopathic preparations have, therefore, mainly tackled non-cancer illnesses. For example, a formulation developed in Argentina and originally called HANSI™ for "homeopathic activator of the natural system immune" and renamed in the US as Enercel™ was strikingly effective in treating childhood diarrhea46. Either the same or modified Enercel™ product has recently also been reported as providing marked benefits to tuberculosis infected AIDS patients[47]. This and other homeopathic products have also proven to be effective in promoting the non-scaring healing of wounds[48].

The Enercel™ tested in the childhood diarrhea study was not strictly a homeopathic since in performing Gas Chromatography:Mass Spectroscopy (GC:MS) on the product, xylocaine (Lidocaine) was unequivocally detected. Lidocaine is related to procaine and in turn to cocaine. The undisclosed inclusion of Lidocaine in HANSI™ was possibly the outcome of a German immigrant to Argentina around the time of the Second World War. Historically, the Huneke brothers (Ferdinand and Walter) in Germany and Ana Aslan (1897-1988) in Romania had long maintained that procaine had exceptional healing and rejuvenating qualities unrelated to its anesthetic actions[49]. A number of Germans took refuge in Argentina and possibly an effort was made to obscure the presence of Lidocaine in the early development of HANSI™ and later Enercel™. Instead, the homeopathic benefits are attributed to a proprietary combination of the various herbal products officially listed as ingredients.

Many other compounds in herbal, tincture, essence, bitters or in more purified chemical forms have been purported to have wide ranging clinical benefits covering many types of illnesses. So too do certain pharmaceutical compounds have benefits well beyond their initial clinical indication; one example being dilantin[50]. While not being diluted to the levels used in preparing homeopathic formulations, it is possible that some or even many of these compounds may be primarily working through their energizing effects on water. The term "enerceutical™" has been proposed for natural products with water energizing properties.

As noted earlier, homeopathy practitioners have helped promote the concept that water can acquire distinctive changes, which they are able to match therapeutically to different diseases. Similarly, those who market various chemical or herbal remedies will typically claim unique benefits for their particular product. Focusing on disease specificity of various products is a useful marketing tool that helps avoid direct cost and efficacy comparisons with other potentially competing products. This approach can, however, lead to an understating of the overall potential clinical benefits of many enerceutical™ products.

Substantial generalized health benefits have also been ascribed to the moderate consumption of alcoholic beverages[51], regardless of the actual kind of beverage being consumed. Again, it is conceivable that the 5-45% ethanol in different alcoholic beverages is adding to the energetic property of the accompanying water. Consuming the beverage can, in turn, potentially assist in energizing the entirety of the body's water.

ii) Physical Energetic Methods

A powerful observation is the apparent capacity of energized water to exert a positive effect on other samples of water, which are simply positioned so as to be in close vicinity with the energized water. This approach formed the basis of the water-energizing device devised by Johann Grander (1930-2012)[52]. Regular water is passed through a metal container, within which there is a sealed sample of energized water derived from an aquifer in Austria. Mr. Grander did not disclose his method of imparting energy to the water and possibly, it was intrinsically energized or vitalized simply by its location. In any event; piping regular water around the inner, sealed collection of more energized water is apparently able to improve (revitalize) ordinary water; as evidenced by the continuing customer support of the Grander device. Reported data on using the Grander device include increased agricultural productivity and confirmed reduced scale formation in heat exchangers[52]. Additional examples of distant effects of energized water include the use of double-walled

mugs, in which activated water is placed between the outer and inner walls of the mug[53]. There is also the use of sealed tubes containing energized water as stirrers for regular water. Water activating sealed stirrers containing zeolite-like material[54], are also being marketed. The energized water and the zeolite-containing water stirring devices are consistent with the presence of water activating energy fields radiating from the devices.

An even simpler physical energetic method of water activation is for the water to flow as a spinning vortex, with a resulting centripetal (inward or implosive) force. The basic concept of water gaining energy by its natural spiraling in streams and rivers is attributed to Viktor Schauberger[55-56] (1885-1958), an Austrian forrester and self-taught water engineer. This effect can be achieved using twisted internal fittings into water pipes. Single or double vortexes generating rotors and nozzle outlets have also been designed[22]. Some of the nozzles simulate a Caduceus coil. It is possible that the coils are exerting a mechanical force on the water so as to disrupt some of the intermolecular hydrogen bonds. Theodor Schwenk (1910-1986) further suggested that energized water retains its dynamic movements by describing microscopic patterns, indicative of internal movements, within drying droplets of energized water[57].

Sound waves have also be used to agitate water and special emphasis has been given to apparent beneficial effects of ~528 Hz. Vibrations of various crystals, e.g. ANCHI™ crystals from Japan are similarly reported as benefiting water, either by being placed into the water or placed in close vicinity to the water or even to the human body[58].

Passing an electrical current through water is commonly used to enhance its energetic activity.

Although, a high voltage electrostatic field can transiently influence the orientation of water molecules, either naturally present or added small amounts of electrolytes are required for electrical transmission to occur through water. As noted above, hydrogen and oxygen gases are then generated during electrolysis at the cathode and anode electrodes, respectively. The water around the cathode acquires an alkaline pH and is commonly collected for drinking. One of the electrodes can be positioned above the water so as to ionize the air and, thereby, create a high voltage plasma (ionized) gas. The resulting "plasma activated water" can be further separated into an acidic (hydrogen ion-rich) and alkaline (hydroxyl ion-rich) water; the latter being recommended for human consumption. William Rhodes (b. 1922) and later Yull Brown (1922-1998) observed that the combination of gases generated from the electrolysis of salt-containing water could be ignited to provide a flame for welding purposes[13]. Water was shown to form on the welded surface. While initially attributed to the recombination of hydrogen and oxygen atoms, the more reasonable recent proposal is that a highly energized form of water vapor is being formed. Indeed, one can often observe a relatively slowly evolving stream of bubbles arising midway between the two electrodes during the electrolysis process. Rather than using the gas mixture (commonly referred to as Brown's gas) for welding, it can, as stated above, be bubbled through regular water as an effective means of enhancing the water's health promoting qualities[13]. A strong magnetic field and even radio waves can seemingly also induce the formation of ignitable water vapor from distilled water, possibly without necessarily any accompanying hydrogen or oxygen gas formation.

Magnets have commonly been used to alter some of the properties of water, including its level of energy. With moving water, the applied magnet field typically has a fixed location, such as having magnets attached to water pipes. Magnets can also be placed within the water line so as to rotate in response to the water flow. Lorentz magnetic forces acting on the moving electrical charges can orient water molecules into a more pronounced dipole (dielectric) arrangement. When the water to be treated is stationary, the magnetic fields are usually pulsed by the repetitive charging/discharging of an electromagnetic device Both approaches have also been used with fuel lines in automobiles and are said to enhance the combustibility of gasoline and to increase mileage[59]. If true, this implies that water is not the only liquid capable of being modified in ways to enhance its energetic properties.

Electromagnetic energies of ultraviolet, visible, infrared and radio wavelengths have variously been reported as improving the quality of drinking water. There is also a striking overlap between electromagnetic devices primarily designed to directly transmit energy into the body and those used to activate water. Many of these medical devices comprise high frequency electrical charging and discharging. They include Royal Raymond Rife's (1888-1971) Beam Ray[60]; Geroges Lakhovsky's (1869-1942) Multiwave Oscillator[61]; Edgar Cayce (1877-1945) Violet Ray[62]; and Pannos Papas (b. 1947) Papimi machine[63]. Their purported beneficial effects in humans are essentially similar to those attributed to consuming energized water. Rife was able to adjust the frequency of electrical charging of inert gases within a glass tube. He argued that the appropriate mortal oscillating frequency (MOR) of charging of the gas was a critical issue in the destruction of specific pathogens, which he further argued were the underlying causes of cancer. His assertion of different MOR's being needed for different types of pathogens may be similar to the untested assumption of Hahnemann that homeopathic formulations have differing disease specificity. Lakhovsky also assumed that healing may be occurring because of a particular frequency, but since the required frequency was not known, he designed his device to simultaneously provide multiple frequencies. Cayce's Violet Ray was intended for localized treatment. Ingesting or applying an "animated ash" made from electrical arcing of bamboo, enhanced the efficacy of his Violet Ray therapy. The Papimi machine employs a pulsed electromagnetic field (pemf) therapy, with its emphasis on its magnetic component. Recent reports using the Papimi-type machines emphasize their capacity to directly activate water[54].

A more provocative suggestion is that energy, which is different from conventional electromagnetism, is also capable of changing the structure of water. Moreover, according to some investigators, this form of energy is freely available from the environment. Dr. Wilhelm Reich (1897-1957) was among the more influential practitioners applying this line of reasoning to water. He postulated that "orgone" energy could be concentrated in chambers comprising multiple layers of electrical conductive and electrical insulating materials[64-65]. Water placed within such a chamber becomes environmentally energized. Various orgone energy collecting and transmitting devices are currently being marketed on the internet. They essentially comprise collections of mixed ceramics embedded in plastic housing, which commonly is in the shape of a pyramid. Another device developed by the late Victor Roehrich (1928-2012) and referred to as "Intrasound," uses a series of seven differently cut aluminum pipes arranged in a circle. Upon heating for 14 days, the pipes are said to resonate with and to amplify a "universal" energy field. Arguably, the aluminum pipes would have become coated with aluminum oxide, a highly dielectric material as a result of the 14-day heating. In any event, the device is said to be forever capable of radiating energy into nearby materials, including water, kaolin (a type of edible clay) and various gels[66].

The concept of an unconventional energy can be traced back to many earlier investigators who were not specifically concerned with water. Prominent proponents of such energies include Franz Mesmer[67] (1734-1815; animal magnetism); Baron Karl von Reichenbach[68] (1788-1869; odic force, which can be collected from sunlight, transmitted through metals and able to ionize air); General A. J. Pleasonton[69] (1801-1894; cobalt blue glass panels significantly enhanced plant growth when used in his greenhouse); Albert Abrams[70] (1864–1924; radionics) and Nikola Tesla[71] (1856-1943; impulse or radiant energy; discussed later). Additional names given to an envisioned, non-conventional energy include; etheric, dark, subtle, zero-point, free, proto, prana, vril and scalar. The term scalar relates to energy without a directional component (vector) and which, therefore, normally has no net movement. It may, however, have an affinity for, or be otherwise influenced by, various devices. For instance, certain pyramids, coils and egg shaped containers, may affect the directional flow of such energy and enhance the capturing of the energy by nearby water.

Humans have considerable electrical activity, especially involving the brain, heart and muscle[72]. There may also be additional energy-related human and animal forces. For example, the nature of consciousness is still poorly understood but is commonly viewed as an energetic process. Of considerable interest, therefore, is the possible mutual relationship existing between conscious or emotional thought and the energy property of water. Dr. Masaru Emoto (b. 1943) is widely cited for demonstrating that water, rapidly cooled to below 0°C, will crystallize into attractive, symmetrical ice structures if it is first exposed to pleasant, harmonious human thoughts[73-74]. By contrast, human thoughts of anger and hostility are said to be disruptive to the crystallization process, leading to misshapen ice particles. It is likely that a reduced freezing point allows for slower and, therefore, more orderly crystallization of water. Dr. Emoto's observations may, therefore, be a manifestation of a consciousness induced lower freezing point of energized water. Dr. Emoto's business associates further state that drinking his energized water will achieve improved emotional wellbeing of consumers.

It can be inferred that if harmonious thoughts can create energized water then practicing this mode of thinking ought to be capable of directly enhancing the functional properties of the person's own body fluids. Moreover, since energized fluids can apparently radiate effects onto nearby fluids, other individuals could benefit by simply being within the vicinity of a highly energized person. These possibilities may explain the health-related benefits of many forms of religious practices and the apparent success of faith healers[75]. The concept of individuals being able to transmit energies to others also forms the basis of Reiki and related therapies[76-78]. Qualified practitioners commonly emphasize the importance of reverence, humility, gratitude, serenity, blissfulness, forgiveness and related virtues in both themselves and in those who are expecting to be healed[75]. Other healers place value on laughter[79], which is interesting because laughter is said to correlate with rapid frontal lobe electroencephalogram readings (e.g. ~30 Hz), typically achieved by experienced Tibetan monks during deep meditation. Laughing, as well as feelings of calm serenity and also yarning are also associated with a preponderance of parasympathetic over sympathetic activity of the autonomic nervous system. There may also be enhanced production of brain acting chemicals, including oxytocin, dopamine and serotonin, associated with certain mindsets. Conversely, stress is consistently associated with the occurrence and progressive deterioration of various illnesses. Underreported clinical studies have consistently confirmed generalized health benefits of participation in stress-reducing activities and especially utilizing positive mental attitudes, including engaging in regular laughter in disease suppression.

Resonance sounds, such as by chanting or the use of Tibetan Bowls; visual sights of beauty, including paintings and geometric designs; dance; and using psychedelic drugs to experience an out-of-body Kundalini-like experience, may also be capable of altering the mind state of certain individuals, such that they become capable of directly energizing their own body's fluids, or even energize water to be consumed by others.

Clergy of Christian, Jewish and other faiths claim to render water "holy" by reciting various prayers and/or hymns. The prayer cited by Catholic priests specifically requests that the holy water be capable of dispelling sicknesses[80]. Again, in performing this ritual, priests seemingly focus their minds away from analytic, deductive reasoning ascribed to the left side of the brain towards more uplifting, spiritual attitudes, attributed to right sided or even non-cortical brain activity. As noted above, effective practitioners may simply have learned to boost the parasympathetic component of the autonomic nervous system, while suppressing the sympathetic component. It is noteworthy that Holy

Water is typically used externally and that stocks are replenished by the periodic addition of regular water to the remaining supplies.

Since insects and rodents are apparently capable of differentiating between plants grown with energized versus ordinary water[21], the sensory element involved may be far more primitive than either analytic or emotional intelligence. Possibly, analytic thinking and other advanced human brain activities act as a barrier to both the receptivity of the brain to an external source of energy and the capacity of the brain to positively influence one's own body fluids.

Sensitive, high-resolution technologies are becoming available to assess the electrical activities of the brain, heart and muscles in various situations. Water may come to be used as a medium to test the body's electrical or other energy-related capacities and, thereby, offer fresh insight into the field of psychic neurobiology.

Proposed Explanation for the Activation of Water

A reasonable conclusion regarding the many diverse methods of activating water is that they cause a quantitative alteration in a specific, unifying functional parameter. In other words, it is proposed that the variously generated types of energized water critically differ from regular water in a common, quantifiable parameter, which directly correlates with the energy level of the water.

A second reasonable assumption stemming from surveying the multiple methods of activating water is that separated electrical charges are integral to the activation process. The separated charges can be in the form of certain dielectric atoms,

molecules or colloids. Separated charges can also be generated on electrodes or repeatedly induced on various electrical devices. Magnetic forces are also capable of inducing electrical charges, as well as altering their orientation and, thereby, further separating existing charges.

The electrical properties of colloids in a fluid warrant further consideration[81]. Surrounding each of the intrinsically charged core particles is an extended collection of opposing charges derived from the fluid. These external shells of similar charges provide the repulsive force preventing aggregation of the inner core particles. A basic question is why the outer fluid charges do not collapse onto the core particles, but rather persist as opposing charges in the fluid surrounding the particles? As discussed in more detail later, it would seem as if some factor is limiting the extent of the electrostatic attraction of opposing charges.

Ordinary water carries a very small proportion of unopposed charges in the form of hydronium (H_3O^+) and hydroxyl (OH^-) ions. Individual H_2O molecules by themselves are also potentially charged in that the oxygen atom is electrically negative in respect to the positive charges present on the two hydrogen atoms. Hydrogen bonding between the oxygen atom and the hydrogen atoms on different water molecules can, however, effectively neutralize the potential dielectric charges on individual water molecules. Indeed, it is these hydrogen bonds, which maintain water in a cohesive or liquid phase between $0^\circ C$ and $100^\circ C$[82]. Some extent of inter-molecular hydrogen bonding probably exists even in water vapor and may act to reduce the theoretically achievable vapor pressure of evaporated water.

A third postulate is that water activation is an ongoing process that is mediated by the absorption of a continuing replenishing supply of an environmental energy. This energy can be both directly received from the environment or via energy- absorbing dielectric materials (including previously energized water), or devices.

This particular postulate can be restated as follows: That the electrical charges on dielectric atoms, molecules, colloids, previously energized liquids and various devices, mediate the collection of the proposed external energy and the transmission of the energy in either the same or a derivative form into water. Rapid on and off switching of electrical or magnetic charges is typically employed when electrically driven devices are used. With regards to dielectric molecules, actual oscillations may similarly occur such that an attracted energy modifies the charged molecules in such a way that the added energy, or a variant of it, is periodically released to do useful work on other molecules, including water. The activating molecules then presumably return to a state, which is again receptive of the external energy, such that a continuing cycling or oscillating phenomenon can occur.

Nikola Tesla has provided an important clue to the possible nature of the water-energizing radiant energy. He described an impulse force accompanying abrupt electrical discharge[83-86]. He personally could sense the force and distinguished it from both regular electricity and from conventional electromagnetic radiation. He initially assumed that his electrically powered equipment was directly generating the atypical energy in the form of very minute particles. He subsequently suggested the probability of tapping into and inducing the movement of an abundant source of preexisting energy within the "ether" element of the environment[87]. Conceivably, therefore, the strong separated electrical charges generated on his equipment were attracting and concentrating the widely available etheric energy[87], which was then being released upon abrupt electrical discharge. Tesla was able to filter (remove) much of the accompanying standard electromagnetic and electric forces using coils. This enabled the remaining unconventional energy to be projected over long distances. Tesla further observed the continuing presence of an atypical energy field long after the final electrical discharge from his device, as if the projected etheric energy had a self-perpetuating quality[85]. Possibly, therefore, dielectric materials within the vicinity of the electrical device became further receivers/transmitters of etheric energy.

Electrical power has additionally been shown not to be essential for certain types of devices to become an antenna of the external, etheric force. For example, Wilhelm Mohorn (b. 1954) has developed, under the trade name "Aquapol," a coil-based water-modifying device[88]. Although not designed to enhance the quality of drinking water, it can nevertheless reverse the upward seepage of water into the foundations of buildings, even though it is placed at the level of the ceiling of a water-damaged building. The device comprises two flat spiral "Tesla-like" coils as energy input antennas, attached to a plate with a downward facing cone shaped coil, possibly as a filter of electrical energy or as an etheric energy projecting antenna. The air around the later coil is said to show increased ionization (personal communication). Similar to oscillating crystals, various arrangements of coils are considered capable of resonating in response to environmental energies. The resonance can possibly also be reflected in rapidly cycling etheric charges.

The basic proposition is, therefore, that the presence of separated (free) electrical charges is the primary means whereby an external, environmental or etheric energy can be captured and transferred to water. The separated electrical charges can be provided by certain dielectric atoms, molecules (including energized water) and colloids and also by specific electrical and other devices. This raises the question of why do some dielectric compounds and devices work far better than others? For compounds, it could be a quantum requirement possibly based on the spatial relationship of the opposing charges. For instance, in many molecules, including in most water, the opposing charges are essentially bonded to, and thereby masked to varying degrees by the corresponding opposing charges on adjacent dielectric molecules. If the intermolecular bonding were to be lessened, the charges would presumably become more prominent and, thereby, may become better absorbers of the presumed environmental energy. As noted above, the absorbed energy may also need to induce a secondary change within the compound, such that at a particular threshold, the added energy is released. With regards to electrically or magnetically driven devices, the efficiency of energy transfer is presumably dependent upon the rapidity and completeness of the charging/discharging cycles.

This scenario can also potentially apply to water itself and allow for a continuing positive feedback mechanism, such that even highly purified water can mediate a continuing activation process. It is also possible that water molecules may actually transform into two energized forms; one that can act as a primary, time-dependent receiver of the environmental energy, as well as becoming a continuing transmitter of a secondary water-modifying energy. The other, less active form could be an absorber and conveyer of transmitted energy from distant water molecules, but not able to tap into the primary environmental

energy source. This distinction may explain why only certain types of energized water are able to retain and to even enhance their activity over time.

The apparent direct transfer of energy to and between water molecules would be somewhat analogous to the spread of electron activating electromagnetic energy in a laser[89]. Lasers allow the broader transfer of a repeatedly absorbed form of energy received by some molecules to a larger set of energy absorbing molecules. In much the same way that energy input is required to both trigger and continue the laser process, efficient water activation is presumably best achieved by continuing input of the natural physical energy from the environment.

What could be the fundamental purpose of such a charge attracting energy? Intuitively, a natural force must exist which prevents the annihilation of charges resulting from the blending of opposing charges. The question of why electrons do not collapse into the nucleus of atoms has indeed been previously raised but never satisfactorily answered. The usual comment is that it is simply not allowed by quantum mechanics or that the electrons transform into a non-compressible spherical waves, when within the close vicinity of the nucleus[90].

Except in the original formation of neutrons from protons plus electrons[91], approaching electrons and protons do not fuse into uncharged entities. With the formation of hydrogen atoms from separated ions, an electron avoids direct contact with a proton because its linear approach to the proton is changed into an orbiting motion. This reflects the imposition of a kinetic (direction altering) force.

A reasonable suggestion is that a kinetic energy limiting electrostatic attraction (KELEA) force counteracts the increasing strength of electrostatic attraction as the distance between opposing charges diminishes. It can be likened to a theoretical kinetic energy of orbiting planets limiting gravitational attraction (KELGA). KELEA may well collect on separated charges well before opposing charges are about to enter the same space. It is even conceivable that the spin or some other quality of the attracted force is different between positive and negative charges, thereby providing a functional barrier preventing the eventual fusion of opposing charges.

KELEA and the Alternative Cellular Energy (ACE) Pathway

Various observations made during the author's studies on infections caused by stealth adapted viruses can now be explained using the KELEA model. Stealth adaptation is an immune system evasion mechanism in which viruses lose or mutate genes coding for the relatively few antigenic components normally targeted by the cellular immune system[92-97]. Consequently, stealth adapted viruses fail to evoke the typical inflammatory response expected of an infectious disease. Yet infected patients and experimentally inoculated animals commonly recover from severe illnesses caused by these viruses. Recovery from the cell damaging (cytopathic) effect (CPE) of these viruses can also occur in tissue cultures[93]. The recovering cultured cells characteristically become pigmented with material, much of which is expelled from the cells. The extracellular material can aggregate via a self-assembly process into numerous particles, ribbons and threads[93]. Cellular repair from virus infection is attributable to the extracellular material since the replacement of the tissue culture medium by re-feeding the cultures leads to rapid (within minutes) reactivation of the CPE[93]. The reactivation does not occur if particles from other repaired virus cultures are included in the re-feeding medium. These particles are presumptively energizing the culture medium and consequently the intracellular water, enabling the cells to continue to resist the virus CPE.

Biochemical analyses of tissue culture media of recovered stealth adapted virus cultures have revealed a remarkable array of novel chemical compounds, many comprising an aromatic ring structure. The aggregated particles and fibers contain differing collections of identifiable minerals[93]. While each particle typically contained relatively few minerals, overall a wide array of minerals is present.

Consistent with their apparent water energizing properties, the materials are electrostatic, which largely explains their capacity to self-assemble into larger particles and fibers. They also fluoresce under both regular light and UV illumination, with lingering phosphorescence[93]. The UV fluorescence is enhanced with certain dyes, including neutral red and acridine orange. Using such dyes, confirms the intracellular production of fluorescing material in virus-infected cells. The smaller extracellular particles also exhibit multidirectional streaming movements, with a tendency to form into temporary clusters, which subsequently undergo abrupt dispersion. The particle movements are well beyond those attributed to Brownian motion.

Some of the larger aggregated structures also respond with movements to electrical, magnetic and sound energies. Indeed, non-attached pigmented clusters of cell can occasionally be easily rotated using a simple hand-held magnet. Ferromagnetism can also be readily demonstrated with some of the cell-free larger particles. Moreover, slowly forming vapor bubbles, likely corresponding to the water vapor component of Brown's gas, can form adjacent to some of these particles in medium, as well as in distilled water[93]. This observation adds further to the suggestion that the particles can directly affect the tissue culture medium.

Both intracellular and extracellular inclusions are present in biopsies of brain and other tissues obtained from stealth adapted virus infected patients and inoculated animals[94-96]. Of major importance is the very extensive disruption of mitochondria (the major producer of ATP from food metabolism), which can be observed in apparently still viable cells[94].

Based on the foregoing information, it was reasonably concluded that the energy-responsive particulate materials, termed ACE pigments, were providing a non-mitochondria source of cellular energy[93-94]. Clearly, ACE pigments have dielectric properties in common with many of the previously described water energizing, enerceutical™compounds. They also share properties with melanins of human, animal, plant and fungal origins. This is noteworthy, since melanins have also been proposed as a naturally occurring non-metabolic source of cellular energy[98-99], and may indeed represent another form of ACE pigments. Fundamentally, it is proposed, therefore, that the ACE pathway is a process for capturing etheric[97] and possibly other forms of environmental energies, leading to a beneficial activation of the body's water.

A dual pathway for generating cellular energy is, therefore, envisioned. One is based on the oxidative phosphorylation of food and primarily results in the generation of ATP. It can be argued that the approximately 2,000 Calories generated by food in a typical human daily diet is really insufficient to maintain body temperature as well as provide for the daily energy expenditures of muscles, brain and other organs. There can also be additional metabolic demands in response to infections or needed for tissue repair. The ACE pathway is viewed as a major back-up system, which can presumably help overcome the deficiencies or increased demands placed on the food generated energy pathway. It may also provide for aspects of consciousness, which may not be fueled by the conventional metabolic process. It may even lead to the synthesis of chemical molecules, as suggested by the formation of lipid-like membranes and pyramid shaped structures in cultures containing ACE pigments, but devoid of any living cells[97].

Of the numerous illnesses linked to stealth adapted virus infections, autism has been given special attention[100]. Anecdotal reports of significant clinical improvements have been received from parents of autistic children, who have followed my research and opted to administer HB-101, humic acids and/or zeolites, to their child.

A fresh approach to the potential energy-based therapy of children with autism stemmed from positive findings treating herpes simplex virus (HSV) infections using ultraviolet (UV) light in conjunction with the application of neutral red dye to the herpes virus induced skin lesions. The neutral red dye illumination with white fluorescent light was originally reported as being successful in the early 1970's[101]. The approach was largely discarded, however, when efforts at reproduction of the results were unsuccessful[102]. It is noteworthy that incandescent light rather than fluorescent white light was used in the seminal study arguing against the procedure, without realizing that fluorescent white lighting has a UV component absent from incandescent lighting. Dr. Jon Stoneburner who had chosen to use UV illumination, rather than fluorescent white light had independently discovered the benefit of using neutral red dye on herpes skin lesions. An incidental observation of Dr. Stoneburner was that the herpes skin lesions would fluoresce during therapy. Dr. Stoneburner and I were able to confirm the fluorescence as well as the expedited healing of HSV, herpes zoster virus (HZV) and human papillomavirus (HPV) skin lesions[103]. The fluorescence often extended beyond the area of the lesion to which the neutral red dye is added and could sometimes be readily seen within the skin surrounding the lesion. In patients with multiple herpes lesions, applying neutral red dye to a single lesion would result in healing of the other lesions, which would also fluoresce if directly exposed to UV light. The primary and secondary fluorescence of skin lesions would typically require a few minutes to fully develop and it would often begins to fade after 15-30 minutes. The skin lesions also typically show significant swelling due to ingress of serous fluid, which could actually proceed to form fluid droplets on the skin.

Because of FDA reluctance to approve the direct application of neutral red dye to herpes virus skin lesions, an effort was made to formulate a Lidocaine and neutral red containing homeopathic solution, which would behave comparably to the fluorescing material within HSV skin lesions. In preparing these formulations, it was observed that neutral red fluoresced in (>40%) ethanol similar to the fluorescence seen when added to HSV lesions. Moreover, in the dissolving process, the neutral red particles underwent marked kinetic movements. Particles of neutral red dye sprinkled onto a plastic dish containing ethanol would typically form linear streams as they partially dissolved. Remaining particles tend to aggregate into a focus from which

an occasional particle will be rapidly expelled, only to slowing return back towards the focus. As noted above, the abrupt dispersion of clustered particles had previously been observed with ACE pigments in long-term cultures of stealth adapted viruses. It had also been seen in freshly mined and finely ground humate particles suspended in water. The intensity of fluorescence and the kinetic activity of sprinkled particles of neutral red dye, could be maintained in lower concentrations of ethanol by using various energizing additives, including Lidocaine, various herbal tinctures and Brown's gas. Enhanced activity could also be noted in various alcoholic beverages beyond what was expected based solely on ethanol content. Diluting Lidocaine-energized ethanol solutions with water can also lead to a striking separation of the ethanol from the water, with Lidocaine forming a flocculent at the interface of the two liquids.

A 5-gallon homeopathic solution of a variety of tinctures based on the stated components in Enercel™, but containing 0.1% Lidocaine and 5% ethanol was prepared. It showed a bright yellow fluorescence when sprayed onto paper towels soaked in tap water containing neutral red dye (~1 mg/ml). This system was tested for its efficacy in treating HSV skin lesions. Specifically, a neutral red dye soaked paper towel was placed onto a Saran™ covered HSV skin lesion. The Lidocaine-ethanol containing formulation was sprayed onto the neutral red dye soaked paper towel and the towels illuminated for 30-60 minutes, with several repeat spraying of the formulation within this time period. This treatment was shown to be effective in expediting the healing of the underlying HSV skin lesion. Moreover, within a few minutes of starting the therapy, the underlying herpes skin lesion and surrounding skin could be shown to fluorescence upon direct illumination with the UV light.

A similar approach was extended to the treatment of children with autism[104], with convincing signs of clinical improvement and inducible areas of direct UV skin fluorescence, occurring even beyond the treated areas. The clinical improvement was not due to the fluorescence property of the formulated solution, since beneficial effects occurred using black plastic instead of Saran™ wrap. In spite of earlier striking successes, the solution lost its healing activity after several months. It was being stored in a large capped carboy and frequently succussed before removing samples for various uses. An explanation for the loss of activity was subsequently discovered and is explained below.

Supplies of regular and purportedly energized water have been obtained from various sources and tested for the dissolving pattern of sprinkled neutral red dye particles. Differences were seen in both the dissolving pattern of the particles of dye and in the repulsion and attraction movements of the remaining non-dissolved materials. With many water sources, stationary particles of neutral red dye would become surrounded by slowly extending concentric red discs of the dissolving dye. This pattern is typically seen using regular water and with some of the supposedly energized water. In other energized water samples, however, the sprinkled neutral red particles moved in a linear manner, with a tendency of the particles to then stop and begin to return back in a slow to-and-fro manner. The speed and extent of motion varied with water of different sources.

A range of water-energizing methods was then evaluated using the neutral red dye assay. These included the use of additives and of external energy sources. Discernable changes occurred with water exposed to various compounds including humic/fulvic acids, zeolites, magnesium oxide prills, Kiko pellets, various herbal tinctures, Brown's Gas and small amounts of previously energized water. The water energizing activity of humic/fulvic acids and zeolites can potentially be increased by procedures, such as high heat and strong acid treatment, which are designed to break intermolecular covalent bonding in favor of hydrogen bonding. Water activating effects were similarly seen with various external energy sources including those providing electrostatic fields, electromagnetic forces, magnetism and sound energy. Of special note, finely embossed aluminum foil, as is available in the form of holographic stickers, was able to transmit a dynamic energy to water. The heat and pressure used in the embossing process may well lead to finely variable level of surface oxidation that may essentially result in complex patterns of charge separation. From the perspective of manufacturing of medical devices with free (separated) electrical charges, embossed metal foil has an enormous advantage in its ease of production and in the simplicity of its potential external use on the body and on containers used to hold liquids.

The different water energizing approaches can be combined with generally additive effects on the linear motility of sprinkled neutral red particles. An exception to this basic observation occurs when the surface tension of the energized water is reduced such that the neutral red dye sinks into the water. Otherwise, this assay provided a reasonably useful semi-quantitative assessment of differences between water processed in various ways. With some of the more effective water energizing procedures, the water would become UV fluorescent upon the addition of the neutral red dye. Another observation with the more energized water was the polymerization of the neutral red dye into small fibers. There was also the formation of numerous vapor bubbles in the water, which tended to remain attached to the base of the dish.

The next major observation was that activated water in capped container, showed a progressive reduction in measured weight. Typically, within 1-2 hours, a reduction in weight of >0.5 mg/ml was observed with energized water compared to <0.1 mg/ml with regular water. Weight reduction occurred very slowly if the container was tape-sealed, suggesting that the weight loss was primarily, but not entirely resulting from the release of vapor; seemingly being more readily produced by the energized water, compared to regular water. This has been a consistently reliable observation, even though it has elsewhere been reported that orgone energy treated water may evaporate slower than untreated water in open dishes[65]. The newer observation argues that so-called homeopathic formulations, but actually examples of energized water, would be better stored in sealed containers. It also provided an explanation of the loss of activity of the formulation being tested on children with autism.

Energized ethanol with or without neutral red dye were placed into Ziploc bags and either the whole bag or smaller portions of the bag was thermally sealed, to yield flaccid, flat, fluid containing bags. An interesting and repeated observation was the progressive expansion of the sealed bags over several weeks. The bags slowly transformed from being flaccid to exerting a strong, outward pressure well beyond the 20 mm Hg, maximum measurement of an ophthalmic tonometer. This expansion was due to a markedly increased vapor pressure overlying the liquid, which clearly exceeded atmospheric pressure. Expansion did not occur using non-energized ethanol and the observation using energized

ethanol clearly defies the proportional relationship of volume, temperature and pressure of ethanol vapor. Thermally sealed Ziploc bags containing powdered dry ice can be expanded to a similar high pressure, but the pressure is not maintained even when residual dry ice is present. This is because the sealed bag is quite permeable to the carbon dioxide gas. In contrast, the greatly increased pressure from the activated ethanol containing bags becomes a stable finding lasting beyond many months, with a slowly decreasing volume of remaining liquid. This suggests that the polyethylene plastic bags probably do allow for the slow passage of ethanol vapor. As such, the maintaining of the high pressure over many months points to a continuing energy input into the ethanol. The increased vapor pressure exists in a form in which it could be potentially converted to mechanical energy, realizing one of Tesla's goals of freely extracting useful energy from the environment. Figure 1 shows an example of the expansion of an originally flaccid portion of a sealed Ziploc bag containing energized ethanol plus neutral red dye. Expansion also occurs with energized ethanol alone.

A degree of transient vaporization of various energized solutions is also apparent in the crackling sound heard upon succussion of a closed container, even when an overlying air layer is not present. When air is present, a container of energized water will commonly respond to vigorous shaking or succussion with a temporary spinning vortex of vapor bubbles.

Yet another striking observation is the rapid bursts occurring in bubbles, which slowly form when a small quantity of a humic acid solution is added to energized ethanol. Individual bubbles abruptly release an outward acting force as the previously slowly expanding process reaches a threshold leading to the bubble nearly completely disappearing. After several seconds, the small residual bubble will again slowly expand and once more collapse upon quickly releasing a repulsive (kinetic) force.

Summary and Conclusions

The basic postulate of this article is that the strength of attraction of opposing electrical charges is subject to a countervailing force, so as to prevent the charges from fusing with each other. This force is potentially breached, as in the energy-dependent formation of neutrons from the fusion of protons (positive charge) with electrons (negative charge). The charge-attracting, anti-fusion force comes from the environment and can exert an influence on other spatially separated opposing charges. In particular, it can reduce the strength of intermolecular hydrogen bonding of both water and ethanol molecules, leading to significantly enhanced vaporization of individual molecules from these liquids. More importantly, the attached energy can add to the kinetic (movement) activity of liquid molecules and that of certain compounds. It can presumably also add to the capacity of the water and compounds to support biological processes.

A fundamental process of even regular metabolism is the splitting of water into separate hydrogen and oxygen atoms. The hydrogen atoms are in turn converted to free positively charged protons and negatively charged electrons. The separation of positive from negative charges requires energy to counteract the attraction between opposing electrical charges. It is possible that KELEA facilitates the separation of the charges, with the added energy being regained as regular metabolic energy, as in the production of ATP from the passage of protons through the ATP synthase enzyme[105].

Denoting KELEA as a kinetic energy is consistent with apparent repulsive force seen in some of its manifestations. Repulsion is clearly seen with the rapid movement of individual neutral red dye particles away from clusters of non-dissolved particles in energized water and especially in energized ethanol. Another example may be the reported prolongation of the discharge times upon repeated use of an electroscope, as first reported by Reich and attributed to orgone energy[106]. Attraction followed by repulsion can also be commonly observed in various simplified electrostatic experiments, yet is not generally discussed. Since, there can be transduction of energy between liquid molecules, an analogy can also be drawn to the working of a laser. A further consideration, is that the spacing of opposing charges in molecules may have a quantum influence on the release and subsequent regaining of KELEA, such that the effect can still be rather remarkable with distantly separated charges in large molecules, colloids, etc.

Certain chemicals, including humic/fulvic acids, zeolites, herbal tinctures, homeopathic solutions, Lidocaine and neutral red dye, appear to be particularly effective in capturing KELEA and mediating its transfer to water. Activated ethanol is a simple and effective provider of KELEA or a derivative energy to water. The transfer of energy to water is not dependent upon direct physical contact, as is also shown by the apparent transmission of energy from various electrical and magnetic devices, occurring upon their rapid, abrupt, periodic discharging. The unifying concept is that a range of both chemicals and electrical devices may operate through the common mechanism of KELEA mediated activation of the body's water. This concept provides an important foundation to well designed clinical trials. Of added interest is to explore the possibility that the body's water can be activated even more directly by utilizing an intrinsic biological capacity, which may exist in all life forms, including humans.

The immediate opportunity exists to assess the beneficial outcomes of working with KELEA activated water or directly with KELEA itself in agricultural applications and in animal and human clinical studies. A collective effort by many participants will greatly facilitate worldwide progress in applying the paradigm espoused in this article. An immediate challenge is the development of simple, reliable assays for individuals to measure the energetic levels of their ACE pathway and that of animals and plants. One approach with humans and animals is to determine the extent to which samples of body's fluids can be further energized, using one or more of the procedures discussed in this article. ACE deficient individuals may also express easily discernable levels of non-fully charged ACE pigments, assessable by their UV fluorescence of body fluids, with and without the addition of neutral red dye solution. Indeed, certain stealth virus infected patients produce many fluorescing ACE pigment particles and fibers, which can be irritating to the skin. This clinical condition is referred to as Morgellon's disease[107] and seemingly represents an attempted healing process that has gone awry. Small numbers of particles and a single fiber from a Morgellon's patient readily activated water, as shown by formation of numerous vapor bubbles, diminution of weight (>1.0 mg/ml in an hour), induced fluorescence and other measures.

Clinical studies are also underway to assess the efficacy and optimal use of KELEA activated water, more conveniently termed ACE™ Water, in medical conditions potentially attributed to an insufficiency of cellular energy (ICE). Certainly, prior successes reported for homeopathy, ought to be easily surpassed using the more highly energized water products, which can now be easily and inexpensively produced. So too, consumption of ACE Water™ may well duplicate if not exceed many of the reported benefits associated with the use of numerous natural products and energy devices. Eating food cultivated with ACE Water™ will also likely lead to substantial health benefits, as well as to increased productivity of the crops, along with less need for pesticides. The diffusion pattern of sprinkled neutral red dye is a reasonable initial screening procedure, except in circumstances in which the surface tension is so low that the particles submerge below the surface. The next easiest assay to perform is fluorescence of the water at an approximately neutral pH. A variety of additional assays are available, including the capacity to energize ordinary water and/or the extent to which the water itself can be further energized.

Finally, the basic concept of KELEA clearly lends itself to major efforts to detect this unconventional form of energy. Many of the simple observations already recorded are amenable to further analysis using more sophisticated equipment. Results from these studies will in turn lead to improved methods of generating and conveying KELEA for biological purposes, as well as understanding its inherent nature. It is even possible that KELEA may relate to the dark energy, which has eluded physicists.

Footnotes and Selected References:

1. Pangman MJ. Evans M. (2011). Dancing with Water: The New Science of Water. Uplifting Press, Coalville, UT. ISBN: 978-0-9752726-2-6.

2. Lower S. (2013). Water science pseudoscience, fantasy and quackery. http://www.chem1.com/CQ/

3. Pollack GH. (2013). The Fourth Phase of Water, Enner and Sons, Seattle, WA.

4. Collings PJ. (2002). Liquid Crystals. Princeton University Press. Princeton NJ.

5. Ellis J. (2013). www.johnellis.com

6. Wiggins P. (2008). Life depends upon two kinds of water. PLoS ONE 3: e1406. doi: 10.1371/journal.pone.0001406.

7. Xia M-S, Hu C-H, Zhang H-M. (2006). Effects of tourmaline addition on the dehydrogenase activity of Rhodopseudomonas palustris. Process Biochem 41: 221-225.

8. http://excelexgold.com/

9. http://en.wikipedia.org/wiki/Cherenkov_radiation

10. http://en.wikipedia.org/wiki/Sonoluminescence

11. http://www.starchamberproducts.com/

12. Shul Yin L. (2012). Stable water clusters, meridians, and health. Forum Immunopathological Dis. Ther. 3: 193-219.

13. http://www.eagle-research.com

14. Batmanghelidj F. (1997). Your Body's Many Cries for Water. Global Health Solutions, Falls Church, VA.

15. http://www.youtube.com/watch?v=RFr7H65rznE

16. http://en.wikipedia.org/wiki/Kirlian_photography

17. Jaffrin MY, Morel H. (2008). Body fluid volumes measurements by impedance: A review of bioimpedance spectroscopy (BIS) and bioimpedance analysis (BIA) methods. Med Eng Phys. 30: 1257-69.

18. Agre P, Preston GM, Smith BL, Jung JS, Raina S, Moon C, Guggino WB, Nielsen S. (1993). Aquaporin CHIP: the archetypal molecular water channel. Am J Physiol. 265: 463–76.

19. Chaplin. (2007). The memory of water; an overview. Homeopathy 96: 143-150.

20. Montagnier L, Issa JA, Ferris S, Montagnier J-L, Lavallee C. (2009). Electromagnetic signals are produced by aqueous nanostructures derived from bacterial DNA sequences. Interdiscip Sci Comput Life Sci 1: 81–90.

21. www.kikotechnology.com

22. http://www.fractalwater.com

23. www.cwt-international.com

24. http://www.plasmaquinton.com

25. Zeolites. Journal published by Elsevier Press. Amsterdam.

26. www.humicsubstances.org

27. http://en.wikipedia.org/wiki/Clathrate_compound

28. Rivest JB, Jain PK. (2013). Cation exchange on the nanoscale: an emerging technique for new material synthesis, device fabrication, and chemical sensing. Chem Soc Rev. 42: 89-96.

29. http://emrousa.com/

30. http://matse1.matse.illinois.edu/ceramics/ware.html

31. Howenstine J. (2007). Improve drinking water with prill beads. http://www.newswithviews.com/Howerstein/james54.htm

32. http://www.hb-101.com/

33. http://en.wikipedia.org/wiki/Fluorescence

34. http://www.edgarcayce.org

35. Makino S. (1999). The Miracle of Pi-Water. IBE Co. Nagoya, Japan.

36. Emmons C. (2009). ORMUS: Modern Day Alchemy. Dreamgate Press, Santa Ana CA.

37. Nakao A, Toyoda Y, Sharma P, Malkanthi Evans M, Guthrie N. (2010). Effectiveness of hydrogen rich water on antioxidant status of subjects with potential metabolic syndrome—An open label pilot study. J Clin Biochem Nutr. 46: 140–149.

38. Xia C, Liu W, Zeng D, Shu L, Sun X, Sun X. (2013). Effect of hydrogen-rich water on oxidative stress, liver function, and viral load in patients with chronic hepatitis B. Clin Trans Sci. 6: 372-375.

39. Hayashi H. (2013). Hydrogen Rich Water Guidebook. http://new-water.org/world/index.html

40. http://www.townsendletter.com/Dec2006/ThreeYears of HCl Therapy.pdf

41. Marcy EE, Hunt FW. (1868). The Homeopathic Theory and Practice of Medicine. William Radde publisher, New York

42. Mandal PP, Mandal B. (2001). A Textbook of Homeopathic Pharmacy. B. Jain. Kokata, India.

43. Mathie RT, Hacke D, Clausen J, Nicolai T, Riley DS, Fisher P. (2013). Randomized controlled trials of homeopathy in humans: characterizing the research journal literature for systematic review. Homeopathy 102: 3e24.

44. Zanasi, A., Mazzolini, M., Tursi, F., Morselli-Labate, A.M., Paccapelo, A., Lecchi, M. (2013). Homeopathic medicine for acute cough in upper respiratory tract infections and acute bronchitis: A randomized, double-blind, placebo-controlled trial. Pulm Pharmacol Ther pii: S1094-5539(13)00125-9. doi: 10.1016/j.pupt.2013.05.007.

45. Altunc U, Pittler MH, Ernst E. (2007). Homeopathy for childhood and adolescence ailments: systematic review of randomized clinical trials. Mayo Clinic Proc 81: 69–75.

46. http://wjohnmartin.blogspot.com/2004/10/synopsis-of-research-by-w-john-martin.html

47. www.enercel.com

48. Oberbaum M, Markovits R, Weisman Z, Kalinkevits A, Bentwich Z. (1992) [Wound healing by homeopathic silica dilutions in mice]. Harefuah. 123: 79-82.

49. http://en.wikipedia.org/wiki/Ana_Aslan

50. Dreyfus J. (1988). A Remarkable Medicine has been Overlooked. Dreyfus Medical Foundation, New York.

51. Norrie PA (2003). The history of wine as a medicine *in* Sandler M, Pinder R. Wine a Scientific Exploration. Taylor and Francis, New York.

52. http://www.grander.com/

53. http://energymugs.com/

54. http://stirwands.com/

55. Alexandersson O. (2002). Living Water: Viktor Schauberger and the Secrets of Natural Energy. ISBN: 978-0717133901

56. Coats C. (2001). Living Energies. Gateway, Dublin ISBN: 978-0717133079

57. Schwenk T, Schwenk W. (1989) Water: The Element of Life. Anthroposophical Press.

58. http://anchicrystals.com/Research.html

59. Faris AS, et al. (2012). Effects of magnetic field on fuel consumption and exhaust emissions in two-stroke engine. Energy Procedia 18: 327–338.

60. http://en-wikipedia.org/wiki/Royal_Rife

61. http://en-wikipedia.Georges_Lakhovsky

62. http://en-wikipedia.Edgar_Cayce

63. http://www.papimi.gr/

64. Mazzocchi A, Maglione R (2010). A preliminary study of the Reich orgone accumulator effects on human physiology. Subtle Energies and Energy Med. 21: 41-50.

65. DeMeo J. (2011) Water as a resonant medium for unusual external environmental factors. Water J. 3: 1-47.

66. http://www.angelfire.com/tn/intrasound/page7.html

67. http://en-wikipedia.Franz_Mesmer

68. http://en-wikipedia.Carl_Reichenbach

69. http://en-wikipedia.Augustus_Pleasonton

70. http://en-wikipedia.Albert_Abrams

71. http://en-wikipedia.Nikola_Tesla

72. Becker RO, Selden G. (1985). The Body Electric. William Morrow New York.

73. Emoto M. (2004). The Hidden Messages in Water. Beyond Words Publishing, Hillsboro, OR.

74. Emoto M. (2004). Healing with water. J Alt Comp Med. 10: 19-21.

75. Gilkeson J. (2000). Energy Healing: A Pathway to Inner Growth. Marlow, New York.

76. http://www.reiki.org/faq/whatisreiki.html;

77. http://healthyogalife.com

78. Teixeira PC, Rocha H, Coelho Neto JA. (2010). Johrei, a Japanese healing technique, enhances the growth of sucrose crystals. Explore (NY) 6: 313-323

79. http://www.laughteryoga.org

80. http://www.newadvent.org/cathen/07432a.htm

81. http://www.substech.com/dokuwiki/doku.php?id=stabilization_of_colloids

82. Chaplin M. (2013). http://www1.lsbu.ac.uk/water/

83. Tesla N. (1900). The Problems of Increasing Human Energy. Century Magazine June

84. Tesla N. (1901). Method of Utilizing Radiant Energy. Patent No. 685,958 Issued November 5, 1901.

85. http://www.teslasociety.ch/info/NTV_2011/free.pdf.

86. Vassilatos G. (2013) http://aetherforce.com/the-true-story-of-teslas-radiant-matter/

87. Wachsmuth G. (1932). Etheric Formative Forces in Cosmos, Earth and Man. Translated by Olin Dantzler Wannamaker, Anthroposophic Press, New York.

88. http://www.aquapol.co.uk/

89. http://en.wikipedia.org/wiki/Laser

90. Mason F, Richardson R. (1983). Why doesn't the electron fall into the nucleus? J Chem. Ed. 60: 40-42.

91. http://www.thenakedscientists.com/forum/index.php?topic=44598.0

92. Martin WJ. (2013). Stealth Adaptation of Viruses. Review of earlier studies and updated molecular analysis on a stealth adapted African green monkey simian cytomegalovirus (SCMV). Omics Group Scientific Reports 2: 794. Doi: 10.4172

93. Martin WJ. (2003). Stealth virus culture pigments: A potential source of cellular energy. Exp. Mol. Path. 74: 210-223.

94. Martin WJ. (2003). Complex intracellular inclusions in the brain of a child with a stealth virus encephalopathy. Exp Mol Path 74: 179-209.

95. Martin WJ, Glass RT. (1995). Acute encephalopathy induced in cats with a stealth virus isolated from a patient with chronic fatigue syndrome. Pathobiology 63: 115118.

96. Martin WJ. (1996). Severe stealth virus encephalopathy following chronic fatigue syndromelike illness: Clinical and histopathological features. Pathobiology 64:18.

97. Martin WJ. (2005). Etheric Biology. Exp Mol Path 78: 221-227.

98. Solis-Herrera A, Arias-Esparza MC, Solis-Arias RI, Solis-Arias PE, Solis-Arias MP. (2010). The unexpected capacity of melanin to dissociate the water molecule fills the gap between life before and after ATP. Biomed Res. 21: 224-226.

99. Adamski AG. (2013). Quantum nature of consciousness and the unconscious collective of Carl G. Jung. Neuroquantology 11: 466-476.

100. Martin WJ. (1995). Stealth virus isolated from an autistic child. J Aut Dev Dis. 25: 223224.

101. Melnikoff RM. (1972). Photodynamic inactivation of herpes simplex infections. Calif Med.116: 51–52.

102. Myers MG, Oxman MN, Clark JE, Arndt KA. (1976). Failure of neutral-red photodynamic inactivation in recurrent herpes simplex virus infections. New Eng J Med. 293: 945-949.

103. Martin WJ, Stoneburner J. (2005). Symptomatic relief of herpetic skin lesions utilizing an energy based approach to healing. Exp. Mol. Path 78: 131-134.

104. Martin WJ. (2013). Alternative cellular energy (ACE) pathway activation in the therapy of autism. J Alzheimers Dis Parkinsonism. (Abstract) doi: 10.4172/2161-0460.S1.002.

105. Sherratt HS. (1991). Mitochondria: structure and function. Rev Neurol. 147: 417-430.

106. Spitzer RL (1953). Wilhelm Reich and Orgone Energy: http://www.srmhp.org/0401/reich.html.

107. Savely VR, Stricker RB. (2010). Morgellons disease: Analysis of a population with clinically confirmed microscopic subcutaneous fibers of unknown etiology. Clin Cosmet Investig Dermatol. 13: 67-78.

Figure 1. A photograph showing the marked expansion and stretching of a thermally sealed portion of a Ziploc bag, into which a solution of energized ethanol plus neutral red dye had been added before sealing. The bag was initially flaccid, but became progressively expanded and markedly turgid over the ensuring several weeks. This particular container has remained expanded for over 9 months with a noticeable slow decrease in the residual fluid levels. Similar expanded bags have burst from their internal pressure, while others have remained expanded without any residual liquid. Equivalent marked expansion occurs to sealed portions of Ziploc bags containing energized ethanol without added neutral red dye. These examples do not photograph as clearly as those containing the dye. No expansion occurs using regular ethanol with or without neutral red dye.

Preface to Section 2

The world is facing a major crisis in the impaired physical and mental health of many of its inhabitants. In the US, over half of adults have one or more chronic medical illness. Psychiatric illnesses are also pervasive and not uncommonly occur in conjunction with medical conditions. Illnesses have a huge negative impact on the economic productivity of Nations by limiting educational achievements and by reducing both the quantity and quality of work performance. Illnesses also divert substantial financial and human resources to the needs of patient care. Mental illnesses render whole communities at risk for irrational, aberrant and criminal behaviors; typically reflecting the perpetrators' disregard for inflicting human suffering. Mentally impaired individuals are themselves at added risk for commercial exploitation in various business dealings, adding to their sense of despair. Parents are especially troubled by mental illnesses occurring in their children; with increasing risks of autism, learning and behavioral disorders, suicide and later life poverty

Several explanations for the deterioration of the Nation's mental health have been offered, including environmental toxins, electromagnetic pollution, poor nutrition, lack of regular exercise, infections with conventional pathogens, etc. A role for genetics has also been emphasized to explain particular susceptibility of certain individuals to various environmental factors. In spite of huge expenditures of time and money, very little progress has been made in stemming the worldwide upward trending of chronic illnesses.

This book is focused on the reality of a stealth adapted virus epidemic being a major contributing factor in the development of many of the chronic illnesses with major neurological and/or psychiatric manifestations. Later publications will address the potential role of stealth adapted viruses in cancers and in other illnesses, including the triggering of autoimmune diseases in humans and in animals.

Brain malfunction can lead to chronic illnesses outside of the normal purview of neurologists and psychiatrists. For example, lowered pain threshold can greatly amplify symptoms ascribed to musculoskeletal disorders. The autonomic nervous system regulates cardiovascular, gastrointestinal, hormonal and other activities of the body. Disruption of this aspect of brain function can, therefore, manifest in diseases occurring in these other organ systems. Altogether, there are many possible consequences of stealth adapted virus induced impaired functioning of the brain. An understanding of the myriad of chronic illnesses potentially caused by stealth adapted viruses may help reduce the somewhat futile endeavors to classify chronically ill individuals into precisely defined and discrete clinical entities.

The book is also focused on a new therapeutic paradigm based on enhancing the alternative cellular energy (ACE) pathway. Cellular energy is not only derived from the metabolizing of food, but also from the ACE pathway. This pathway clearly has a role as an adjunct to the immune system in the suppression of conventional infectious diseases and is uniquely able to suppress infections caused by stealth adapted viruses. It is also especially suited to suppress infections when the immune system is rendered inactive, as in the case of AIDS and in patients on chemotherapy.

The potential therapeutic applications of the ACE pathway extend well beyond infectious diseases. Many chronic illnesses can be directly ascribed to an insufficiency of cellular energy (ICE). This can occur from restricted delivery of oxygen, blood supply and/or nutrients to tissues as well as from various intrinsic metabolic anomalies. In a preliminary study, the blood oxygen level of a patient with emphysema was significantly increased using the same approach as earlier described for treating children with autism.

The ACE pathway can be conveniently enhanced using KELEA absorbing enerceutical™ foods and liquids, including water. Moreover, the mind may be capable of delivering KELEA to the body. If so, it would validate many beliefs in faith healing and in the importance of a positive mental attitude in the prevention and suppression of illnesses.

Extended clinical testing is certainly needed to validate and to optimize the application of ACE pathway based therapies; while also providing further support to the underlying concepts. Agricultural practices can also be upgraded to bring KELEA into the cultivation of food and animal husbandry.

The enormity and urgency of moving forward with these and related endeavors is daunting. Rather than continuing at the usual pace of making incremental advances, far more could be accomplished with the widespread involvement of many of the existing institutions devoted to human welfare. This entire book is intended as a motivating tool for others to engage in a worldwide cooperative endeavor to utilize and extend ACE pathway based science.

Public health authorities have been quick to besmirch the research since it has a negative bearing on the safety of vaccines. Even more powerful has been the opposition to the linkage of stealth adapted virus contamination of an experimental poliovirus vaccine to the origin of AIDS. Bringing this provocative topic into the open will hopefully fuel added interest in the research and encourage more people to become involved.

The additional material in the second Section of the book is, therefore, intended to appeal to a wide collection of various activists and patient support groups. The diversity of topics is in keeping with creating a major coalition to help translate the research to everyday practical applications. Including somewhat repetitive articles written at different time periods is a testament to both the persistence of the author's efforts and the obstinacies of public health authorities.

Topic 1 is the proposed role of cytomegalovirus contamination of the CHAT polio vaccine in the transformation of chimpanzee simian immunodeficiency virus (SIV) to the human immunodeficiency virus (HIV). The included sketches outlining the manufacturing of poliovirus vaccines and of an SIV infected chimp receiving a contaminated polio vaccine were prepared in 1996. When shown at conferences, the evoked response has typically been a warning to anticipate retaliation. This is a long way off honest scientific inquiry.

While the tragic emergence of HIV was clearly inadvertent, the unwillingness of public heath authorities to review possible past errors is more protective of themselves than the public at large. Public health authorities have also publicly stated that stealth adapted viruses do not exist and have rejected any suggestions for pursuing vaccines as a possible source for certain stealth adapted viruses. Topic 2 addresses the Public Health nonfeasance in this matter. It includes documents illustrative of early efforts to have the Food and Drug Administration (FDA) and the Centers of Disease Control and Prevention (CDC) seriously consider SCMV-derived stealth adapted viruses. The first of these documents is a Handout, which was provided at a "Twentieth Century Plagues Symposium" held in Los Angeles March 1st, 1996. This document also details additional information regarding the development of polio vaccines. It records the indifferent responses of both FDA and CDC to unsolicited proposals for extended research on SCMV. The document has been online since the time of the Symposium at several web sites including www.s3support.com

The next document "Political and Economic Compromises Affecting Public Health: Lessons from Contaminated Polio Vaccines" is a more personalized account of some of the public health hurdles encountered with the research. This article has also been openly available on the Internet. It is followed by a copy of an unanswered e-mail sent to CDC following a discussion relating to a poster presentation at a year 2000 conference. A copy of poster presented at a CDC sponsored conference in March of 2002 is included under this topic.

The research was also presented at a 2002 Institute of Medicine (IOM) Workshop entitled "The Infectious Etiology of Chronic Diseases." The conference was attended by and partly organized by the Director of the Infectious Disease Division of the CDC. Following the Workshop, I submitted an article for inclusion in the Proceedings. The article was conveniently omitted from the Proceedings of the Workshop. It has been online since and is appropriately also included in this book.

The third topic covered in this part of the book is related to vaccines. It begins with a broad review of vaccines first published in 2004. A transcript of a more recent presentation given in July 2013 at Las Vegas follows. This presentation addresses the unnecessary use of excessive amounts of adjuvants when multiple vaccines are administered and raises the specific issue of adjuvants potentially provoking cellular immunity against some residual minor antigens expressed on stealth adapted virus infected brain cells. This adverse event can explain the encephalitis-like illness occurring in some adolescents receiving the human papillomavirus (HPV) vaccine. (An earlier Internet published comment I made on this topic is included).

The remaining topics are more clinically oriented. They deal with the limitations in the practice of psychiatry. A description of multiple illnesses occurring in a family is then provided. This article was also rejected for publication in a CDC sponsored Journal of Emerging Infectious Diseases.

A selection of specific clinical conditions is then discussed. These include CFS; autism; epilepsy; Morgellon's disease (in which the skin fibers are identified as ACE pigments); so called "chronic Lyme disease" (most cases of which are due to stealth adapted viruses); post traumatic stress syndrome (PTSD); myofascial disorders; Alzheimer's disease; drug addiction; criminal behavior and multiple myeloma. A speculative article on obesity/diabetes is added to emphasize the wide range of rethinking about diseases that can be inspired by the concept of there being stealth adapted viruses.

While somewhat repetitive in their message and in listed references, the articles and comments reinforce the basic messages of the book. "Stealth adapted viruses exist and may explain many chronic illnesses." The second and far more promising message of the book relates to the enormous potential for improving world health by enhancing the ACE pathway. Again, while the immediate focus is on illnesses caused by stealth adapted viruses, the ACE pathway is in many ways better than the immune system in combating conventional virus, bacterial and fungal infections in humans and in animals. For example, on the negative side, the immune response can help in disseminating certain infectious diseases. It can also lead to undue cellular damage of neighboring cells by the accompanying inflammation, which can also result in permanent scarring. The immune system is also capable of causing cellular and tissue damage through auto-immunity.

As noted earlier, the potential value of the ACE pathway extends well beyond infectious diseases to the many medical conditions, which can be viewed as due to ICE, as well as a means of providing scarless healing of wounds. Enhancing the ACE pathway also has widespread applications in agriculture. Not all of these topics could be covered within the confines of this book.

The book ends with a call for action to implement the scientific findings and for the formation of groups to help coordinate international efforts and document progress. The work really deserves a "Manhattan" style international empowerment endeavor.

Stealth Adapted Virus Contamination of Experimental Polio Vaccines and the Origin of AIDS

The discovery of stealth adapted viruses hit its first major political hurdle when the initial isolate was identified as originating from an African green monkey simian cytomegalovirus (SCMV). This finding implied that the virus was probably a contaminant of polio vaccines produced using cytomegalovirus (CMV) infected cultured kidney cells from this species of monkeys (1). An even bigger political hurdle arose when the research further indicated a potential linkage between CMV contaminated experimental polio vaccines and the origin of the AIDS virus.

Human CMV was identified from the beginning of the AIDS epidemic as being an important co-pathogen with HIV and responsible for many of the more serious consequences, such as retinitis and encephalitis (2). Some researchers, notably Dr. Peter Duesberg of the University of California Berkeley, actually argued that HIV alone was non-pathogenic and illness was dependent upon other factors, such as CMV (3). Primate studies conducted by Dr. Peter Barry and colleagues of the University of California Davis, convincingly showed synergism between simian immunodeficiency virus (SIV) and CMV (4-5). One of the possible mechanisms of synergy was that CMV coded for a protein (called US28 in human CMV), which can be used as a cell entry receptor by HIV (6).

I first became interested in a possible association between stealth adapted viruses and HIV when I learned that a close, but not intimate friend of the patient from whom I first isolated a stealth adapted virus, shared many of her symptoms. I cultured his blood and observed that he too was strikingly stealth adapted virus culture positive. Additional testing by his physician showed that he was also HIV positive. I concluded that stealth adapted viruses might be important co-pathogens for HIV and may certainly account for some of the neuropsychiatric symptoms encountered by AIDS patients. I also surmised that stealth adapted CMV might have participated in the initial development of HIV from the simian immunodeficiency virus (SIV) of African primates and prepared slides (Figures 1 and 2) to illustrate this possibility.

In his book, "The River," Edward Hooper specifically argued that HIV arose from the area in the Belgium Congo, where Dr. Hiliary Koprowski had tested his experimental polio vaccine, termed CHAT (7). Molecular studies had confirmed that HIV arose from simian immunodeficiency virus (SIV) of chimpanzees. Mr. Hooper wrongly assumed that Dr. Koprowski's CHAT vaccine was produced using cultured kidney cells of chimpanzees. His overall efforts were undermined when it was later shown that the CHAT vaccine was produced, not using chimpanzees, but using cultured kidney cells of rhesus monkeys (8).

Both recorded in Mr. Hooper's book and confirmed by first hand discussions with an individual working in Dr. Koprowski's Congo facility, chimpanzees were widely used in the testing of the CHAT vaccines. I was also informed that minced kidneys from chimpanzees were routinely cultured and inoculated with polio vaccine virus. Of special note were the illnesses developing in the chimpanzees with many unexplained deaths. A contractor, Mr. Ricky Mann, had to go to enormous efforts to continue the supply of chimpanzees, obtaining them from all parts of Africa. Several of the chimpanzee animal handlers also became sick with a wasting disease referred to as "thin man" syndrome. Some of them developed rhinoscleroma of the nose and had other indications of severe Klebsiella bacterial infections, consistent with immunodeficiency.

The probable contamination of the CHAT vaccine was described by Dr. Albert Sabin in one of his published articles from the time (9). He observed a cytopathic effect (CPE) from the vaccine not attributed to poliovirus, yet it was a difficult virus to culture. In his correspondence Dr. Koprowski took offense at the criticism from Dr. Sabin and in reply Dr Sabin stated "Farewell my one time friend."

In 1958, Dr. Kopowski arranged for the collection of some 340 sera of Congolese children before and after inoculation with the CHAT vaccine. The sera were provided to the United States Public Health authority for confirmation of the production of anti-polio antibodies. Aliquots were said to be stored at Fort Dietrich, Maryland. My early suggestion to public health authorities that these sera be retrieved and tested for anti-RhCMV antibodies fell on deaf ears.

This is no longer necessary, since it has been confirmed that the CHAT vaccine contains DNA of rhesus cytomegalovirus (RhCMV). This information was provided in molecular studies conducted by the British Bureau of Standards in a 2005 publication (10).

Public health officials have not openly addressed the issue of illnesses developing in chimpanzees' and their handlers. Rather, efforts have been made to highlight publications, which unconvincingly suggest an earlier formation of HIV. For example, a specious argument was raised in a publication from Los Alamos Laboratory that on the basis of the current rate of mutation and the known divergence of quasispecies of HIV, that the original virus must have arisen prior to the 1950's (11). I inquired from one of the authors if this conclusion would hold if a second virus, for example, CMV was present during the early emergence of HIV. I was bluntly told that this would invalidate their conclusion but they did not want to even raise the issue. As it stands powerful objective arguments can be made for HIV as an inadvertent consequence of efforts to develop a polio vaccine.

References

1. Smith KO, Thiel JF., Newman, JT., Harvey E., Trousdale, MD., Gehle, WD., Clark, G. (1969). Cytomegaloviruses as common adventitious contaminants in primary African green monkey kidney cell cultures. J Natl Can Inst. 42: 489-496.

2. Jacobson MA, Mills J. (1988). Serious cytomegalovirus disease in the acquired immunodeficiency syndrome (AIDS). Clinical findings, diagnosis, and treatment. Ann Intern Med. 108: 585-94.

3. Duesberg PH. (1989). Human immunodeficiency virus and acquired

immunodeficiency syndrome: Correlation but not causation. Proc. Natl. Acad. Sci. USA, 86: 755-764.

4. Barry PA, Pratt-Lowe E, Unger RE, Marthas M, Alcendor DJ, Luciw PA. (1990). Molecular interactions of cytomegalovirus and the human and simian immunodeficiency viruses. J Med Primatol.19: 327-37.

5. Sequar G, Britt WJ, Lakeman FD, Lockridge KM, Tarara RP, Canfield DR, Zhou SS, Gardner MB, Barry PA. (2002). Experimental coinfection of rhesus macaques with rhesus cytomegalovirus and simian immunodeficiency virus: pathogenesis. J Virol. 76: 7661-71.

6. Rucker J, Edinger AL, Sharron M, Samson M, Lee B, Berson JF, Yi Y, Margulies B, Collman RG, Doranz BJ, Parmentier M, Doms RW. Utilization of chemokine receptors, orphan receptors, and herpesvirus-encoded receptors by diverse human and simian immunodeficiency viruses. J Virol. 71: 8999-9007.

7. Edward Hooper, The River: A Journey Back to the Source of HIV and AIDS (Harmondsworth: Penguin; Boston: Little, Brown, 1999.

8. Sabin AB. (1959). Present position on immunization against poliomyelitis with live virus vaccines. Brit Med J. 1 (5123): 663-80.

9. Berry N, Jenkins A, Martin J, Davis C, Wood D, Schild G, Bottiger M, Holmes H, Minor P, Almond N. (2005). Mitochondrial DNA and retroviral RNA analyses of archival oral polio vaccine (OPV CHAT) materials: evidence of macaque nuclear sequences confirms substrate identity. Vaccine. 23: 1639-48.

10. Baylis SA, N. Shah N, A. Jenkins A, Berry NJ, Minor PD.(2003). Simian cytomegalovirus and contamination of oral poliovirus vaccines. Biologicals, 31: 63–73.

11. Korber B, Gaschen B, Yusim K, Thakallapally R, Kesmir C, Detours V. (2001). Evolutionary and immunological implications of contemporary HIV-1 variation. Br Med Bull 58: 19-42.

Polio Vaccines

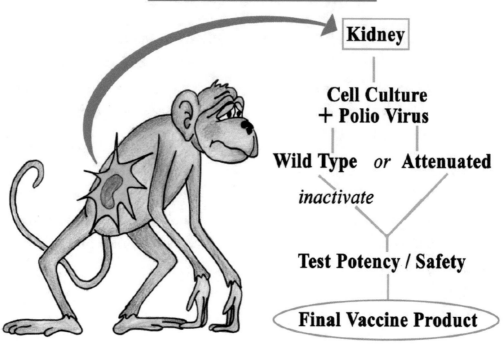

Kidney

Cell Culture
+ Polio Virus

Wild Type   *or*   Attenuated

inactivate

Test Potency / Safety

Final Vaccine Product

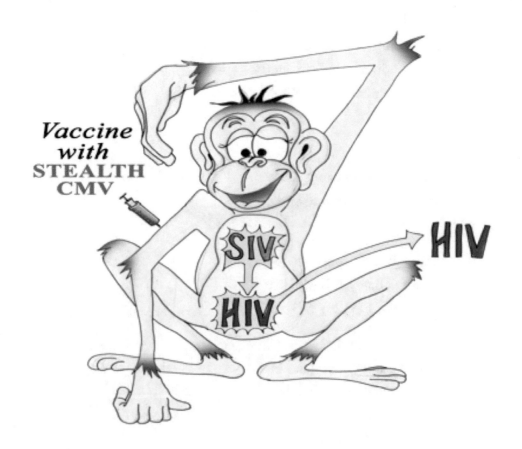

Vaccine
with
STEALTH
CMV

SIV

HIV

HIV

(Posted on the World Trade Organization Discussion Group 12/12/02)

Stealth Adapted Rhesus Cytomegalovirus and the African Origin of AIDS

(Posted by Dr. Stephen B. Palmer, a colleague)

According to Dr. W. John Martin, M.D., Ph.D., Founder of the Center for Complex Infectious Diseases, Rosemead California, cytomegaloviruses contaminating experimental poliovirus vaccines in Central Africa may have provided the genetic push allowing a relatively harmless simian immunodeficiency virus (SIV) of chimpanzees to evolve into the human immunodeficiency virus (HIV) that causes AIDS.

Potential contamination of polio vaccines has been a smoking gun since a 1972 study showed cytomegalovirus in kidney cultures from African green monkeys. Recent testing by the United States Food and Drug Administration (FDA) showed that monkey cytomegalovirus DNA was still detectable in some of the vaccine lots produced after the 1972 discovery. Contamination certainly would be expected in some early experimental lots of polio vaccines extensively studied in Congo and other parts of Africa.

Cytomegaloviruses promote the growth of retroviruses, such as SIV and HIV, by providing a receptor molecule allowing for the entry of the virus into cells. The cytomegaloviruses of monkeys used to make polio vaccines have more SIV and HIV receptor molecules than do the cytomegaloviruses of either chimpanzees or humans. Martin showed this in an African green monkey cytomegalovirus, and others have shown it with rhesus monkey virus.

Pre and post immunization sera were collected from 340 Congolese children immunized with an experimental polio vaccine in 1958. Testing of these stored sera for vaccine induced anti-cytomegalovirus antibodies would address the issue of cytomegalovirus being present in the vaccine.

Martin has also been told of the use of kidneys taken directly from chimpanzees to grow polio vaccine virus. This information did not come to light during the discussions relating to Ed Hooper's book "The River," that drew the public's attention to the use of Africa as a testing ground for polio vaccine. Africans, and indeed all of mankind, have a reasonable right to know the details of these studies.

As further reported to Martin, there apparently was an illness among the vaccine workers that caused wasting. It was referred to as the African equivalent of "thin man." Many of the captive chimpanzees also became sick. In spite of these findings, chimpanzee kidneys were still being shipped to the United States for hepatitis vaccine and other studies.

Martin's own work has pointed to a process whereby viruses evade immune recognition by simply deleting genes that code the relatively few viral components targeted by the cellular immune defense. The process has been termed "stealth-adaptation." Stealth-adapted African green monkey simian cytomegaloviruses have been detected in patients with illnesses involving the brain.

Rather than seizing the opportunity to learn more about these viruses, Public Health officials seemingly continue to maintain the "don't want to know" attitude. A technologist working at the Centers for Disease Control and Prevention (CDC) even told Martin of a simian cytomegalovirus being grown from a person who received an experimental rubella virus vaccine. This vaccine was also produced in African green monkey kidney cells. As Martin recently concluded, "I feel the public should expect more from those entrusted with the Nation's health, including a willingness to more openly address the question of vaccine safety. Possibly by placing the issue in terms of World Heath and, in particular, the role of polio vaccines in the origins of AIDS in Africa, other organizations might respond to the challenge posed by stealth-adapted viruses."

The members of the WTO should ask their respective governments to look into the vaccine safety issue through their own institutions and pharmaceutical houses. I doubt that it is the intent of the pharmaceutical industries of the World to slowly kill all of their customers. The issue is as important as any other we face today.

Handout of Presentation at the "Twentieth Century Plagues" Symposium

Embassy Suites Hotel, Los Angeles International Airport, March 1, 1996

Overview: Stealth viruses are defined as cytopathic viruses able to establish a persistent infection because of deletion and/or mutation of specific genes, which, if expressed, would evoke effective anti-viral cellular immunity (1-5). This presentation will mainly focus on stealth viruses derived from African green monkey simian cytomegalovirus (SCMV). Federal health authorities have been slow in addressing the potential transmission of stealth viruses in live polio viral vaccines.

Brief History of Poliomyelitis: Epidemic paralytic polio began to appear towards the end of the 19th century (6). The first report of a major outbreak was made by the Swedish pediatrician Dr. Carl Medin in 1889. Polio epidemics soon appeared throughout Europe. Attempts at animal transmission failed until 1908 when Dr. Carl Landsteiner (of blood group fame), inoculated two monkeys with a spinal cord extract obtained at autopsy from a 12 year-old boy. One monkey died acutely while a Rhesus monkey became paralyzed.

A major outbreak of paralytic polio occurred in the United States in 1916 affecting nine thousand individuals in New York City (7). Each summer there were reported clusters of polio in various parts of the United States. An unusual polio-like illness affected California in 1934. One hundred and ninety

eight medical personnel at the Los Angeles County Hospital became sick (6). They were not paralyzed, but instead developed a chronic fatigue syndrome (CFS)-like illness. Attempts to isolate polio virus were unsuccessful. A class action suit was settled in 1938 for several million dollars, supposedly with the stipulation that there be no publicity to suggest that caring for polio patients could possibly be hazardous.

Basic studies showed that (i) polio virus initially infected the alimentary tract; (ii) Only a small percentage (<1%) of infected individuals actually developed paralysis; (iii) there were 3 serologically distinct Types (1, 2 and 3); (iii) most cases of paralytic polio were due to Type I, as exemplified by the highly virulent Mahoney strain; (iv) serum from a previously infected individual contained protective type-specific antibodies; (v) polio virus could be passed in monkey and mouse brains but "brain adapted" polio did not grow well in non-nervous tissues (6).

A major breakthrough occurred in 1948 when Dr. John Enders showed that by using antibiotics, it was possible to culture intestinal cells on which polio virus could be propagated and could produce a quantifiable cytopathic effect (CPE).

Development of Polio Vaccines: There were two competing approaches to vaccine development.

The influenza vaccine model involved the use of inactivated (killed) virus. The yellow fever vaccine model involved the use of a weakened or attenuated strain as a live viral vaccine. Several unsuccessful efforts were made in the 1930's using these approaches with the polio virus grown in monkey or mouse brain.(6).

Following the work of Dr. Enders, Dr. Jonas Salk was commissioned in 1953 by Basil O'Connor, Director of the National Foundation for Infantile Paralysis (March of Dimes Foundation), to produce an inactivated polio vaccine (IPV). Dr. Salk grew virulent polioviruses on Rhesus monkey kidney cells and inactivated the harvested viruses using a 1:4,000 dilution of formaldehyde. Serious questions concerning (i) the necessity of using the Mahoney isolate for Type 1 polio, (ii) the dynamics and completeness of formaldehyde inactivation and (iii) statistical analyses of the early trial data using experimental vaccine lots were smothered by O'Connor's powerful Public Relations campaign and by the public's eagerness for a vaccine. The Federal Laboratory of Biological Safety (a component of the National Institutes of Health) licensed polio vaccine on April 12, 1954. This occurred in spite of Dr. Bernice Eddy notifying her supervisors that several monkeys developed paralysis from the vaccine made at Cutter Laboratories.

Soon after the launching of the commercial vaccine lots in April 1955, there were 10 deaths and 192 cases of vaccine induced paralytic polio. Most of the cases were traced to the Cutter vaccine. Vaccine lots from Park Davis and Eli Lilly were also found to contain live virus and the Industry knew that Dr. Salk's assumptions about formaldehyde were invalid. Efforts were made to single out Cutter as the only defective vaccine, while at the same time

improve upon the inactivation protocol. The vaccine program was halted on April 27th and resumed on May 27th, hopefully still in time for the anticipated July onset of natural polio cases. By November, new manufacturing safeguards were introduced, such as an additional filtering step, without any recalls of existing lots. The Laboratory of Biological Safety was reorganized and placed under the Food and Drug Administration (FDA) as the Division of Biological Standards. For her good work, Dr. Eddy was removed from polio vaccine testing!

Although the efficacy of IPV was exaggerated by changing the clinical definition of paralytic polio and other maneuvers, it led to a significant reduction in polio cases. It still probably caused several cases from residual live virus. Public Health critics of the program were shunned and their protests nullified by legislating compulsory vaccination.

Dr. Albert Sabin, Dr. Hilary Koprowski and others were more interested in developing an attenuated polio virus which could be administered as a live vaccine. Dr. Koprowski conducted extensive human and animal trials in Central Africa. A 1956 trial in Ireland was marred by the occurrence of polio in vaccine recipients. Dr. Sabin had more success, mainly because he made use of Dr. Renato Dulbecco's technique of plaque purifying attenuated viral isolates. By 1960, Dr. Sabin had convinced much of Europe that his vaccine was less expensive and created more rapid and longer lasting immunity than did IPV. Moreover, excretion of live vaccine virus indirectly vaccinated others within a community and helped out-compete virulent viral strains.

The subtle nucleotide changes, mainly within the 5' non-coding region, that distinguish attenuated from wild type virus have recently been identified. Much of the difference involves a single nucleotide substitution. The attenuated strains can revert to more pathogenic variants, and even approved vaccine lots contain some revertants. This problem occurs more frequently with the least altered, Type 3 polio vaccine strain.

SV40 Contamination: From 1957 Dr. Eddy was involved in pioneering work with Dr. Sarah Stewart on the ability of a mouse Polyoma virus to induce tumors in hamsters. Dr. Eddy was suspicious that monkey kidney cells might have a polyoma-like virus. In April 1960, she reported to her supervisor, Dr. Joseph Smadel that extracts from Rhesus monkey kidneys induced tumors in hamsters similar to those induced with Polyoma virus. Her work was dismissed by Dr. Smadel. Moreover, she was severely chastised for subsequently mentioning her findings at a Polyoma virus conference held in New York in August 1960. She was instructed that all future utterances had to be written and submitted for approval by Dr. Smadel.

Dr. Maurice Hillerman of Merck was also concerned that the Rhesus monkeys being used for polio vaccine were becoming increasingly infected with unknown viral agents. He asked the Director of the Washington D.C. Zoo to arrange a shipment of a different species of monkey, imported separately from the usual routing. He received African green monkeys from Africa via Madrid and Baltimore. Extracts from Rhesus monkeys produced a strong vacuolating cytopathic effect on the African green monkey kidney cells. The responsible virus was the 40th virus isolated from Rhesus monkeys and called SV40. It was subsequently shown to be the same virus as found by Dr. Eddy to cause cancer in hamsters.

SV40 is endemic in Rhesus monkeys, producing very little CPE. It does not occur naturally in African green monkeys and when present, it produces a strong CPE. With suppressed publicity, FDA undertook a major effort to switch the polio vaccine seed lots from Rhesus to African green monkeys using anti-SV40 antibodies to neutralize the contaminating virus. Brief and belated mention of the problem occurred in the press; even Dr. Eddy was subsequently allowed to publish her findings. She had, however, lost her laboratory and her research support.

The SV40 issue was initially seen as a problem for the live viral vaccine. Unfortunately, the virus was not inactivated by 1:4,000 formaldehyde and parental inoculation created more infections than oral administration. However, since there were no apparent ill effects, the concern over this virus subsequently waned. Yet, recently there have been reports of SV40 virus in both childhood choroid plexus brain tumors (8) and in mesotheliomas (9). The children were too young to have received SV40 directly from a vaccine and in all likelihood this virus is now circulating within the human population.

Live polio virus vaccine was licensed to Lederle (American Cyanimid) as Orimune(R) in 1961 and has since largely replaced inactivated vaccines. Occasionally (8-10 cases per year in the United States), the Type 3 vaccine polio reverts to virulence and induces disease.

Viral Contamination of Live Polio Virus Vaccines: During early vaccine developments almost 50% of the kidney cultures established for polio vaccine production were discarded because of obvious viral contaminants (mainly foamy retroviruses but also adenoviruses and non-identifiable agents). Few if any of these adventitious agents were thought to be pathogenic for man. An exception was the monkey B virus, a herpes simplex-like virus known to cause fatal encephalitis. Another major scare arose in 1968 in Marburg, Germany when monkeys intended for vaccine production transmitted an acute hemorrhagic illness to their caretakers. Renewed efforts went into screening for unknown viruses. In 1968, Dr. Kendall Smith, an electron microscopist at FDA Division of Biological Standards, detected atypical viral-like agents in some batches of polio vaccines. He unsuccessfully argued for the use of a fluorescence focus assay using sera from monkeys tested on their own kidney cell cultures. Consideration was given by Lederle to switching from monkeys to human cells for vaccine production. This approach was taken by Pfizer, which had established a polio vaccine production facility in England. Their product was called Diplovax and was derived from human diploid cells.

In a joint Lederle/Bureau of Biologics study conducted in 1972 "All eleven monkeys studied demonstrated the presence of CMV-like agents. These monkeys all originated from Kenya over a short period of time. Seven of these monkeys would have passed our existing test standards." In a "Cytomegalovirus Contingency Plan", dated August 4th, 1972, Dr. Vallancourt from Lederle argued that "Unless and until Pfizer's Diplovax is in abundant supply, the BB [Bureau of Biologics] cannot risk Lederle being off the market." On March 16th, 1973, Dr. Vallancourt wrote to the President of American Cyanamid, "I do not believe our problems with the slow release of specific lots of Orimune(R) are a result of a Pfizar influence..... Furthermore, if the Bureau wanted to restrict us they could bring up the subject of CMV (Cytomegalovirus) in our substrate (i.e. African green monkey kidney tissue) which they have not done, even though they have told us the monkeys in the collaborative study performed in 1972 were all positive for this agent."

In 1977 assays for retroviral reverse transcriptase activity were applied to some polio vaccines. Certain vaccine lots tested positive and this activity could be transmitted to cell cultures, which developed an atypical CPE. It is now known that many of the African green monkeys used were infected with simian immunodeficiency virus (SIV). Interestingly, African green monkeys brought to the Caribbean earlier this century were not infected with SIV, raising the possibility that SIV is a more recent infection than generally thought and possibly of experimental vaccine origin. Discussion at the time of possible Type C retroviruses in polio vaccines led to more detailed electron microscopy. Particles were seen in polio batch 3-444, but were considered more likely to be cell debris than actual viruses. I worked at the time as Director of the Viral Oncology Laboratory at the Bureau of Biologics. In 1978, Dr. G. Aulakh and I confirmed that a bulk monopool of polio vaccine contained a considerable amount of DNA, not all of which could be accounted for as simple cellular debris. I was not provided any information on the earlier concern about CMV, but was told that the monopools were submitted only to determine if they met the mutually approved and mandated tests for polio vaccines. In other words, they were not available to continue this line of experiments. I still remember the project terminating statement from the Bureau's Director: "Stop worrying about it, every time you eat an apple you ingest foreign DNA."

The issue of probable contamination of polio vaccines with adventitious viruses was difficult for anyone to contradict. Maintaining support for the established vaccine program was, however, variously argued: (i) The continued use of primary cultures from kidney cells would help avoid any suggestions that established cell lines may have progressed towards cancer and may contain "oncogenes". (ii)

The polio vaccine had been administered to many millions of individuals without apparent adverse effects. It was, therefore, unnecessary and anyhow too late, to correct the situation; (iii) The viruses detected were not the dreaded Marburg agent or the much feared Type C cancer associated retroviruses and were, therefore, of little significance and were probably destroyed when passing through the stomach; (iv) The pharmaceutical industry had to be encouraged to manufacture vaccines and the public had to be encouraged to take vaccines for the government-sponsored immunization programs to be successful. (v) The programs were predicated on existing costs and profit margins and could be threatened if additional testing was imposed. (vi) Many companies had ceased vaccine manufacturing within the United States and any further reduction could make the country more vulnerable to threatened germ warfare, etc., etc.

Neuropsychiatric Illnesses: It is generally believed that the incidence of chronic brain dysfunctional disorders has steadily increased over the last several decades. Conditions such as chronic fatigue syndromes, fibromyalgia, migraine, attention deficit hyperactivity disorder, autism, mental depression, dementia, schizophrenia, etc., are being increasingly diagnosed. Our society has somewhat complacently accepted these conditions as inevitable consequence of "progress". Moreover, compared to efforts aimed at clinical management of these diseases, relatively little effort has gone into identifying their etiology. The involvement of basic researchers has been especially hampered by the imposition of confusing clinical case definitions. Medical specialists have somewhat arbitrarily segregated various facets of neuropsychiatric illnesses into a wide array of nosologically distinct, territorially protected, clinical entities. The ever-increasing multiplicity of named

dysfunctional brain syndromes may simply reflect the complexity of the brain, rather than imply the presence of innumerable etiologic processes.

Search for Viral Infections in Neuropsychiatric Diseases: I initiated a project in 1986 to see if patients diagnosed as having the chronic fatigue syndrome (CFS) were virally infected (1). Early findings using the polymerase chain reaction (PCR) with primer sets cross-reactive with all of the known human herpesviruses gave weak positive signals with one third of CFS patients, but not with any of several laboratory controls (10). I later found that many CFS patients would also give low, but detectable PCR responses using primers designed to amplify the tax gene of human T lymphocytotropic viruses.

The potential significance of the marginal responses seen in CFS patients became clearer in 1990 when strikingly positive PCR responses occurred with cerebrospinal fluid (CSF) samples from two patients and on a brain biopsy of a third patient. All three patients had unexplained severe encephalitis-like illnesses. The brain biopsy came from a schoolteacher who had gradually lost her capacity for written and oral communication, but was otherwise alert with no localizing neurological signs. She did, however, have an abnormal MRI with periventricular lesions. The brain biopsy showed mild gliosis with no inflammation. Complete and incomplete forms of herpesvirus-like particles were present in several of the glial cells. Some of the virus positive cells showed marked vacuolated changes and lipid accumulation (11). The PCR positive CSF samples were acellular, indicating an absence of inflammation. One had come from a newborn infant with delayed neural development. The other from an adolescent with residual severe brain dysfunction, following what was called atypical herpes simplex encephalitis.

From about 1987, I had devised several approaches to try to grow human herpes virus-6 (HHV-6) from CFS patients using cord blood lymphocytes. Routine human fibroblast cultures were also established. The latter cells frequently showed a delayed, transient CPE somewhat suggestive of human CMV. Buoyed by the PCR findings in the severe encephalopathy cases, efforts were again undertaken to culture a virus from a PCR positive CFS patient (D.W.). After 6 weeks of culture, enlarged, rounded, vacuolated cells suddenly began to appear in both human and monkey derived cell lines. Cell syncytia were easily recognized. The CPE could be passed to many other cell types and to cells of multiple species. By electron microscopy, numerous complete and incomplete herpesvirus-like viral particles were seen. The foamy, vacuolated cells did not, however, stain with antisera specific for the immediate-early gene of human CMV. Nor did the cells stain with antisera specific for HSV, HHV-6/7, VZV or adenoviruses. PCR assay using the HTLV tax gene primer gave two well defined products. Sequencing of one of the PCR products (GenBank accession # U09212) identified a region of partial sequence homology to the UL34 gene of human CMV. The sequence of the other PCR product (GenBank accession # U09213) was suggestive of a possible herpesvirus origin but could not be aligned with any of the known herpesviruses (3). Viral DNA was isolated and cloned. A region of sequence homology to African green monkey simian cytomegalovirus (SCMV) was first found in early 1995. Unrelated to our work, Dr. Gary Hayward of Johns Hopkins University submitted additional sequence data on SCMV to GenBank in December 1994. From his sequence data, it was unequivocal that the CFS patient's virus had originated from SCMV (5).

In 1991 I coined the term "stealth" to describe the viruses I had been seeking. I used this term primarily because it carried the connotation that the viruses could occur in the absence of inflammation. The SCMV-derived virus was designated stealth virus-1. A closely related virus (stealth virus-2) was isolated in early 1991 from the CSF of a patient with an acute severe encephalopathy following a 4 year history of a manic depressive illness. The patient (B.H.) has remained in a vegetative state since 1991. DNA sequencing of a PCR amplified product from her infected cultures has identified the virus as being similar but not identical to the SCMV-derived stealth virus-1. Numerous other stealth viral isolates have been obtained and are awaiting resources to perform the necessary sequencing analyses.

Animal Studies: Stealth virus-1 induces an acute neurological illness when inoculated into cats (12). The early manifestations developed within a week and included gingivitis, bloody ocular and nasal discharge, lymphadenopathy, pupil dilatation with photophobia (squinting in response to light) and nuchal hair loss from rubbing against the cage. There was a reduction in body temperature beginning in the second week which averaged 0.6°F at 2 weeks and 0.8°F at weeks 3 and 4. Most striking was the marked behavioral changes in all of the virus-inoculated cats. They lost the playfulness that was present prior to injection and became reclusive and irritable. They resisted being handled and the animal caretakers resorted to wearing leather gloves. By palpation, the enlarged lymph nodes and various muscle groups were identified as being painful for the animals. The severity of the illness peaked at around 4 weeks with definite improvement noted in the cat necropsied at 6 weeks. A cat that was maintained to 15 weeks had resumed normal activities by week 10 and appeared to be symptom free. Histological examination of animals' brain tissue showed foci of cells with cytoplasmic vacuolization and an absence of any inflammatory reaction. Electron microscopy on several cats confirmed the presence of occasional herpes-like viral particles and, more commonly, the presence of variable patterns of accumulations of viral-like granular and other membranous structures suggestive of subgenomic viral expression.

Unstable, Fragmented Nature of the Stealth Virus-1 Genome: Restriction enzyme digests of stealth virus-1 have been cloned and partially sequenced. The genome appears to comprise multiple fragments of approximately 20 kilobase pairs. The virus displays an unusually high degree of sequence microheterogeneity suggesting infidelity of DNA replication or an alternative replication strategy (13). The concept of a fragmented, unstable viral genome is consistent with electron microscopic observations (14). It has important implications for the approaches to be taken in using molecular techniques to screen for these viruses. It also raises concerns about the possible oncogenicity of stealth viruses (16).

Stealth Viruses Isolated from Patients with Severe Encephalopathy: Since the beginning of these studies, a number of patients have been identified with strongly positive cultures and complex clinical case histories. While the initial focus was on CFS patients, the more recent emphasis has been on patients with severe encephalopathy. In many of these cases, a preceding CFS-like illness was present. The clinical observations support the concept of potentially infectious neuropsychiatric illnesses attributed to varying degrees of dysfunction of different regions of the brain (2). Brain biopsies have been obtained in five patients and have shown many of the characteristic changes noted in the infected cats. In some of the biopsies, there is an additional vasculitis component. Possibly, the most promising clinical aspect of these illnesses is the recovery process seen in several, but not all, severely ill patients. The mechanism of recovery is the most important goal of stealth virus research. In a step towards this goal, I am hoping to obtain sufficient sequence data on isolates from severely ill adult patients and from autistic children, to determine which, if any, have originated from African green monkeys. I realize that while monkey cells have been used for polio and adenovirus vaccines; dog and duck kidney cells have been used for rubella vaccines; chicken cells have been used for measles and mumps vaccines; and a very wide range of animals cells have been used for animal vaccines. Unfortunately, only limited sequence data are available on most animal viruses. Some stealth viruses may simply have originated by the down sizing of known human viruses, especially human herpesviruses. In these endeavors, I have so far unsuccessfully sought the help of both the FDA and the CDC.

Notification to CDC, FDA and Lederle: CDC was notified in early 1991 that a repeat positive culture was obtained on patient D.W. and a similarly positive culture was obtained on the comatose patient B.H. An invitation to visit my laboratory to see the cultures was declined. CDC was again notified in mid 1995 when the sequence data on stealth virus-1 showed a greater relatedness to SCMV than human CMV. FDA Center for Biologics Evaluation and Research (CBER) was contacted in March of 1995 and the President of Lederle sent a letter in July, 1995. Prompted by FDA personnel, unsolicited requests for collaborative and financial assistance were sent to

CDC and to FDA in June 1995, followed by visits to both agencies. The unsolicited proposal format is designed to protect an agency from the accusation that it has given a competitive advantage to someone seeking government funds. The proposals included detailed clinical descriptions of 12 cases of severe, otherwise unexplained, encephalopathy. In the FDA proposal, I asked to do a surveillance study to determine the prevalence of simian CMV-derived stealth viruses in humans and in the monkeys used for polio vaccine production. I also wanted to test some vaccine monopools. In the CDC proposal, I asked for help in sequencing the viral isolates from the 12 patients.

Response from CDC Received 5 Months After Submission: "Reviewed by the National Immunization Program (NIP) and the National Center for Infectious Diseases (NCID) of CDC. It was determined that the proposal did not address NIP research needs at this time." There was also the comment from NCID that "The twelve patients reported have variable clinical histories with no obvious epidemiological links between them. The evidence that a virus can be isolated from these cases is unconvincing. No independent confirmation has been reported in the results described in this proposal and the evidence for the existence of an infectious agent which the offeror (sic) calls a 'stealth virus' remains unconvincing."

Unofficially, I was told that Ms. Joanne Patton, now working in Dr. John Stewart's laboratory at NCID, recalled hearing of a patient becoming infected with SCMV, which had contaminated a rubella virus vaccine grown on African green monkey cells. From discussions with Ms. Patton and Dr. Stewart, there seems to be some regret that CDC had not made efforts to pursue this event. A request

for a brief statement in the Mortality and Morbidity Weekly (MMWR) on atypical viral encephalopathy was also denied.

Response from FDA Received 7 Months after Submission: "Request for support to screen for the prevalence of simian cytomegalovirus derived stealth viruses in humans is best sought from an agency such as the NIH, rather than the FDA. Requests to screen the monkey colonies that are used in vaccine production for the presence of stealth viruses. These are studies that would need to be conducted with the manufacturers.

Request to screen polio samples (the bulk mono-pools) for stealth viruses. Screening of these samples, if done, should be carried out in CBER's own laboratories. We are evaluating the types of studies that might be best done using PCR-based techniques. However, even if such sequences are found in a PCR-based assay, the question whether transmissible, replication competent viruses are present would still need to be addressed.

We appreciate your informing us and others, not only in your submitted proposal but also at the FDA Workshop on cell substrates and at the IOM meeting on vaccine safety, about your concerns vis-a-vis simian cytomegalovirus and potential related viruses in the polio vaccine. We will be addressing this concern in our Laboratories".

Other Relevant Actions: The Advisory Committee on Immunization Practices (ACIP) had made a recommendation to move from a series of four injections of live polio vaccine to a split protocol of two injections of inactivated vaccine (IPOLTM, produced by Pasteur Merieux-Connaught Laboratories) followed by two injections of live vaccine. This was ostensibly to help reduce the occurrence of the approximately 8-10 cases a year of poliomyelitis from Type 3 vaccine revertants. Although not cited as a reason, such a move would expand the production of inactivated vaccines should a subsequent total switch to inactivated vaccine be recommended. A decision to delay the implementation of this plan was made at the February ACIP meeting. There will be opportunity for public input at the June meeting.

FDA has held open meetings to discuss possible hazards of interspecies bone marrow therapy in AIDS patients and interspecies liver and other organ transplants. FDA was unable to muster support to restrict the first baboon bone marrow trial.

Summary: The concepts that certain stealth viruses have arisen as contaminants of live viral vaccines and that vaccinations may have untoward consequences, have not been embraced by either vaccine manufacturers or Public Health agencies. As the experimental data unfold, however, the existence and the clinical importance of stealth viral infections are less easily reputed. Indeed, stealth viral infections may explain a wide range of neurological and neuropsychiatric diseases, serving as a common thread linking these diseases, and possibly accounting for their ever increasing prevalence.

If a vaccine program were to be initiated today, one would surely not import wild monkeys from Africa, create short term primary kidney cultures, add a human virus and administer the crude gamish derived from the virally infected cells to virtually every child in the country. Nor would one want to withhold applying the many molecular biological techniques developed over the last 30 years to assess vaccine purity. Yet this is essentially the situation with live polio vaccine and comparable arguments can be made for other human and animal viral vaccines.

Various meetings, phone calls and document exchange have uncovered a sense of frustration within the Federal Public Health System, Industry, and the general public with what appears to be a resistance of those in authority to face the issue of prior, if not also present, vaccine contamination. If animal viruses have been inadvertently introduced into humans, the sooner we find out the better. I would very much appreciate continuing to hear about patients with seemingly complex infectious and/or neuropsychiatric illnesses. Volunteers are needed to help bring these patients to the attention of the Public Health System. Work on the in vitro efficacy of stealth virus inhibitors needs support so that clinical therapeutic trials may soon become a reality. 1634 Spruce Street, South Pasadena CA 91030 and telephone number 626-616-2868. The email address is wjohnmartin@hotmail.com. The internet web sites are www.s3support.com and www.ccid.org

References

- Martin W.J. Viral infection in CFS patients. in "The Clinical and Scientific Basis of Myalgic Encephalomyelitis Chronic Fatigue Syndrome." Byron M. Hyde Editor. Nightingdale Research Foundation Press. Ottawa Canada pp 325-327, 1992.

- Martin W.J. Stealth viruses as neuropathogens. CAP Today 8 67-70, 1994.

- Martin WJ, Zeng LC, Ahmed K, Roy M. Cytomegalovirus-related sequences in an atypical cytopathic virus repeatedly isolated from a patient with the chronic fatigue syndrome. Am J Path. 145: 441-452, 1994.

- Martin WJ. Stealth virus isolated from an autistic child. J Aut Dev Dis. 25:223-224, 1995.

- Martin WJ, Ahmed KN, Zeng LC, Olsen J-C, Seward JG, Seehrai JS. African green monkey origin of the atypical cytopathic 'stealth virus' isolated from a patient with chronic fatigue syndrome. Clin Diag Virol. 4: 93-103, 1995.

- Paul JR. A history of poliomyelitis. Yale University Press, New Haven, 1971.

- Gould T. A summer plague. Polio and its survivors. Yale University Press. New Haven, 1995.

- Lednicky JA et al. Natural simian virus strains are present in human choroid plexus and ependymoma tumors. Virology 212: 710-717, 1995.

- Cristaudo A, et al. Molecular biology studies on mesothelioma tumor samples: preliminary data on H-ras, p21 and SV40. J. Environ Path 14: 29-34, 1995.

- Martin WJ: Detection of viral related sequences in CFS patients using the polymerase chain reaction. In Hyde B ed. The Clinical and Scientific Basis of Myalgic Encephalomyelitis/ Chronic Fatigue Syndrome. Nightingale Res Found. Ottawa Canada, pp278-282, 1992.

- Martin WJ. Severe stealth virus encephalopathy following chronic fatigue syndrome-like illness: Clinical and histopathological features (submitted)

- Martin WJ, Glass RT. Acute encephalopathy induced in cats with a stealth virus isolated from a patient with chronic fatigue syndrome. Pathobiology 63: 115-118, 1995.

- Gollard RP, Mayr A, Rice DA, Martin WJ. Herpesvirus-related sequences in salivary gland tumors. J Exp Clin Can Res.15: 1-4, 1996.

- Martin WJ. Genetic instability and fragmentation of a stealth viral genome. Pathobiology (in press).

Political and Economic Compromises Affecting Public Health: Lessons from Contaminated Polio Vaccines

W. John Martin, MD, Ph.D.
(Initially Written in 2003 with some revision in 2005)

There are growing health problems being faced by many Americans. The autism rate has doubled in California over the last 4 years and special educational needs of school children are soaring. Severe childhood behavioral problems necessitating residential care are financially forcing many parents to relinquish custody of their children to the state. In 2001 a sampling of only 19 mid-sized states revealed the astounding number of 12,700 children who were orphaned to the state because of their parents' inability to cover institutional costs. This heart rendering loss of legal custody does not even ensure medical therapy, but merely protective restraint. Nearly half of the elderly will experience the memory loss and emotional fragility of Alzheimer's disease. Published figures from the Centers for Disease Control and Prevention (CDC) confirm that over 20,000 patients are hospitalized each year with a brain illness diagnosed as encephalitis. Yet in over 60 percent of such patients, even detailed laboratory studies fail to reveal the underlying cause. Many more individuals who experience an alarming decline in brain function are not hospitalized. Some are driven to such tragedies as drug addiction, suicide or mindless criminal behaviors. It has become commonplace to hear of individuals with chronic fatigue syndrome (CFS), diabetes, arthritis and debilitating mental illnesses.

Individually, any of these increasingly prevalent diseases ought to trigger a serious investigation for the possibility of an infectious cause. When lumped together, and with the added knowledge of virus contaminated vaccines, it is difficult to comprehend the smug indifference to reports of stealth adapted viruses by those entrusted with protecting the public health.

This article is written primarily to document major lapses in public health decision-making. It is also an effort to increase public awareness of a probable cause of our Nation's deteriorating health. The article begins with an accounting of events surrounding the development and testing of poliovirus vaccines. It then explores the unwillingness of current public health officials to openly address inadvertent mistakes possibly made by their predecessors. Preserving the image of a faultless system is used as justification for inaction and even active suppression of new information. This is a personal account prompted by my own experiences with public health officials and their reactions to my repeated attempts to speak on behalf of those afflicted with stealth-adapted viruses.

I have previously summarized the early history of polio as a disease entity and the subsequent development of both inactivated (Salk) and attenuated or weakened (Sabin) poliovirus vaccines. This information has been available on the web site www.ccid.org since being presented at various public health conferences since 1995. Primary cultures of Rhesus monkey kidney cells were initially chosen to grow poliovirus for vaccine production. The decision to use freshly cultured cells was based on an understandable concern that established long-term cell lines may have acquired some genetic changes leading towards the formation of cancer cells. The potential alternative risk expressed by several prominent scientists was that freshly harvested monkey kidney cells might be carriers of unknown viruses. This legitimate concern was largely dismissed.

The 1960 finding that Rhesus monkey kidney cell cultures were commonly contaminated with an animal cancer-causing virus, termed SV40, should have been a wake up call; especially since the formalin inactivation process used for Salk vaccine was essentially ineffective against this virus. Previously approved SV40 contaminated polio vaccine lots were never recalled from the marketplace. A switch was, however, made to using kidneys of African green monkeys. Rather than reestablishing SV40-free polio viral stocks, use was made of an anti-SV40 animal antiserum (itself a potential source of contaminating viruses), as a way of clearing the contaminant.

The other major change in the early 1960's was the reluctant admission that the Salk vaccine was inferior to Dr. Sabin's live attenuated vaccine in the rapidity of providing immunity and cost. Moreover, the Sabin vaccine was transiently excreted by those who were vaccinated leading to secondary infection and presumed vaccination of others within the community. The omission of a formalin inactivation step should have led to a redoubling of efforts to screen the donor monkeys for non-apparent infections. This was never done. Rather a political conflict arose over the Division of Biological Standards (a forerunner to the Bureau of Biologics component of the Food and Drug Administration, FDA) decision to license Dr. Sabin's vaccine. The pharmaceutical company, Lederle, a Division of American Cyamamid had been working with an attenuated poliovirus developed by Dr. Hilary Koprowski, who later became the director of the Wistar Research Institute. Unlike Dr. Sabin vaccine, which was based on plaque purified polioviruses, the polio strains employed by Dr. Kroprowski were less purified and prone to cause breakthrough polio in some recipients. More serious was evidence obtained by Dr. Sabin that Dr. Koprowski's vaccine was contaminated with an unknown virus. The government never pursued the apparent contaminant in Dr. Koprowski's vaccine. Rather it was facing a threatened lawsuit by Lederle to have their vaccine licensed at the same time as that of Dr. Sabin. The government appealed to Dr. Sabin to forego patenting his vaccine virus and to freely, if reluctantly, provide it to Lederle. Dr. Koprowski left Lederle to become the director of the Wistar Institute.

Continuing concerns regarding other possible contamination of polio vaccine grown in African green monkeys persisted throughout the 1960's. Up to half of all cultures were being rejected because of apparent viral contamination. Common sense suggestions such as screening the monkey's serum for antibody reactivity against their cultured kidney cells were dismissed since it would undoubtedly lead to rejection of even more cultures. Dr. Leonard Hayflick, also working in the Wistar Research Institute, had successfully cultured a human lung derived cell line that lacked the longevity or chromosomal abnormalities of cancer cells. This cell line, designated WI-38, was a potential substitute for monkey kidney cells. Dr. Hayflick had provided the cells to Pfizer for successful polio vaccine production in England. Seemingly not wanting to provide support to the Wistar Institute or Pfizer, both Dr. Sabin and Lederle argued strongly for the continued use of monkeys, including the need for monkeys in safety studies.

Nevertheless, Lederle and the FDA's Bureau of Biologics addressed the issue of potential virus contamination of polio vaccines in 1972. Kidneys from eleven monkeys were set aside from vaccine production to see what, if any, contaminating viruses might be present. All eleven monkeys grew out African green monkey simian cytomegalovirus (SCMV). Only four of the isolates would have been detected using the then mandated government screening tests. Lederle had a more sensitive indicator cell line, even better than WI-38.

The 1972 findings were never made public by FDA authorities. Lederle prepared a Contingency Plan in the form of an internal memo. In it, they argued that the Bureau of Biologics would not have the courage to take their product off the market in favor of Pfizer's vaccine. They also stated that they could contest the findings; repeat the studies; contend that any orally administered contaminating virus would be destroyed as it passed into the stomach; etc. These arguments would provide added time to continue making thousands of doses of the existing vaccine. Alternatively, they could begin to treat the monkeys with an anti-viral agent. In making their case, Lederle's main assertion was that if SCMV was infectious for humans, it would have resulted in acute illness, yet none had been seen over the years of using the vaccine. Even at this time, many investigators, including some FDA officials, were well aware of data linking viruses to chronic illnesses, including cancer. Human CMV had also been linked to congenital illness including mental retardation in children born to asymptomatic mothers. Mankind was not served well by the failure of FDA to publicly discuss the 1972 study.

From 1976 to 1981 I was employed as a Medical Officer within the Bureau of Biologics, (now the Center or Biologics Evaluation and Research) of the FDA. I served as Chief of the Viral Oncology Branch, within the Division of Virology. While primarily a research position, my mission was to help ensure that vaccines were not contaminated with viruses, which could potentially cause cancer. Of most concern were retroviruses that are able to switch between RNA and DNA forms using an RNA dependent DNA polymerase (reverse transcriptase). Almost under cover, Dr. John Petricanni a more highly placed official began to test concentrated lots of poliovirus vaccines for reverse transcriptase activity. Of several vaccine lots tested, lot 3-444 yielded definitely positive results. Moreover, even after the poliovirus was neutralized with antibodies, the vaccine was still able to induce damage to cultured cells, with the further production of reverse transcriptase activity. I was informed of the work and asked if I could become involved. Electron microscopy of the vaccine revealed several particles, which some had suggested could be retroviruses. In my opinion, the particles were too variable to be identified as typical retroviruses. I was more impressed and indeed dismayed by the enormous quantity of cellular debris in the supposedly purified vaccine. With the assistance of Dr. Gurmit Aulak, DNA was isolated from a sample of the bulk vaccine. While, the vast majority of the DNA hybridized with African green monkey DNA, a residual portion of the DNA was seemingly not of monkey origin.

I informed the Bureau Director of the apparent presence of non-monkey tissue derived DNA in the polio vaccine. Instead of disclosing the earlier study, the Director simply shrugged the finding with a dismissive statement "every time you eat an apple you ingest foreign DNA." In spite of the aberrant findings, the vaccine lot was approved for human use. The Bureau Director was subsequently appointed President of the Medical Research Division of American Cyanamid. American Home Products later acquired American Cyanamid and the corporate name subsequently changed to Wyeth, one of its subsidiaries. Pfizer later acquired Wyeth.

I asked another senior FDA official to explain the apparent secrecy regarding the 1972 study. He responded that data obtained on regulated products were "proprietary." He went on to emphasize how important it was to be supportive of vaccine manufacturers, especially since they might be needed in case of biological warfare. I further asked about doing more studies on the polio vaccine. "You really can't do studies on products unless they are agreed to by the manufacturer."

Studies at the Los Angeles County Hospital and University of Southern California (LAC+USC) Medical Center

I moved from FDA to the National Institutes of Health (NIH) and subsequently joined the Uniformed Services University for Health Sciences. This allowed me to work at the National Naval Medical Center and to obtain Medical Boards in Anatomic and Clinical Pathology, with sub-specialty certification in both Immunopathology and Medical Microbiology. In 1985, I joined the University of Southern California (USC) as a tenured professor of pathology. Along with other activities, I began to search for a virus cause of CFS. Following several years of suggestive culture findings and clearly positive findings using the polymerase chain reaction (PCR), I eventually isolated a virus from a CFS patient. Based on the foamy appearance of the cells developing in the viral cultures, I initially suspected a spumavirus (spuma is the Latin for foam). I

communicated this opinion to Dr. Paul Cheney who was being supported by the CFIDS Association of America. Dr. Cheney visited with me in Los Angeles and recommended that the CFIDS Association provide financial support. Unfortunately, they also substituted "fact" for "opinion" as they claimed their success to the Centers for Disease Control and Prevention (CDC). Dr. Walter Gunn from CDC responded that Dr. Thomas Folkes from the CDC was also isolating spumaviruses. Another CDC researcher, Dr. Brian Mahy, was asking who was my virologist, as if I had no clue of what I was doing.

I agreed to a study in 1992 organized by Dr. Gunn soon after he retired from CDC and entered into a consultative agreement with the CFIDS Association. The study did not distinguish fatigued patients from poorly selected controls. Rather than focusing on why anyone should be testing positive, the conclusion was I had failed to provide the prized "diagnostic test for CFS." I well knew this to be true since even at the time of Dr. Cheney's 1991 visit, a similar virus had been isolated from a comatose patient with a history of a severe bi-polar illness. This second virus isolate was provided to the Los Angeles County Public Health Laboratory, which reportedly sent a portion to the California State Health Department. Both laboratories had difficulties maintaining the virus and dismissed it as a contaminant.

By 1992, I also knew that the viruses had characteristics of a herpes viruses, although I still considered the possibility of a herpes-retrovirus hybrid. DNA sequence data became available in 1994. It indicated a herpes-like virus in that the sequences were related to, although distinct from human cytomegalovirus. In 1995, the cytomegalovirus-related sequences in this virus were unequivocally identified as being derived from SCMV, as was the second virus isolate. Most of the viruses being cultured from other patients were seemingly unrelated to SCMV, suggesting that there were potentially many sources of atypical viruses. The common feature was an apparent failure to evoke an inflammatory response, supporting the earlier designation of stealth or stealth adapted viruses.

The unequivocal SCMV origin of the two stealth adapted viruses clearly implicated the probable source as being from contaminated polio vaccines. I conveyed this important information to the FDA, CDC, Los Angeles County Health Department, Lederle and USC officials.

It was also of interest that an atypical virus had been isolated in 1972 from a brain biopsy from a 6-year-old boy from Birmingham, Alabama. In 1976 it was also shown to be SCMV. I contacted Dr. Charles Alford, the boy's pediatrician. It became clear that he had been discouraged from suggesting a vaccine origin of the virus and felt pressured to agree it must have been a laboratory contaminant. This was clearly not so with the repeatedly isolated stealth adapted virus cultured from the CFS patient; because her blood was also positive for the virus by PCR. Moreover, the virus had growth characteristics unlike any likely contaminant.

I visited with the FDA and was provided an opportunity to briefly speak at a vaccine workshop. Through efforts by an anti-vaccine support group, I was also invited to speak at an Institute of Medicine's Vaccine Safety Forum held in Washington D.C. on November 6, 1995. I soon learned that in the next's day closed session that Industry representatives were furious that I had been invited. Still, I had made my point since a subsequently published summary stated:

"A final important question is whether there are long-latency adverse events following vaccinations. A vaccine researcher has studied several adults who have experienced deteriorating brain function over a period of years that in some cases has left them permanently comatose (Martin et al., 1994, 1995). Tests of their brain tissues, blood, and the fluid circulating in their brains and spinal cords have ruled out known causes of brain deterioration. According to the speaker, genetic studies suggest that their illness is caused by a newly identified cytomegalovirus closely related to a type that infects African green monkeys. Although there is no evidence of the possible source of the genetic material identified, because the kidney cells of these monkeys are used to culture the viral strains used in the live-virus polio virus vaccine, the speaker hypothesized that at least some lots of this vaccine might harbor a strain of cytomegalovirus that can cause severe brain deterioration decades after vaccination. Although the monkey kidney cells currently used in oral polio virus vaccine production come from laboratory monkeys that are not thought to be infected with viruses and that are screened for viral contamination, previously unsuspected (and therefore not screened for) viruses or viruses present at very low levels could theoretically contaminate the vaccine."

Unbeknown to me at the time, Mr. Walter Kyle Esq., a lawyer involved in polio vaccine litigation quickly filed suit against Lederle, demanding on the basis of my talk that they provide him with samples of the polio vaccines received by his clients, for stealth virus testing within my USC Clinical Laboratory. I had previously agreed with a patient advocate to test a blood sample from Whitney Williams, a child with an AIDS-like illness. Her parents had suspected that her illness had come from the polio vaccine she had received. A successful Court Order, opposed by both

Lederle and the FDA, had been issued to authorize testing of a very small aliquot of the bulk vaccine for African green monkey simian immunodeficiency virus (SIV). The Court Order specifically forbade the testing for HIV or any other virus. It was reasoned that if I could culture a stealth adapted virus from the child, a further application could be made to the Court for direct stealth virus testing of the polio vaccine.

I had resisted prior efforts from USC to move away from controversial testing of CFS patients. The Dean expressed to me his concern that I was polarizing some of the University's alumnae. I also knew that American Cyanamid had funded the Department's Chairman. Not too surprisingly, therefore, my Clinical Laboratory at USC was closed nine days after the IOM meeting. I was told that an inspection showed certain chemicals were not being properly stored. I was also informed that the medical technologist was dismissed and that the University had to provide him with back pay and pay-in-lieu-of-notice. Other accounts were said to be overdue. All of the approximately $20,000 of donated funds quickly disappeared from my USC gift account. I refused to authorize the disbursement, especially since the technologist had kindly agreed to reduce funding, as did several volunteers. Without hesitation, I also refused to sign a statement that I would do no more patient related stealth virus testing. I set about correcting the chemical storage problems, but was confronted in early January 1996 with the demand for an additional $15,000 to supposedly cover overhead costs. The money was required before my laboratory would be reopened. I took leave from USC knowing that as a tenured professor, I could not be easily fired.

I located suitable space outside of USC, which was initially offered at $1 per month. I was able to obtain sufficient equipment to meet the requirements for a Federal and State Licensed Clinical Laboratory. I also spent time visiting the Los Angeles County Health Department Virology Laboratory. This allowed me to maintain several of the stealth adapted virus cultures. Interestingly, the head technologist recalled the culture she had received five years previously. I demonstrated how to better maintain the cultures and explained how she was misinterpreting cellular changes in some of her cultures as being due to toxicity. The full expression of the cytopathic effect (CPE) required frequent re-feeding of the cultures.

I was able to further improve upon the culturing technique in the new laboratory and also able to derive additional DNA sequence data. Not only did the data confirm the SCMV origin of the virus, but also it showed the virus to be genetically prone to mutations. Moreover, it had acquired cellular genes and seemingly also some genes from bacteria.

A Johns Hopkins' virologist let it slip that the FDA wanted his help to disprove my virus theory. I found this disappointing since my unsolicited requests to FDA and to CDC for collaborative support had been denied. I was, however, invited to attend a small closed meeting at NIH later in 1996. No report of the meeting was ever issued. Instead of an open inquiry, biased individuals expounded unreasonable and contradictory statements to discredit the research. i) You don't have any evidence for a virus; ii) yes, there may be a virus, but it can't have come from a human; iii) we all know that CFS is not infectious; etc. I sensed little concern or compassion for the patients for whom I was describing positive culture results. My hope for some official to culture the blood from the patient from whom I had repeatedly obtained positive cultures remained unmet. I also considered discussing recent findings of some bacteria-related sequences in the virus cultures, yet I knew this information would be misrepresented as proof of contamination. It was, however, an indication of potential transfer of stealth adapted viruses through bacteria, which has major public health implications. I coined the term "viteria" for viruses able to replicate bacterial sequences.

Over the next several years, I learned of many memorable patients for whom their clinicians had requested stealth adapted virus testing. This led to an ever-broadening perspective of the illnesses potentially attributed to stealth adapted viruses. For example, I saw these viruses as a cause of what was being called "chronic Lyme disease." The positive test for Borrelia bacteria was probably attributed to virus incorporated bacterial sequences.

I also became interested in the Gulf War Syndrome (GWS) after being contacted by a conscientious administrator at the Office of Naval Research. She requested that I should look for stealth viruses in the gamma globulin injections given to troops to reduce their risk of contracting hepatitis A. She said she could not raise the issue internally because it would create a ceiling restricting her career. I submitted a grant on this topic to the Department of Defense. I also had the occasion, along with Dr. Tom Glass of the University of Oklahoma, to make a presentation on GWS at a symposium organized by Purple Heart veterans at the Oklahoma State House. The Governor of Oklahoma, a high level official from the Veterans Administration (VA) and a spokesperson for the CDC, attended the symposium. At the start of the meeting, the VA

official proclaimed that the "Government will not let any stone go unturned in finding the cause of the Gulf War Syndrome." He balked, however, when I suggested it could be infectious with family members being at risk. He also recoiled when I began linking the possibility of contaminated poliovirus vaccines to the emergence of AIDS. The CDC representative was also indifferent to my talk and to the data provided by Dr. Glass of illness occurring in cats inoculated with stealth adapted viruses. The grant was rejected.

On another occasion, I accompanied a delegation to discuss GWS with Congressional staff members. The morning presentations dealt with potential movement of toxic chemical clouds from Khamisiyah towards the troops. The delegation celebrated at lunchtime upon learning they had secured a funding commitment to study the effects of low level chemical exposure. "By the way John, they don't want to hear anything about viruses and we have agreed to pull you off the agenda for this afternoon." It was reminiscent of a tactic I had seen played out at the NIH where dissident visiting guests were offered grants in return for their silence.

I had previously tried to communicate with Congress about stealth adapted viruses. I naively did so through Ms. Beth Clay, a senior staff member of Congressman Dan Burton. I doubt if any of the material I provided to Ms. Clay was ever read. I also was prompted to contact the Congress after being told in 1997 that I could not view tissue sections of spinal cords inoculated with licensed lots of polio vaccines in routine safety studies. The reason again given was that all studies done on regulated products are essentially proprietary to the industry providing the product. I described the problem to staff members of the United States Government

House and Senate Commerce Committees. A legal counsel for the House Committee was willing to insert a provision in an upcoming FDA Reform bill that would "require Industry to agree to waive its proprietary restrictions in the event that a safety issue was identified on a regulated product." This would allow the safety concern to be freely disseminated to the scientific community. I was tasked with seeking support from the American Medical Association (AMA) and also advised to contact someone at the Hover Institute at Stanford University. The AMA political liaison officer in Washington, D.C., could see no advantage of the public learning "that doctors were not in the knowledge loop." Senior FDA officials told the legislative counsel that to be a "sieve" of such information about competitor's products would lessen their prospects for high salaried post government "service" employment. The Hover Institute expert was unwilling to confront the Pharmaceutical Industry. Thus, the matter was dropped.

Reports of family centered and even community wide outbreaks of complex illnesses kept coming to my attention. Dr. Donovan Anderson reported that he had seen over 100 patients from the Mohave Valley region of Arizona, who initially presented with a gastrointestinal illness. Many went on to develop a chronic neurological and fatiguing illness. He reported both the occurrence of the gastrointestinal symptoms and the subsequent neurological symptoms to the Arizona State health authorities. He told them that I had found many of his patients to be stealth virus infected. Not only did the Arizona State Public Health department dismiss the reports, but also Dr. Anderson soon found himself the target of probes by both the Arizona State Medical Board and the Federal Drug Enforcement Agency.

A hairdresser in Joelton, TN, realized she, two staff members and a number of her clients were experiencing general malaise, muscle aches and pains, and impaired brain function. When asked if she knew of other patients, she took the initiative of describing her illness in the local Shopper magazine. Well over 100 people responded. As soon as the issue of stealth viruses arose, the local health authorities, supposedly on advice from CDC, let the matter die.

I continued to publish on cases from the Mohave Valley and elsewhere and to present the findings at various scientific conferences. I found it difficult to engage CDC and FDA researchers in scientific discussion. I attended an NIH conference on SV40. Over morning coffee, I asked a group of technicians from Lederle if they were still identifying SCMV in kidney cell cultures. I was told that they did so in some 10% of the cultures. I went straight to an FDA representative at the same conference suggesting the importance of using PCR assays to further screen bulk polio vaccine submissions for SCMV. The straightforward answer was "We would not know what to do with a positive result."

In 2002 I was told that my Clinical Laboratory license was coming up for early re-inspection of its Federal license. The California Department of Health Laboratory Field Services performs the inspection on behalf of the Centers for Medicare and Medicaid Services (CMS). The laboratory had been inspected on 3 earlier occasions, 1996, 1998 and 2000, without any deficiency or shortcoming ever reported. Because of inexperience of the local inspector, I was informed that the Health Department had decided to send a senior inspector from the Berkeley office. My early expectation of a serious effort to understand the research soon gave way to disappointment that the clinical testing was being scrutinized with the foregone conclusion of finding something non-compliant with Federal regulations. The essential premise was that since others had not confirmed the work, it must not be valid. Yet the inspector knew of no one who had actually tried to duplicate the work or who could factually criticize any of the prior publications. To conclude that what I had been observing for over 10 years was nothing but the overgrowth of normal cells was striking in its dishonesty.

Earlier that year, I had performed an Institutional Review Board (IRB) approved study to test blood samples from individuals donating blood on the University of California Irvine (UCI) campus. The primary intention was to obtain more negative control blood samples for reference purposes. Probably not surprising, however, 10% of the samples tested positive. The Blood Bank Director at UCI was strongly criticized for agreeing to the study. In order to disregard the finding, it was necessary for the inspector to conclude that the assay procedure was not detecting viruses and as a CDC employee was once quoted "There's no such virus that has been validated or shown to be true." Another public health official more clearly declared that stealth virus "doesn't exist."

I was required to provide the State inspector with complete copies of all of the procedure manuals, etc. In so doing, I was hopeful of rapid validation by the State Public Health Laboratory of how relatively easy it is to culture stealth adapted viruses. I also made a request to the Laboratory Head of the Los Angeles County Department of Health to do independent testing. This seemed reasonable since I had previously shown their chief virology technologists several examples of typical stealth adapted virus CPE and knew that she was certainly capable of performing

the work. This request was rejected. Instead, the State and Federal authorities concluded that testing for stealth adapted viruses was placing the Nation's health in "Immediate Jeopardy."

I spoke with Ms. Karen Nickel of the Laboratory Field Services. She informed me that the inspection and subsequent action were taken at the behest of the CDC. I still thought some good might come from the inspection and began correspondence with Mrs. Karen Fuller of CMS. Five letters are attached to show the tenor of my unsuccessful appeal to CMS.

I was required to write a letter to every physician who had ever requested virus testing stating the Health Department's conclusion. The State had also justified their inspection on a complaint received from a disgruntled patient. While at USC, I had tested her blood, that of her dog and of a companion. She was angry at my reluctance to retest her blood in the new laboratory. She then joined the chorus of patients being diagnosed as having chronic Lyme disease, arguing that my initial stealth virus report had delayed her arriving at this supposedly now correct diagnosis. The laboratory inspection process started to get out of hand, when the woman requested from the Health Department the name of every tested patient for a class action lawsuit. Karen Nickel apologized to me for this unforeseen outcome. Nevertheless, the woman went ahead with a frivolous lawsuit against USC and me. At most, she may have given $200 to my USC Gift Account, yet claimed she had paid thousands. A local television station ran a derogatory report based on her ridiculous assertions. Far more damaging has been an internet posting from the National CFIDS Association about stealth viruses being bogus. This prominently displayed posting has done enormous harm by discouraging potential supporters of my research.

The Director of the National Center for Infectious Diseases at CDC helped organize an October 2002 conference by the Institute of Medicine on "The Infectious Etiology of Chronic Illnesses." I presented my findings and subsequently submitted a paper for inclusion in the proceedings. The paper was never published and my name removed from the Speakers Biographies, although it still can be seen on the Agenda. When I later queried why the paper was not published, the now ex-CDC employee offered the paltry excuse that he had to accept the opinion of others. I posted the actual paper that I submitted on the Internet and it is included later in the book.

In 2005, CDC was confronted with an individual and his father who both tested positive for SCMV using a PCR assay performed in a commercial primate-testing laboratory. The individual developed a fatiguing illness with cognitive impairment shortly after partying on a business trip. Both his son and his father became fatigued soon after his return. He was concerned about a transmissible disease, including the possibility of HIV. While he repeatedly tested negative for HIV, he did learn of my research and had his blood tested for SCMV in the commercial primate laboratory. He shared the positive results with me. I visited the laboratory and was assured the results were definitive and not due to any other type of CMV. I again explained the research to a CDC official. All CDC did was to request the University of Georgia to do cursory viral culture on a blood sample from the individual. Their result was said to be negative and the issue was dropped. I was at least able to constructively advise the individual on certain approaches to combating his stealth virus infections.

I also submitted an article to the CDC's managed "Emerging Infectious Disease Journal". It described a striking family history of an apparently transmissible illness involving three generations. A 65-year-old man developed what was initially considered Alzheimer's disease with additional features of Parkinson's disease. His wife cared for him for a while but she became fatigued and cognitively impaired. The man's daughter offered to accommodate her parents in her house but soon after the arrangement was made, she became sick, as did her husband and 4 children. The husband's diagnosis is ALS while the 3 younger children all have severe learning disorders. The CDC declined publication of the article and, it too, is included in this book.

Conclusion

It is clear that CDC and FDA have not been willing to accept the existence of stealth adapted viruses as a potential cause of the increasing incidence of various chronic illnesses, including those with prominent neuropsychiatric manifestations. Indeed, there are many indications of both agencies stifling potential research on these viruses. A major reason is that the available data bear on the question of the agencies diligence in helping ensure the safety of vaccines. Had the initially isolated stealth adapted virus been a derivative of a human virus, less political and scientific opposition would likely to have been asserted.

It is commonly argued that questioning vaccine safety will result in non-compliance with public health policies and lead to widespread outbreaks of infections. This approach avoids any accumulated tally of the many adverse events, which have been suppressed in the name of retaining the public's confidence in vaccines. Having to prescreen individuals for possible susceptibility to an adverse effect would add to the cost and inconvenience of vaccination. Publicizing errors could also lead to massive lawsuits restricting continuing efforts to conquer diseases. The suggestion that vaccines may have inadvertently caused AIDS is less of a transgression than not addressing an ongoing virus cause of widespread illness, especially those occurring in children. This consideration more than the former is likely to undermine the claim that the United States is the greatest nation ever to have existed. Considering the possible damage that has already occurred, it is little wonder that some of those in control have chosen to close ranks. It is not for a lack of well-qualified staff or resources that the research has been hindered, but rather it is protection of their agency image, as dictated by some of its leaders. Yet over time, the errors become known and the image becomes tarnished.

Through post-government employment opportunities, Industry has acquired an increasingly influential role in the agencies promotion process. Senior leadership will typically go to those willing to compromise regulations in favor of corporate profitability. The balance of power is now such that there is a perceived need among senior government officials to incentivize industry. In the big picture there should be an absolute ban on government regulators ever subsequently working within the industry they regulate. Furthermore, industry should agree to waive proprietary restrictions on government officials' ability to communicate safety concerns on regulated products to the scientific and medical communities.

By being able to award grants, government agencies as well as industry can capture the allegiance of universities and prominent researchers. Money can also subdue politicians' willingness to confront special interests, even when the burden of their influenced decisions falls onto the public.

The government's handling of illnesses, such as CFS and autism is especially shabby with the appointment of various committees unable to creatively think outside of the accepted paradigms. Even if they could, their tenure would be at risk if they were to criticize their hosts. Many of the specific examples described in this report may serve as models of government malfeasance.

Yet there is hope for a change, especially since like everyone else, most politicians and senior government officials are parents. Many are beginning to experience the ravages of the current epidemics of mental illnesses. Increased public awareness of the events surrounding the development and testing of polio vaccines and of the unwillingness to pursue the concept of stealth adaptation will make it easier for such politicians to demand a full accounting of the past and current practices of those entrusted with the nation's health. Well-supported, focused research will very likely provide therapeutic answers whereas continuing to ignore the issues will simply result in an ever-increasing national and international tragedy.

Example of Attempts to Engage CDC

The following e-mail sent to Dr. Phil Pellett on 8/14/2000, following a Herpesvirus Conference. At that time, Dr. Pellett was Head of the Herpesviruses Branch of CDC. As noted above, I never did receive a reply.

E-mail to Dr. Pellett:
Topic Stealth-adapted viruses

Dear Phil,

Thank you for the discussion during the last evening of the International Herpesvirus Workshop. You were willing to talk bluntly, yet in a constructive manner, regarding CDC shunning of my research. As you said, CDC administrators look to you for scientific judgment on matters relating to herpesviruses. Without your support, there is little chance of any response to my requests that CDC pursue what I perceive to be a serious Public Health problem. As I recall, the major points of our discussion were as follows: You spent approximately 45 minutes at my poster and came away with the impression that some of the sequence data must be incorrect. You were concerned that sequence homology matchings should be more uniform and not differ along a stretch of nucleotides. You asked if all of the sequencing had been fully double-stranded and whether I had reviewed all of the primary data. I indicated that most of the extended sequencing had been performed by Lark Technologies, at Houston Texas. Although there was some internal overlapping, the sequences were primarily derived from one-way reactions. While there is, therefore, the possibility of an occasional nucleotide error, this would have had no effect on the conclusions that I was drawing from the data. I understand that you hold your own sequence-related studies up to a particularly rigorous standard, but this has more to do with the types of conclusions that you are trying to draw, rather than justifying dismissing any sequences that are not verified by double-stranded confirmation. Most of the sequences that I have obtained have been on GenBank for a long time and can be reviewed directly by anyone who's interested. The irregular matching that you noted is indeed interesting. It goes along with an earlier publication on the genetic instability of the virus. You also suggested that real science ought to be obvious to any intelligent scientist and that it was my responsibility to present the work so that it would be more widely accepted. Again I disagree. Most scientists are pretty fixed in their belief system, and historically any shift in a prevailing paradigm has been met with resistance. The average scientists cannot be expected to plow through loads of someone else's raw data, or as you said "interpret my data for me." The CDC is something of an exception, however, since its mission is to be vigilant for possible threats of emerging infections. The poster provided a good opportunity for a scientific discussion, but you chose to view it in my absence. I, therefore, do not know if you fully understood and appreciated the significance of what was being presented. I was surprised by your suggestion that I sought a chance to discuss my work at CDC as a "cheap" way of claiming CDC recognition. I view my challenge as primarily to get CDC to listen and to take some action. I hope you will continue to provide some assistance by engaging in more meaningful discussions of actual sequence data and patients' histories. In particular, I would like to know you responses to the following issues that I have raised.

1. Do you doubt that the virus for which I have extensive sequence data was derived from an African green monkey simian cytomegalovirus.

2. Do you doubt that the virus most probably originated in a live poliovirus vaccine; or that it came from a patient (who is still living); or that it can induce severe illness when inoculated into cats.

3. Are you convinced that the virus has some unusual sequences that at least qualifies it as being an atypically structured virus.

4. Do you feel the apparent absence of UL83 and UL55 related genes could provide the virus a way of evading recognition by the cellular immune system.

5. Are you willing to accept that the virus has recombined with cellular sequences, including the CXC chemokine coding gene, melanoma growth stimulatory activity, a potential oncogene.

6. Do you see any significance in the apparent amplification of the US28 chemokine receptor coding gene.

7. Do you accept the presence of bacteria-derived sequences within the viral genome.

8. Are you aware of the high proportion of patients with unexplained encephalitis-like illnesses, including cases in which brain biopsies have been submitted to CDC for review.

9. Given our positive tissue culture findings in several such patients, can you dismiss the probability of widespread infections with atypically structured viruses.

10. Don't you think we owe it to those responsible for the Nation's Public Health to have some of these topics more openly discussed?

Enough questions for now. Most of the papers dealing with stealth-adapted viruses are on the web site www.ccid.org I hope we will continue to dialogue and thanks again for the time provided in Portland. Kind regards, W. John Martin, M.D., Ph.D.

End of e-mail.

Comments on some items referred to in the e-mail sent to Dr. Phil Pellett:

Issue of DNA sequence differences between stealth adapted virus and published DNA sequence of SCMV. Genetic instability and variability is a feature of stealth adapted viruses and probably relates to the mode of virus replication. A manuscript on this topic had been published.[1]

Item 4. The proteins coded by UL55 and 83 provide 2 of the 3 major targets for immune recognition of hCMV.[2] No evidence has yet been found for these genes in the prototype stealth adapted SCMV derived virus.[3] The third gene comprising an effective target for T cell immunity (UL123) is mutated in the stealth adapted virus.

The relevance of amplification of US28 genes (item 6) is that the product of this gene acts as a receptor for HIV. Both African green and rhesus

177

monkey CMV have 5 copies of this gene as opposed to the single copy in human and chimpanzee CMV.[4-5] The British study confirms that the polio virus vaccine used in the Congo in the 1958 was contaminated with rhCMV. It is very plausible that chimps receiving rhCMV contaminated polio vaccine could have become persistently infected with rhCMV, especially if it was stealth adapted. This virus could have facilitated the growth of simian immunodeficiency virus (SIV) and its evolution to become HIV. Individuals working on the vaccine project have described illnesses in both chimps and a number of African workers (thin man syndrome). The CDC is said to have retained pre- and post-polio virus vaccine inoculation sera from a 1958 study performed in the Congo. I have suggested that these sera be retrieved and tested for antibodies to rhCMV.

The issue in item 5 of stealth viruses recombining with possible cancer causing genes (oncogenes) is of major concern, especially when linked to data showing the probable passage of stealth adapted viruses through bacteria (item 7).

Item 8. Unexplained encephalitis cases occurring within the United States is not given much emphasis. Over 19,000 cases of encephalitis are diagnosed annually. Yet in spite of intensive studies the cause of over 50% of the cases remains undefined. If severe cases go undefined, how many milder cases could be caused by the same or similar undefined agents. With diseases such as autism on the rise, it is inexcusable for CDC and other public health agencies not to investigate stealth adapted viruses.

Cytomegalovirus (CMV), itself, should be of major concern to CDC. Between 0.5 and 2% percent (average 1%) of all children are born infected with human CMV.[9] From 10-20% of these children will show clinical signs of infection, especially indications of brain damage. A current estimate for the United States is 400 neonatal deaths occur from congenital CMV with another 8,000 children left with hearing, visual and/or mental handicap. CMV has also been linked to other illnesses, especially in immune suppressed individuals. Regular CMV is, therefore, not a trivial infection. Realization that SCMV and presumably hCMV can undergo stealth adaptation is clearly important and its study well within the mandate of CDC.

References

1. Martin WJ. Genetic instability and fragmentation of a stealth viral genome. Pathobiology 64: 9-17, 1996.

2. Martin WJ. Stealth adaptation of an African green monkey simian cytomegalovirus. Exp Mol Pathol. 66: 3-7, 1999.

3. Martin WJ. Chemokine receptor-related genetic sequences in an African green monkey simian cytomegalovirus-derived stealth virus. Exp Mol Pathol. 69:10-6, 2000.

4. Sahagun-Ruiz A, Sierra-Honigmann AM, Krause P, Murphy PM. Simian cytomegalovirus encodes five rapidly evolving chemokine receptor homologues. Virus Genes. 28: 71-83, 2004.

5. Sequar G , Britt WJ , Lakeman FD, et al. Experimental coinfection of rhesus macaques with rhesus cytomegalovirus and simian immunodeficiency virus: pathogenesis. J Virol. 76: 7661-71, 2002.

CENTER FOR COMPLEX INFECTIOUS DISEASES

3328 Stevens Avenue, Rosemead, California 91770

Phone: (626)572-7288   Fax: (626)572-9288

July 24, 2002

Ms. Karen Fuller,
CMS, DSO/CLIA Division
75 Hawthorne Street, Suite 408
San Francisco, CA 94105

Dear Karen,

I had a pleasant and constructive telephone conversation with Ms. Karen Nickel, Head of the Laboratory Field Services of the California Department of Health. She kindly explained that repeated complaints had been received by the Health Department concerning CCID and in particular the testing for stealth viruses. The complaints came from CDC, certain State Health Departments and individual physicians. The complaints were essentially that patients were being misled into believing they were infected with viruses that did not exist. The California Department of Health was in no position to question the opinion of CDC or to assess how the testing was being used, or possibly abused. Staff shortage prevented action being taken last year, but with continuing complaints, closure of the laboratory was given a higher priority this year. Consequently, the California Department of Health undertook a Laboratory Field Service inspection with a predetermined outcome.

I was not surprised by this forthright explanation. Clearly I need to address the concerns of CDC. For several years, I have been trying to engage CDC in a scientific evaluation of my work. I believe the data are irrefutable, but they obviously can be ignored. In March of this year, I presented a Poster at a CDC sponsored International Workshop on Emerging Infectious Diseases. Disappointedly, it did not evoke further inquiry. I had prepared more data for an International Herpesvirus Workshop that I was unable to attend because of the limited time allowed to respond to the inspection report. At this stage, I am seeking the opportunity for an informal seminar at CDC at which I could present the data and answer questions. I would also like to demonstrate the culture techniques to laboratory technologists.

CMS is in a justified position of helping me by asking for CDC input as to whether my data "establish the existence of atypically structured cytopathic viruses." I really want to get CDC, FDA, NIH and others involved so that I can pursue hopeful leads regarding potential therapies. I am including a copy of the poster that I had presented last March.

Sincerely,

W. John Martin, M.D., Ph.D.

cc. Karen Nickel

CENTER FOR COMPLEX INFECTIOUS DISEASES

3328 Stevens Avenue, Rosemead, California 91770

Phone: (626)572-7288 Fax: (626)572-9288

July 27, 2002

Ms. Karen Fuller,
CMS, DSO/CLIA Division
75 Hawthorne Street, Suite 408
San Francisco, CA 94105
Fax 415-744-2692

Dear Karen,

In good faith, I provided the California Department of Health and your Office a comprehensive response to a Statement of Deficiencies. My response was received by the State on Monday afternoon. Yesterday, I was notified by Fax that my response was unsatisfactory and a recommendation for sanctions has been forwarded to your Office. I trust your Office will not be so quick to judgment.

I would like to work within the system to "establish the existence of atypically structured cytopathic viruses and to demonstrate a culture method for their detection." In this regard, I have reviewed the Inspector General Report on CLIA Regulation of Unestablished Laboratory Test. I realize that CMS does not validate tests and I am not asking you to do so. What I would like, however, is to have your support and guidance in my interactions with Los Angeles County Public Health Laboratory, CDC and CMS.

The Nation's health is our common goal. Before you take any action, I would appreciate an opportunity for an introductory phone conversation. I would also like to contact the author of the Inspector General's report. Some years ago it was suggested that I apply for a "Demonstration Project" with HCFA. I will also see if I can present the work at CDC and NIH. It would be encouraging if some of these efforts were to become part of your determinative process. I look forward to contacting you later today.

Sincerely,

W. John Martin, M.D., Ph.D.
Laboratory Director

180

CENTER FOR COMPLEX INFECTIOUS DISEASES

3328 Stevens Avenue, Rosemead, California 91770

Phone: (626)572-7288   Fax: (626)572-9288

August 7, 2002

Ms. Karen Fuller,
CMS, DSO/CLIA Division
75 Hawthorne Street, Suite 408
San Francisco, CA 94105

Dear Karen,

A serious criticism expressed by the State Examiner related to her interpretation that what she observed in some culture tubes was overgrowth of the cell sheet, or inoculum or reagent toxicity rather than viral cytopathic effect (Tag D6115). This deficiency was highlighted in another citation referring to the lack of confirmation of a positive result with acridine orange or secondary and tertiary culture or other method (Tag D6087). It also related to the issue of verification in Tag D6086, in which the examiner distinguishes between a result being reproducible and being accurate. She cited the laboratory "for not ensuring that testing methods (serology, PCR, electron microscopy, etc), that were described in the patient report, were used to differentiate 'Stealth Viruses' from conventional viruses and to individually characterize different 'Stealth Virus' isolates. Also, that those tests were (not also) performed by an outside laboratory to verify the patient 'Stealth Virus' result."

I fully appreciate the distinction between claiming isolation and characterization of a specific virus, as opposed to describing the development of a cytopathic effect, presumably caused by a virus. I have shown the characteristic cytopathic effect that I routinely observe to various expert virologists, such as Zaki Salahuddin (the discoverer of HHV-6), Luc Montagnier (the discoverer of HIV) and others. They agreed with my own opinion that the cytopathic effect was presumably viral. I have included photographs of positive cultures in several peer reviewed publications and meeting presentations without pre or post publication objections. I have submitted several isolates to the American Type Culture Collection (ATCC). They confirm viral growth prior to placing material in long term storage. I have retrieved several aliquots from ATCC and have experienced no difficulties in re-establishing positive long term cultures.

There are important Public Health implications if a viral cytopathic activity can be summarily dismissed by qualified medical technologists, such as the Examiner. In response to the Inspection Report, I proposed having at least six positive cultures periodically reviewed by an experienced outside virologist. I am also seeking the opportunity to demonstrate the culture techniques to CDC and other Public Health officials. I was hopeful that CMS would retain an open opinion on this critical issue until I could provide input from such authorities.

For some cultures, I do have extensive DNA sequence data and numerous electron micrographs. I did feel that these data justified the statement that techniques such as serology, PCR, electron microscopy, etc can (or could) be used to differentiate among different stealth virus isolates. I have never inferred however that these sophisticated techniques were anything other than research methods. I have never issued a clinical report with the results of serology, PCR, or electron microscopy. Nevertheless, I am more than prepared to delete all reference to such techniques in subsequent reports.

The extent of cellular abnormalities seen in positive cultures is far beyond any of the changes ever seen in an uninoculated culture. I have taken a series of illustrative photographs as part of an ongoing validation study. I will send you and the State copies by regular mail for review. As stated in my previous correspondence, uninoculated cultures were included in each experiment with written documentation of a lack of cytopathic activity included in the sterility quality control that was performed on each batch of MRC-5 cells received.

Sincerely,

W. John Martin, M.D., Ph.D.
Laboratory Director

cc California Department of Health

CENTER FOR COMPLEX INFECTIOUS DISEASES

3328 Stevens Avenue, Rosemead, California 91770

Phone: (626)572-7288   Fax: (626)572-9288

August 11, 2002

Ms. Karen Fuller,
CMS, DSO/CLIA Division
75 Hawthorne Street, Suite 408
San Francisco, CA 94105

Dear Karen,

The argument for Immediate Jeopardy was the conviction that patients were being harmed. This conviction was prompted in part by complains received by the California Department of Health from CDC and other Public Health authorities. These authorities doubted the existence of atypical viral infections that I have been describing over the last several years. The immediate threat to Public Health was explained to me as resulting from possible toxicity of medicines prescribed on the basis of a positive test result.

I was observing and reporting on a cytopathic effect with the qualification that it was presumably viral. I had been assured during prior CLIA inspections that this was a reasonable and legitimate statement. Nevertheless, in the most recent CLIA inspection, the point was made by the examiner that the "findings were consistent with overgrowth of the cell sheet, reagent or inoculum toxicity, rather than viral cytopathic effect indicating that test procedures were not properly verified" (Tag D6115). This opinion naturally led to concerns that I had not documented a lack of toxicity in the uninoculated tubes on each occasion that I had read a positive culture. It also raised a question regarding the possible toxicity of the Ficoll-Paque reagent. Finally, details of the double blind assays certified by Robert Gan and Russ Colins were brought into question, as was the subsequently provided affidavit of Zaki Salahuddin. To establish her case, a wide range of other issues were mentioned in the Examiner's report.

I have responded by restating the argument that I was indeed reporting on a cytopathic effect that, based on an extensive amount of data and on the opinion of acknowledged experts, can reasonably be ascribed as presumably being viral. I have re-examined a number of specimens to demonstrate the reliability and reproducibility of the assay. I have shown features in positive cultures, such as the formation of lipid crystals, ribbon-like structures and pigmented inclusions, that can not be dismissed as artifacts of simple toxicity. I have also sought the opportunity to share samples and to demonstrate culture techniques with State and Federal Public Health laboratories.

I know of no patient who has received therapy solely on the basis of a culture result. The medications that are being prescribed are based primarily on the patients' symptoms and the experience of the prescribing physicians. None of the various medications being used is considered particularly toxic. Moreover, patients will continue to be treated whether or not pre and post therapy cultures are performed. There is really no immediate jeopardy to the nation's health.

I trust you will allow me the opportunity to continue efforts to engage CDC in a scientific dialogue regarding the existence of atypically structured cytopathic viruses. As requested in my previous communications, I would appreciate if either CMS or the State could second my request for an expedited meeting to discuss the scientific data that I have compiled and to demonstrate the cytopathic effect. While I cannot force the federal government to join me in the pursuit of a viral cause of many complex neurological and neuropsychiatric illnesses, I can expect protection from a capricious and unjustified stifling of the work. I firmly believe that imposing sanctions that will prevent the continued examination of patient samples is an inappropriate exercise of the CLIA authority entrusted to CMS. I hope that I can continue to work co-operatively with you and your staff.

Sincerely,

W. John Martin, M.D., Ph.D.
Laboratory Director

cc California Department of Health

CENTER FOR COMPLEX INFECTIOUS DISEASES

3328 Stevens Avenue, Rosemead, California 91770

Phone: (626)572-7288   Fax: (626)572-9288

August 11, 2002

Ms. Karen Fuller,
CMS, DSO/CLIA Division
75 Hawthorne Street, Suite 408
San Francisco, CA 94105

Dear Karen,

I am pleased to enclosed both the protocol and the results of a study that validates the existing cytopathic assay conducted in my laboratory. I have maintained that processed blood from certain patients is able to produce a characteristic cytopathic effect (CPE) in MRC-5 cells. The CPE is characterized by the normal spindle shaped, translucent, closely packed MRC-5 cells becoming enlarged, rounded and fusing into small, and later into larger, three dimensional cell syncytia and clusters. The cellular cytoplasm displays a vacuolated, lipid filled appearance, often accompanied by fine, darkly pigmented inclusions. In many stealth virus cultures the excess production of lipid is so marked that it deposits as extra-cellular crystals. The pigmented material can also take on the form of ribbons and large aggregates. Neither the formation of extra-cellular lipids or other structures can be easily explained on the basis of a simple toxic reaction of the cells.

The validation test consisted of 41 culture tubes. Ten of the tubes were inoculated with stored aliquots of processed blood samples that were previously shown to give a positive CPE. Ten of the tubes were inoculated with stored aliquots of processed blood samples that did not previously cause a positive CPE. Ten samples came from previously untested blood samples provided by the University of California Irvine Blood Transfusion Service. Ten tubes were designated as no-sample or uninoculated controls. Finally, as in all routine studies, a uninoculated control, designated as a Blank tube was included. X Vivo-15 medium was used for all of the tubes. Incubation, feeding and viewing of tubes were performed throughout the experiment in an identical manner without regard to the original inoculum. The tubes were examined daily by myself and independently looked at by Russ Collins, CLS at day 5. I also took photographs of all of the cultures. The results can be summarized as follows:

There was no difficulty in distinguishing a positive from a negative culture
All of the samples that previously tested positive again showed the development of the characteristic CPE
None of the samples that previously tested negative showed the development of a CPE
Three of the blood donor samples did develop a positive CPE, one of which (tube #17) was quite striking
None of the uninoculated tubes (without an added processed blood sample) showed any CPE
The control blank tube also did not show a CPE.
There was a 100% concordance with my day 5 readings and those of Mr. Russ Collins, CLS
None of 20 previously left over tubes from earlier time points showed any signs of the characteristic CPE.

I have typically saved an aliquot from many of the samples that I have tested over the last several years. I can repeat the study at any time intervals that you feel is appropriate. I can also send both positive and negative aliquots of previously tested processed blood samples to an outside laboratory. Among the selected photographs of positive cultures, several show the types of lipid crystals commonly seen in positive cultures. Other photos show the ribbon-like structures and the pigmented aggregated inclusions. I hope this information will forestall the imposition of sanctions, so that I can learn to better understand the clinical importance of such striking findings.

Sincerely,

W. John Martin, M.D., Ph.D.
Laboratory Director

cc California Department of Health

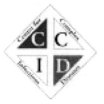

Stealth-Adapted Viruses and Viteria: Insights into
Based on DNA Sequence Analysis of an African Gre

W. John Martin; Center for Comple

Abstract

Stealth-adaptation is a mechanism that allows cytopathic viruses to evade immune elimination through the deletion of genes coding the major antigens targeted by the cellular immune system. A prototype stealth-adapted virus, repeatedly cultured from a patient with chronic fatigue syndrome (CFS) was cloned and partially sequenced. It has a fragmented, genetically unstable, genome. It has retained numerous viral sequences that can be aligned to various regions of the genome of human cytomegalovirus (HCMV). Where the comparison can be made, the sequences match much more closely to those of African green monkey simian cytomegalovirus (SCMV) indicating an unequivocal origin from SCMV. Kidney cells from cytomegalovirus seropositive African green monkeys were, until recently, routinely used to produce live poliovirus vaccine. The SCMV-derived stealth-adapted virus has five adjacent, but divergent, open reading frames that potentially code for molecules related to the US28 CC chemokine receptor protein of HCMV. In addition, the virus has acquired cellular sequences from infected cells, including a set of three divergent genes that potentially code for proteins related to the putative oncogenic CXC chemokine known as melanoma growth stimulatory activity (MGSA/Gro-alpha). The genes in the prototype SCMV-derived stealth-adapted virus, supports current experimental therapeutic approaches based on chemokine suppression. Interestingly, the MGSA-related genes generally lack introns and were, therefore, presumably assimilated into viral DNA from cellular RNA through reverse transcription. The virus has also acquired genetic sequences from various bacteria. This finding has led to the secondary designation of this type of novel microorganism as viteria. Molecularly heterogeneous viruses, inducing similar cytopathic effects in culture (and when examined, non-inflammatory vacuolating cellular damage in brain and tissue biopsies), have been cultured from numerous patients with severe neurological, psychiatric, immunological and neoplastic diseases. In controlled, blinded, studies, cytopathic effects were recorded in 9% of healthy individuals donating blood for transfusion; in contrast to the positive results recorded in virtually all blood samples from patients with various illnesses. The differing clinical manifestations in infected patients may reflect the assimilation of different cellular and other sequences in various stealth-adapted viruses. Stealth–adapted viruses (and viteria) pose a major threat to Public Health. Further information is available on the internet at www.ccid.org.

Background Information

A. There is an increasing incidence of diseases with accompanying signs and symptoms of brain damage. These include neurological and psychiatric illnesses, childhood behavioral disorders, and such common conditions as chronic fatigue, Gulf War Syndrome, so-called "chronic Lyme disease", and many cancers. Altogether, these diseases have an enormous social impact.

B. An infectious cause of many of these chronic illnesses has not been considered primarily because there is no inflammation in the involved tissues.

C. Brain biopsies do, however, show cells with damaged mitochondria, lipid vacuoles, and irregular inclusions. Examples are shown in the figures 1-5.

D. Viral cultures from patients with neuropsychiatric and other illnesses, regularly develop clusters of foamy vacuolated cells. These cellular changes are consistent with infection by actively cytopathic viruses. Figures 6-7.

E. The cultures are also remarkable in the production of large quantities of lipids, including cholesterol esters, and pigmented, protease-resistant, aggregates and ribbon-shaped materials, some of which incorporate metals. Figures 8-15.

F. Viral cultures can induce severe, non-inflammatory, widespread illness when inoculated into cats. The cytopathic effect (CPE) seen in tissues of infected animals is comparable to that seen in the tissue cultures.

G. While the viruses causing CPE in viral cultures differ in different patients, one viral isolate was unequivocally derived from an African green monkey simian cytomegalovirus (SCMV). The issue of probable SCMV contamination of live polio virus vaccines produced in kidney cells of African green monkeys was identified by Industry and FDA in 1972. Unfortunately, this potential problem with live polio virus vaccines was not publicly disclosed, nor scientifically addressed.

H. Continued sequencing of DNA isolated from this cultured virus shows intriguing genetic modifications. Apparent loss of critical viral genes can explain how the virus evades the cellular immune system. Sequencing also reveals the surprizing presence of an assortment of bacterial genes, including genes very closely related to those of Brucella, Mycoplasma, Streptococcus, and other bacterial species. This finding shows the capacity of such viruses to pass, and possibly, to have been passed, through bacteria. Stealth viruses can also potentially incorporate cancer causing cellular genes, as shown by the presence of a cancer-related chemokine gene in the SCMV-derived stealth-adapted virus.

Brain Biopsies From Stealth Virus Infected Patients

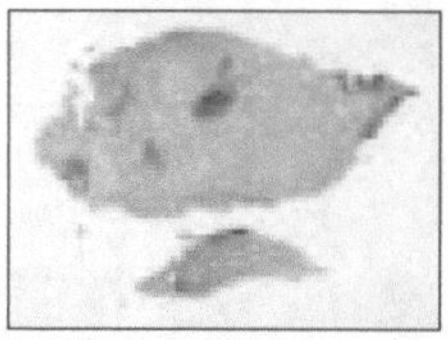

Figure 1. Brain biopsy obtained in 1991 from a stealth virus culture positive school teacher. Her illness began as a chronic fatigue-like syndrome and progressed to a more severe cognitive disorder. Conventional neurological examination was, nevertheless, essentially normal. Periventricular white matter changes were detected using MRI. The pink color of the biopsy is an indication of the absence of inflammatory cells, (lymphocytes and macrophages) that stain blue.

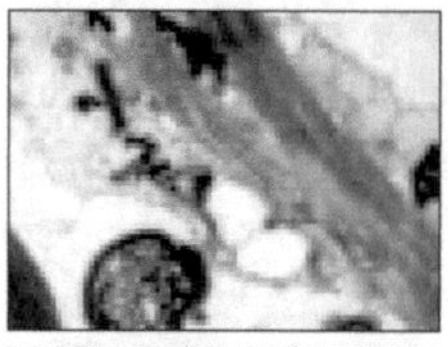

Figure 2. Electron micrograph of an abnormal cell seen in the above brain biopsy. The pale staining material in the vacuoles is lipid. The striking irregularly shaped dark staining materials (inclusions) do not correspond to any normal cellular structures. The long fibers are filaments, typical of glial cells. The round structure at the bottom of the photo is an axon surrounded by a myelin sheath. Mitochondria, seen elsewhere in the biopsy showed degenerative changes.

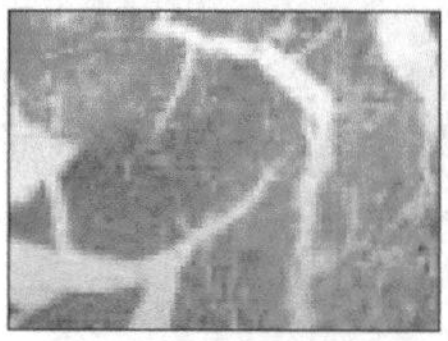

Figure 3. Brain biopsy obtained in 1998 from a stealth virus culture positive 8 year old boy from the Mohave Valley. His illness began as an attention deficit, behavioral problem. Even when gross abnormalities were detected on MRI, there were no clinical signs of motor, sensory or autonomic nervous system dysfunction. The markedly vacuolated appearance, without signs of inflammation, is reminiscent of the changes seen in diseases attributed to prions, such as mad cow disease. The child's mother had previously been shown to be infected. She continues to have repeated bouts of a severe personality disorder.

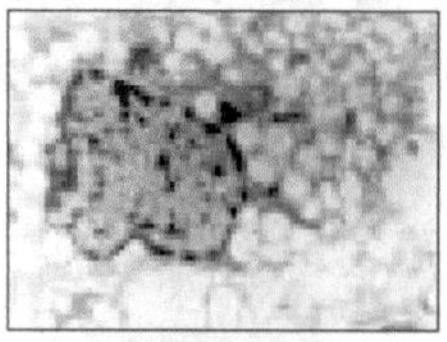

Figure 4. Electron micrograph of a foamy, vacuolated cell in the boy's brain biopsy. Viral particles were not seen.

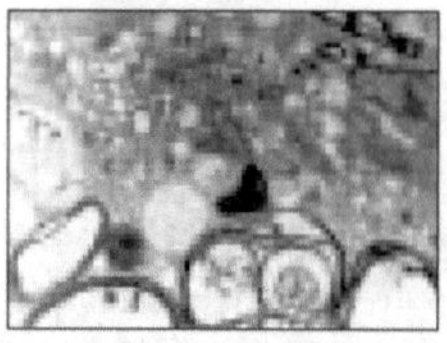

Figure 5. Electron micrograph of another cell showing marked disruption of mitochondria and the presence of an unusual, irregularly staining inclusion. The myelin sheaths show extensive blebing. The child showed a clinical response to ganciclovir, but subsequently died.

Stealth Virus Cultures

Figure 6. Normal MRC-5 fibroblasts seen under phase contrast microscopy. Note the rather bland appearing, closely interdigitating spindle shaped cells. Many cell types can be used to demonstrate the cytopathic effects of stealth viruses.

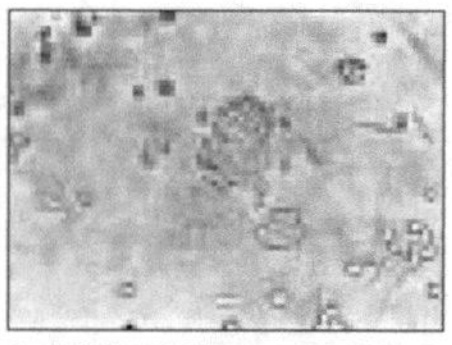

Figure 7. MRC-5 fibroblasts following exposure to mononuclear cells from a stealth virus infected patients. The cells become enlarged and rounded, with a tendency to form large clusters. The cytoplasm develops a foamy vacuolated appearance. Intracellular pigmented material will commonly develop, especially in long term cultures.

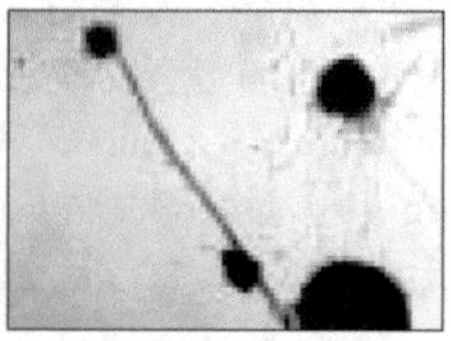

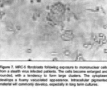

Figure 8. Cell clusters in a long-term stealth virus culture showing the accumulation of dark pigmented material and formation of long ribbon shaped structures. Relatively normal appearing cells, that grew out from the top left cluster, can be seen.

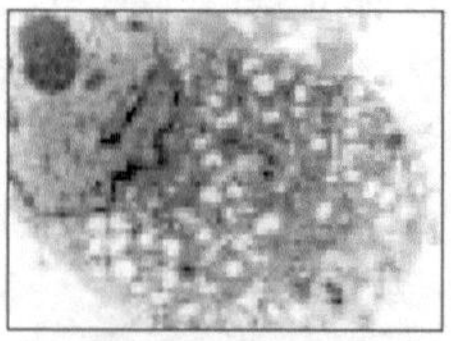

Figure 9. Electron microscopic appearance of a stealth virus infected MRC-5 cell. The culture was from a patient involved in a1996 outbreak of stealth virus infection in the Mohave Valley region of the US. The marked vacuolization is similar to that shown in Figure 4.

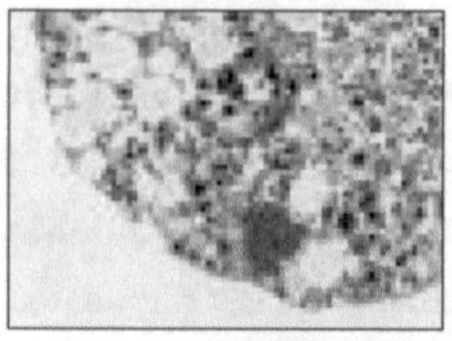

Figure 10. Electron microscopy of a stealth virus infected cell showing widespread accumulations of particulate materials and lipid-filled vacuoles. Unlike infections with many conventional viruses, it is rather uncommon to see intact viral particles.

Structures Developing in Stealth Virus Cultures

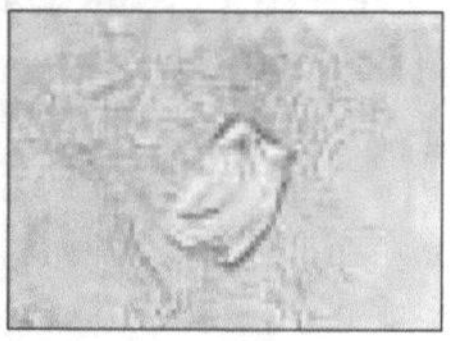

Figure 11. Formation of free-cholesterol-like solid crystals in a stealth virus culture.

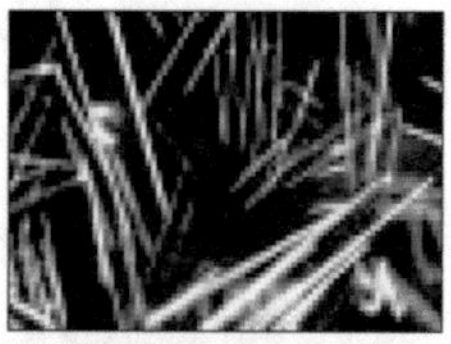

Figure 12. Massive formation of cholesterol ester-like needle shaped crystals in a stealth virus culture. The crystals are best seen under dark field illumination. The culture was obtained from the patient known to be infected with the SCMV-derived stealth virus.

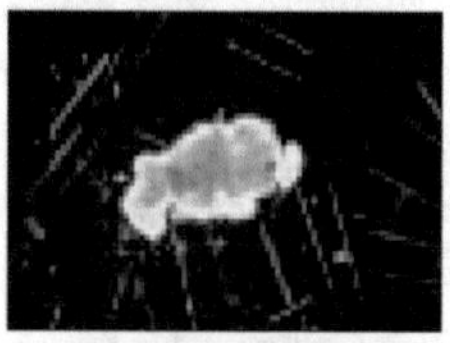

Figure 13. Stealth virus culture examined under dark field illumination showing a floating aggregation of insoluble material and the presence of cholesterol-ester-like needles.

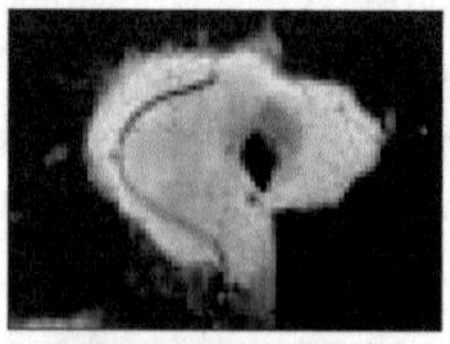

Figure 14. Complex cell cluster developing in a stealth virus culture. It shows a blue colored thread, fine reddish intracellular material and an irregularly shaped dark deposit.

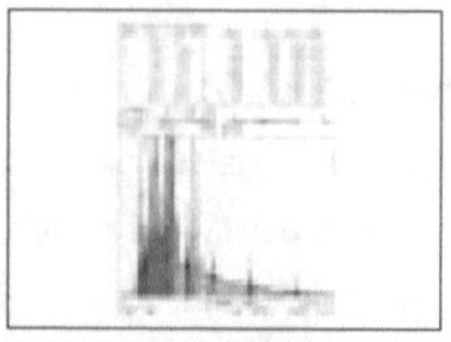

Figure 15. X-ray spectroscopic analysis of aggregated particulate matter in a stealth virus culture. It has accumulated several minerals, including aluminum, titanium, iron and zinc. Other aggregates from the same culture showed different spectroscopic patterns.

Methods a
Sec

DNA isolated from the ste
Nucleotide sequences were

While the complete genome
data are available for Africa
rhesus monkey (RhCMV) cy

Most of the clones aligned to C

Several clones contained aty

Other clones partially ma
associated chemokine, and
human genome.

The sequence data clearly es

Structure of
(>2

UL (Unique
||||-----------------------------------
Genes UL

Note: The majority of anti-CM
product. Other major antige
these genes would enable a

SCMV Relat
Steal

CMV-Related Contiguous Sequences		
Genes	**Length**	
UL 14	1,458	
UL 28-48	28,199	
UL 48-54	8,407	UL
UL 56	767	
UL 57 + ori	2,748	UL
UL 61-69	4,459	
UL 70	1,729	
UL 71-76	6,328	
UL 77-78	1,884	
UL 84-104	25,023	UL
UL 104-105	1,464	
UL 111-112	1,955	UL
UL 115-132	8,628	
UL 141-144	5,820	
US 20-29	16,011	*5 copie
US 30-32	3,978	

Analysis based on >300 clor
evidence that the virus has
absence of UL 83 and UL 5
products of these three ger
recognized by anti-CMV cy
establish that this particular
most likely arose from an SC

Construction, Replication and Potential Therapies
...nkey Simian Cytomegalovirus-Derived Stealth Virus
...ious Diseases, Rosemead, CA 91770

Contact
Telephone
(626) 572-7288
wjohnmartin@hotmail.com
FAX
(626) 572-9288

...mmary of ...udy

...as cloned and sequenced. ...Programs at NCBI.

...known, only partial sequence ...CMV), Baboon (BaCMV) and

...ientity of some clones to SCMV.

...acterial origin.

...teins, including to a cancer ...erated genes present in the

...f atypically structured viruses.

...omegalovirus ...les)

US Region
--|||||||--------------------||||
US 1-32

...e directed against UL 83 gene ...L 123. Deletion/mutation of ...ve CTL mediated immunity.

...the Cultured ...Virus

...stealth Virus Sequences to ...CMV RhCMV HCMV ...matching (Expect value)

...2/465	384/453	159/185
...180)	(e-95)	(e-38)
...9/577	Not avail.	370/492
...0.0)		(e-62)
...5/647	368/406	266/317
...0.0)	(e-137)	(e-57)
...9/901	74/89	75/89
...171)	(e-7)	(e-15)

...chemokine (and also HIV) receptor.

...virus culture. There is good ...able genome. Note apparent ...ed significant mutations. The ...provide the major antigens ... nucleotide homology data ... was derived from SCMV. It ...tch of live polio virus vaccine.

Analysis of the Bacteria Derived Genes

Clones obtained from stealth virus culture that match to bacterial genes		Bacterium showing the highest level of sequence homology	Nucleotide (amino acid) identity Expect value	Bacterial proteins/genes identified by sequence homology to the cloned DNA obtained from the SCMV derived stealth virus culture
Clone	Size			
3B43	3,620	Brucella melitensis	0.0	ribosomal RNA operon C
3B23	8,916	"	0.0	sorbose dehydrogenase, hippurate hydrolase
3B313	7,985	"	0.0	ABC transporter, metal chelation, secreted protein
3B614	5,062	"	0.0	acetyl-CoA carboxylase, dehydroquinate synthase, shikimic acid kinase, recombinase
3B534	612		3e-101	enolase
3B315	4,495	"	(4e-97)	glutathione S-transferase, transcriptional regulator
C16134	4,142	"	0.0	dihydrodipicolinate synthase, guanosine pyrophosphohydrolase, DNA to RNA polymerase
C1616	4,626	"	e-120	UDP-epimerase, 3-demethylubiquinone-9-3 methyltransferase, sodium/bile cotransporter, transposon from plasmid of Agrobacterium (0.0)
3B41	2,869	Agrobacterium tumefaciens	(2e-155)	sensory transduction histidine kinase
3B47	2,024	Sinorhizobium meliloti	(8e-139)	ABC transporter,
C16122	4,915	Zymomonas mobilis	(5e-63)	dihydrodipicolinate synthase, cystathionine gamma-synthase
3B513	8,106	Alpha-proteobacteria sp.	(e-30)	Fe binding, nitrogen fixating, invertase, others
3B512	2,345	Mycoplasma	(5e-98)	ATP-transporter p115-like
3B520	2,797	"	(e-37)	ABC transporters p29 and p69
3B35	2,142	"	(5e-81)	unknown function
3B632	1,396	Streptococcus	(8e-33)	ribonuclease H
3B528	2,049	"	(3e-33)	glucuronyl hydrolase, sugar transporter

Conclusion: Modified bacteria-derived sequences present in stealth virus culture. For many proteins there are structural similarities with related proteins from other bacterial species, leading to the probability of antigenic cross-reactivity.

Example of Cellular Derived Gene

Clone 3B516: (5,820 nt). Codes for UL141, UL144, truncated UL145, and 3 divergent copies of gene matching to human MGSA/Gro-alpha. This is a chemokine with potential oncogenic activity (in melanomas and other tumors). Two of the 3 copies of the gene lack a major intron present in genomic DNA. This suggests recombination with RNA rather than DNA.

Alignments of Predicted Amino Acids of Stealth Virus Genes With MGSA/Gro-alpha

```
Stealth Gene A 4938   NPRFLGVTLLLMSLIAY------CQSTTELRCQCTQTVQGIHPKNIQSVSIKDKGPNCPN 5099
                      NPR L V LLL+ L+A       TELRCQC QT+QGIHPKNIQSV++K GP+C
Human MGSA Gene 12    NPRLLRVALLLLLLVAAGRRAAGASVATELRCQCLQTLQGIHPKNIQSVNVKSPGPHCAQ 71

Stealth Gene A 5100   QEVIATLKNGQKVCLNPTAPMVQKILKKTITDN 5198
                      EVIATLKNG+K CLNP +P+V+KI++K + +
Human MGSA Gene 72    TEVIATLKNGRKACLNPASPIVVKKIIEKMLNSD 104

Stealth Gene B 5469   LLVATLLGTLLASTMVFADK----EERCLCPKTIQGIHPKNIQSVELHEPRDMCPNVEVM 5636
                      L VA LL  L+A+   A      E RC C +T+QGIHPKNIQSV + P C   EV+
Human MGSA Gene 16    LRVALLLLLLVAAGRRAAGASVATELRCQCLQTLQGIHPKNIQSVNVKSPGPHCAQTEVI 75

Stealth Gene B 5637   *VCWYCVIIGKLAHEITYNSLYFSYLHSAKLKNGNEVCLNTEGPMVKKIIEKM 5795
                                       intron                A LKNG + CLN  P+VKKIIEKM
Human MGSA Gene 76    --------------------------ATLKNGRKACLNPASPIVKKIIEKM 100

Stealth Gene C 4583   SPRFLAVALLIVSLIAYSESSQG------IRCECKKGTQKIPENKIVVKKMKRPSGPNHP 4744
                      +PR L VALL++ L+A+      G      +RC+C+ Q I   I    +K P GP+
Human MGSA Gene 12    NPRLLRVALLLLLLVAAGRRAAGASVATELRCQCLQTLQGIHPKNIQSVNVKSP-GPHCA 70

Stealth Gene C 4745   RTEVKDSTKQPGRDPMGRPVS 4807
                      +TEV +T + GR     P S
Human MGSA Gene 71    QTEV-IATLKNGRKACLNPAS 90
```

Additional cellular genes have been identified in several clones obtained from viral DNA isolated from the stealth virus culture. Many of the genes have highly reiterated/repeat sequences. Some of the sequences match to endogenous reverse transcriptase, suggesting a possible mechanism whereby recombinations involving cellular RNA can be back translated into viral incorporated cellular DNA.

Conclusions

Atypically structured, non-inflammation inducing cytopathic viruses definetely exist.

Some of these viruses were derived from simian CMV and have presumably entered the human population from SCMV contaminated batches of live polio vaccines.

Non-inflammatory cytopathic viruses are grouped under the term "stealth." They can be regularly cultured from patients with complex multi-system illnesses, including various cancers. Positive stealth virus cultures were found in approximately 10% of University students donating blood for transfusion. Community outbreaks do occur.

Stealth-adaptation is considered to be a generic process that can involve many types of cytopathic viruses. It presumably occurs through the loss of genes coding for major antigens normally targeted by the cellular immune system.

Tissue culture provides the best method to screen for stealth-adapted viruses. Viral cultures can also provide useful insights into pathology, including formation of lipids, and of protease-resistant protein complexes.

The production of lipids and pigmented materials is viewed as a reparative process helping to maintain cell viability. There is a marked reduction in the intensity of the CPE if the culture medium is not frequently replaced.

Bacteria and cell-derived genes are present in the SCMV-derived stealth virus culture. This important finding indicates the potential intermixing of cellular, viral and bacterial genes in the creation of new highly pathogenic organisms. Viteria is used to define viruses with bacterial sequences. Atypical bacteria can commonly be cultured from stealth virus infected patients.

Stealth viruses are found in cancer patients, many of who have symptoms of an underlying neuropsychiatric illness. The prospect of bacteria transmitting cancer causing viruses is a very serious and urgent public health concern.

Bacterial genes can help explain partial and inconsistent serological and/or PCR diagnostic findings for mycoplasma, (in CFS, Gulf War Syndrome): Borrelia in "chronic Lyme disease"), streptococcus (in PANDAS), etc.

Apparent expansion of chemokines and chemokines-receptor genes provide an adjunctive approach to anti-stealth virusl therapy. Many therapeutic agents are available that can lead to cytokine/chemokine suppression.

Ongoing Research Program

- Complete the sequencing of SCMV and SCMV-derived stealth virus
- Test for passage of this virus through bacteria using molecular methods
- Characterize lipids and proteins synthesized in stealth virus cultures
- Survey patient populations for evidence of stealth virus infections and for any disease-related characteristics of their positive cultures
- Determine if vaccines can activate a stealth virus infection and/or pathology
- Sequence additional stealth virus isolates, especially from cancer patients
- Conduct clinical trials on substances shown to inhibit stealth virus CPE
- Educate clinicians on the multi-system nature of stealth virus infections

Guiding Quotes:

One can only see what one observes, one observes only the things that are already in the mind. *Alphonse Bertollin, 1853-1914.*

The more I look, the more I see, and the more I see, the more I look for. *Teihard de Chardin, 1881-1914.*

PUBLICATIONS

1. Martin WJ, Zeng LC, Ahmed K, Roy M. Cytomegalovirus-related sequences in an atypical cytopathic virus repeatedly isolated from a patient with the chronic fatigue syndrome. Am. J. Path. 145: 441-452, 1994.
2. Martin WJ. Stealth virus isolated from an autistic child. J. Aut. Dev. Dis. 25:223-224,1995.
3. Martin WJ, Ahmed KN, Zeng LC, Ghen J-C, Seward JG, Seehrai JS. African green monkey origin of the atypical cytopathic 'stealth virus' isolated from a patient with chronic fatigue syndrome. Clin. Diag. Virol. 4: 93-103, 1995.
4. Martin WJ, Glass RT. Acute encephalopathy induced in cats with a stealth virus isolated from a patient with chronic fatigue syndrome. Pathobiology 63: 115-118, 1995.
5. Gollard RP, Mayr A, Rice DA, Martin WJ. Herpesvirus-related sequences in salivary gland tumors. J. Exp. Clin. Can. Res. 15: 1-4, 1996.
6. Martin WJ. Genetic instability and fragmentation of a stealth viral genome. Pathology 64:9-17, 1996.
7. Martin WJ. Severe stealth virus encephalopathy following chronic fatigue syndrome-like illness: Clinical and histopathological features. Pathobiology 64:1-8, 1996.
8. Martin WJ. Stealth viral encephalopathy: Report of a fatal case complicated by cerebral vasculitis. Pathobiology 64:59-63, 1996.
9. Martin WJ. Simian cytomegalovirus-related stealth virus isolated from the cerebrospinal fluid of a patient with bipolar psychosis and acute encephalopathy. Pathobiology 64:64-66, 1996.
10. Martin WJ, Anderson D. Stealth virus epidemic in the Mohave Valley. Initial report of viral isolation. Pathobiology 65:51-56, 1997.
11. Martin WJ. Cellular sequences in stealth viruses. Pathobiology 66:53-58, 1998.
12. Martin WJ. Bacteria related sequences in a simian cytomegalovirus-derived stealth virus culture. Exp Mol Path 66: 8-14, 1999.
13. Martin WJ. Stealth adaptation of an African green monkey simian cytomegalovirus. Exp Mol Path. 66:3-7, 1999.
14. Martin WJ. Melanoma Growth stimulatory activity (MGSA/GRO-alpha) chemokine genes incorporated into an African green monkey simian cytomegalovirus (SCMV)-derived stealth virus. Exp Mol Path. 66: 15-18,1999.
15. Martin WJ, Anderson D. Stealth Virus Epidemic in the Mohave Valley: Severe vacuolating encephalopathy in a child presenting with a behavioral disorder. Exp Mol Path. 66:19-30 1999.
16. Martin WJ. Chemokine-related sequences in an African green monkey simian cytomegalovirus (SCMV)-derived stealth virus. Exp Mol Path. 69: 10-16, 2000.
17. Martin WJ. Stealth viruses. Explore 10: Number 4, 17-21, 2001.
18. Martin WJ. Chronic fatigue syndrome among clinicians: A potential role of occupational exposure to stealth viruses. Explore 10: Number 5, 7-10, 2001.
19. Martin WJ. Chemokines and stealth viruses. A blueprint for therapy in infected humans and animals. Explore 11, Number 1, 7-11, 2002.

Updated Examples of Reluctance of Public Health Officials to Consider Stealth Adapted Viruses:

As demonstrated in the responses, or lack of response, to e-mails sent September 2013, senior Public Health officials have steadfastly refrained from showing any willingness to discuss stealth adapted viruses. While some officials clearly do not grasp the concept, it is difficult not to believe that certain officials are intent upon defending what they know is a flawed and occasionally dishonest public health system. They become comfortable with the perks of office and with the promises of further recognition and advancement. Also implied in holding such a position is the possibility of future bonuses for compliance with industry wishes. Yet unnecessary harm has been inflicted by efforts to avoid and probably to have actively suppressed meaningful dialogue regarding stealth adapted viruses. Among those most affected are children living an unfilled life due to autism and related learning and behavioral disorders and by the parents having to witness the calamities of their children.

The following four e-mails were sent in advance of an opportunity to visit the NIH in September 2013.

1. E-mail sent 9/19/2013 to Dr. Thomas Insel, Director of the National Institute of Mental Health and also Head of the Interagency Autism Coordinating Committee; The e-mail was sent following a phone call to Dr. Insel's office.

Title: Request to Meet Briefly
Dear Dr. Insel,

As you may be aware, I have long proposed that autism is a virus infection acquired from the mother during pregnancy. The viruses causing autism are derived from regular viruses but differ in having lost or mutated the relatively few components normally targeted by the cellular immune system. I refer to this immune evasion mechanism as "stealth adaptation." While potentially any virus can undergo stealth adaptation, the best characterized were derived from cytomegalovirus from African green monkeys.

As you can guess, this conclusion drawn nearly 20 years ago, has met with political resistance. I am now confident that there is a therapeutic answer to these viruses and would like to see the issue discussed at the Federal Health level. I will be in Washington D.C. next Monday and Tuesday and would appreciate at least a minute or so to help facilitate future communications. I am familiar with NIH, having worked there for several years. I look forward to an e-mail response. also my cell phone number is 626-616-2868. Kind regards, John.

No Reply

2. E-mail sent 9/19/2013 to Dr. Phillip Krause, Deputy Director, Office of Vaccine Research and Review, Center for Biologics Evaluation and Research, FDA. (I have met with Dr. Krause on several occasions prior to 2002.

Title: Visiting NIH Campus and Request for Possible Brief Meeting‏

Dear Phil,

You may recall my interest in viruses, which evade immune recognition by deletion/mutation of the relatively few genes targeted by the cellular immune system. The work hit a stumbling block when some of these stealth adapted viruses were shown to be derived from African green monkey simian cytomegalovirus. Nevertheless, the work has progressed and has provided new insights into a potent non-immunological anti-virus defense mechanism.

I will be in Washington DC, Monday and Tuesday of next week. I think it is important to keep this issue open since I feel confident the work will eventually be duplicated by others.

Ideally, FDA would take the initiative in doing so. I am happy to help with culturing the viruses from patients, which really isn't very difficult. Its all pretty exciting research and really needs to be more effectively applied. Kind regards, John.

No Reply

3. E-mail sent 9/19/2013 to Dr. Kathy Zoon, formally Director of the Center for Biologics Evaluation and Research, FDA and now Director of Intramural Research at the National Institute of Allergy and Infectious Diseases, NIH.

Title: Visiting NIH next Monday and Tuesday

Dear Kathy,

You may recall my interest in viruses, which evade immune recognition by deletion/mutation of the relatively few genes targeted by the cellular immune system. The work hit a stumbling block when some of these stealth adapted viruses were shown to be derived from African green monkey simian cytomegalovirus.

Nevertheless, the work has progressed and provided new insights into a potent non-immunological anti-virus defense mechanism.

I will be in Washington DC, Monday and Tuesday next week. I was planning to request a visit with you at FDA but see that you are now heading the NIAID intramural program (congratulations). If you could kindly free up a few minutes, I would like to discuss the possibility of working with someone within the intramural program to: i) Confirm how relatively easy it is to culture these viruses; and ii) move forward with NIAID in optimizing simple therapies.

I feel it is important to make up for lost time. Manuscripts summarizing much of the past work will slowly become published, but the real excitement will be to see the research applied. My cell phone number is 626-616-2868. Kind regards, John.

Response sent 9/20/2013

John

Your work sounds very interesting but unfortunately I am really tied up next week. I hope that all works out for the best.

Kathy

4. E-mail sent 9/19/2013 to The Chronic Fatigue Syndrome Advisory Committee (CFSAC). Specifically to OS OPHS CFSAC (HHS/OPHS); Lee, Nancy (NIH/NIMH) [E]; Clayton, Janine (NIH/OD) [E]. Note Dr. Lee is the Designated Federal Officer for CFSAC.

Title: Visiting Washington DC - Virus Cause of CFS‏

Dear Dr. Lee or other CFSAC Official,

As a pathologist, I have had a long term interest in a virus cause of CFS. The viruses causing CFS are somewhat atypical, however, in not being recognized by the cellular immune system.

The work hit some political difficulties with indications that some of the responsible viruses were derived from the cytomegalovirus of African green monkeys. On a brighter side, the work has identified a novel and potent non-immunological defense mechanism.

Since I will be in Washington next Monday and Tuesday, I thought a brief face to face meeting with someone from CFSAC might facilitate further communications. If there is a possibility to drop by either the Hubert H. Humphrey Building or NIH, I would be pleased to do so. Kind regards, John. My cell phone number is 626-616-2868.

Response sent 9/20/2013

Mr. Martin

The CFSAC Support team has no expertise in research around CFS, or in viral research. Let me refer you to Dr. Susan Maier, who is the NIH representative to CFSAC. She is with the Office of Research on Women's Health, but is not herself involved in viral research or CFS research. She may be able to refer you to someone at NIH.

Thanks for your interest.

The CFSAC Support Team

Reply received from Dr. Susan Maier, PhD, Chair, Trans-NIH Myalgic Encephalomyelitis/Chronic Fatigue Syndrome (ME/CFS) Research Working Group, Office of Research on Women's Health (ORWH); DHHS/NIH/OD/DPCPSI on

9/20/2013

Dear Dr. Martin:

Dr. Nancy Lee forwarded your email message from the CFSAC email address regarding your request to meet with someone in DC regarding chronic fatigue syndrome.

Can you tell what type of information you are seeking? I would like to be able to direct your email to the appropriate people rather than having you not find what you need. We are interested in providing you with service that exceeds expectation.

Thanks much.

Susan

E-mail example of an earlier attempt at communicating with CDC.

Copy of an e-mail sent to Ms. Katherine Lyon Daniels, Associate Director for Communication, CDC, on Sept. 26, 2012. The e-mail was sent on the strong recommendation of a close relative. No reply was received to the e-mail or to a request for a follow-up telephone call.

Title: Stealth Adapted Viruses and the ACE Pathway
Dear Katherine,

Nice to speak with you. Possibly the timing is better for CDC to address the issue of stealth adapted viruses, even if some of them arose from African green monkey simian cytomegalovirus (SCMV). I have really been disappointed in my prior contacts with CDC, as reflected in an unanswered e-mail sent in August 2000 to Dr. Phil Pellett, then Head of CDC Herpesvirus Branch. I will also include Abstracts of two prior research articles for you to help assess current scientific interest and/or political reluctance to learn more about these viruses.

The issue is worth pursuing since practical, inexpensive non-immunological approaches to the therapy of stealth adapted and conventional virus infections are becoming available. Kind regards, W. John Martin, MD, PhD. (626) 616-2868.

Included:
Copy of e-mail sent to Dr. Pellett in 2000
Abstract of articles:
Cytomegalovirus-related sequence in an atypical cytopathic virus repeatedly isolated from a patient with chronic fatigue syndrome.
African green monkey origin of the atypical cytopathic 'stealth virus' isolated from a patient with chronic fatigue syndrome.

Submitted but Not Published Paper Presented at the Institute of Medicine's Meeting "Linking Infectious Agents and Chronic Diseases: Defining the Relationship, Enhancing the Research and Mitigating the Effects." Washington DC, October 21-22, 2002. Title: DNA Sequence Analysis of a Stealth-Adapted Simian Cytomegalovirus

W. John Martin
Center for Complex Infectious Diseases
Rosemead CA 91770

Abstract

Stealth-adaptation is a mechanism that allows cytopathic viruses to evade immune elimination through the deletion of genes coding the major antigens targeted by the cellular immune system. A prototype stealth-adapted virus cultured from a patient with chronic fatigue syndrome (CFS) was readily transmissible to cats in which it induced an acute encephalopathy without localizing neurological signs. Vacuolating cellular damage was observed in many tissues of the animals, including the brain, in the absence of an accompanying inflammatory response. The cultured virus was cloned and partially sequenced. It comprises a fragmented, genetically unstable, genome. It has viral sequences that can be aligned to various regions of the genome of human cytomegalovirus (HCMV). Where the comparison can be made, the sequences match much more closely to those of African green monkey simian cytomegalovirus (SCMV), indicating an unequivocal origin from SCMV. In addition to SCMV-derived sequences, the cytopathic virus has acquired both cellular and bacteria-derived DNA. Cellular sequences include putative oncogenes and sequences related to those of endogenous retroviruses. Many of the bacterial sequences match closely to known genes of Brucella, while other genes show greater homology either to mycoplasma or to Streptococcus. The presence of bacteria-derived sequences has led to the secondary designation of this type of novel microorganism as viteria. Molecularly heterogeneous viruses, inducing similar characteristic cytopathic effects in culture (and when examined, non-inflammatory vacuolating cellular damage in brain and other tissues), have been cultured from numerous patients with severe neurological, psychiatric, immunological and neoplastic diseases. The differing clinical manifestations in infected patients may reflect the assimilation of different cellular and other sequences in various stealth-adapted viruses. Tissue culture provides a valuable screening method for the detection of stealth-adapted viruses and for characterizing novel virus-associated products. Cultures can also help avoid the misidentification of stealth viruses for conventional viral and bacterial pathogens purportedly associated with chronic diseases.

Introduction

Chronic illnesses are typically categorized according to the predominating clinical manifestation. Accordingly, they come under the purview of different types of medical specialists, such as psychiatrists, neurologists, rheumatologists, endocrinologists, cardiologists, gastroenterologists, oncologists, etc. While this specialization fosters expertise in diagnosis and therapy, it tends to deemphasize the overlapping clinical features shared by many chronic debilitating illnesses. Symptoms such as fatigue, insomnia, impaired mood and cognition, widespread aches and pains, etc., typically occur in most chronically ill patients regardless of the underlying diagnosis. Individual patients will not uncommonly qualify for having two or more chronic diseases while family histories often reveal a constellation of medical problems among different family members. While most acute infections tend to give rise to a consistent clinical pattern of illness, this is not the general rule for chronic viral or bacterial infections. It is conceivable, therefore, that a common chronic infectious process could be involved in the currently widely recognized increases in many types of chronic illnesses. This consideration does not exclude important additive contributions of genetics, psychosocial, other environmental factors, autoimmunity, etc., in individual patients leading to their particular complex disease process. To help stem the ever increasing prevalence of chronic illnesses, it is vitally important not to overlook an infectious and, therefore, potentially contagious component. This paper reviews evidence that implicates chronic non-inflammatory, stealth-adapted viruses in many forms of human and animal diseases.

Stealth-adaptation refers to the ability of a conventional cytopathic (cell damaging) virus to lose portions of its viral genome that includes genes coding for components normally recognized by the cellular immune system. Although not widely appreciated among most virologists, only a relatively few number of virus coded components provide the target antigens for the vast majority of anti-viral cytotoxic T lymphocytes (CTL). Deletion of these few select genes can essentially enable a virus to avoid effective immune recognition. The best studied stealth-adapted virus is a derivative of an African green monkey simian cytomegalovirus (SCMV). This virus presumably arose from an SCMV contaminated vaccine, such as a live polio virus

vaccine. These vaccines were routinely produced in fresh kidney cell cultures obtained from African green monkeys. In a 1972 joint Lederle-Bureau of Biologics cooperative CMV study[1], all eleven monkeys studied demonstrated the presence of CMV-like agents. Seven of the monkeys would have passed [Lederle's] existing test standards; only one of these monkeys would have passed test methods using Lederle's 130 human diploid cell strain. Lederle proposed that it would begin treating monkeys with DNA antagonists during their isolation period, then remove the kidneys and test exhaustively for the presence of CMV in the kidney tissue. Even assuming that this was done, it is disturbing that 3 of 8 tested lots of live polio virus vaccine released later in the 1970's were recently confirmed as containing SCMV DNA[2]. The present paper will provide a brief summary of studies leading to the isolation and subsequent molecular characterization of a stealth-adapted SCMV-derived virus.

Stealth Virus Detection and Isolation

The research program began in 1986 as a search for a viral cause of the chronic fatigue syndrome (CFS). It was based on the recent description of human herpesvirus-6 (HHV6), and on the availability of a highly sensitive molecular diagnostic assay, known as the polymerase chain reaction (PCR).[3] The PCR assay comprises reacting a sample with relatively short pieces of synthesized DNA (primers) that selectively bind to the flanking sequences of a relatively small section of double stranded DNA present in the particular sample being tested. The DNA is subjected to repeated rounds of heating to separate the DNA strands, cooling to allow the primers to bind and to be extended by DNA synthesis mediated by a heat resistant

DNA polymerase enzyme. This process leads to exponential amplification of the targeted section of DNA, which greatly facilitates its detection and sequencing. Although PCR assays can be exquisitely sensitive, they can yield misleading information. For example, if a pathogen has a mutation or deletion in the targeted section of DNA, a negative PCR assay can result even though much of the remaining parts of the microorganism may be present. Conversely, if the primers are directed against sequences that are partially shared by different microorganisms, spurious identifications can be made based on a positive PCR. In spite of these limitations, PCR assays have proven extremely useful in the search for unexpected pathogens, including as detailed below the detection of stealth-adapted viruses.

The initial PCR assay employed was based on limited known DNA sequences of HHV6. The assay generally yielded negative results in patients with CFS. The PCR assay was modified so as to be cross-reactive with other known human herpesviruses. Other PCR primer sets were also designed to detect various retroviruses. The PCR assays were shown to be even broadly cross-reactive than anticipated, with for example, amplification occurring with adenoviruses by the herpesvirus-reactive primers. The assays were applied to blood samples of CFS patients. Depending on the conditions used, positive responses were seen with approximately a third to a half of the patients tested.[3-5] Parallel testing of blood samples from the vast majority of apparently healthy individuals gave negative results. Some of the positive patient responses were quite striking in their intensity. The importance of the work became apparent when very strong PCR reactivity was seen using a cerebrospinal fluid (CSF) sample from both a newborn child with hepatosplenomegaly, seizures, and impaired neurological development. This finding was followed shortly by a similar strong

reactivity using CSF obtained from an adolescent with major neurological damage resulting from what had been considered a missed diagnosis of herpes simplex encephalitis. In both cases the CSF was totally devoid of any cellular inflammatory reaction and routine virus culture assays were reported as being negative. A positive PCR was also seen using a portion of a stereotactic brain biopsy performed on a school teacher with periventricular lesions identified using magnetic resonance imaging (MRI).[4] Again, the striking feature of the brain biopsy was the lack of any inflammatory reaction. The brain cells did, however, show lipid filled vacuoles, damaged mitochondria and accumulations of irregularly shaped pigmented inclusions. These observations were consistent with non-inflammation-inducing cell damaging (cytopathic) viruses causing a spectrum of neurological illnesses. The term "stealth" was introduced to convey the apparent ability of the viruses to avoid effective immune recognition that would trigger an anti-viral inflammatory reaction. The task was to isolate such a "stealth-adapted" virus.

Many types of viruses can be detected by observing for a cytopathic effect (CPE) on cells grown in tissue culture. Using standard viral culture techniques suitable for the detection of human cytomegalovirus (HCMV), highly suggestive, but non-persisting cellular damage had previously been observed in cultures performed on numerous CFS patients. Clearly positive PCR assays were obtained on blood samples of a patient who experienced an encephalitis/meningitis-like illness in July 1991. A determined effort was made to culture her blood. After some 6 weeks delay, a strong sustainable cytopathic effect was observed.[6] This individual has remained cognitively impaired with personality changes and marked fatigue. She has remained on disability with a diagnosis of CFS.

The striking tissue culture finding was the formation of foamy vacuolated cells that were comparable to what were seen in the brain biopsy. Repeated cultures from this patient provided similar positive results with the additional observation that the CPE would develop much earlier if the culture medium was frequently (daily or every 2nd day) replaced with fresh medium. The positive culture did not react in standard serological typing assays using antibodies specific for conventional human herpesviruses (CMV, HHV6, Varicella zoster, Herpes simplex or Epstein Barr), nor for enteroviruses or adenoviruses. It was also negative for human CMV and HHV-6 using specific PCR-based assays. The patient-derived cytopathic virus was successfully cultured in multiple cell lines from human and animal sources[6] and could even replicate in insect-derived cells. Virus aliquots were submitted to the American Type Culture Collection (ATCC) for long term storage and public access (Accession number VR 2343).

Soon, thereafter, a slightly differing appearing, but otherwise very comparable, strongly positive culture was obtained in a physically different laboratory. The CPE developed in human fibroblasts three weeks after being inoculated with a CSF sample from a patient with a 4 year history of a bipolar, manic depressive illness. The woman had deteriorated clinically and developed seizures prior to admission to L.A. County Hospital. An aliquot of the second sample was sent to the Los Angeles County Public Health Laboratory. Following confirmatory culturing, the County laboratory reportedly sent a sample to the California State Laboratory. This virus was not identified by either the County or State laboratory and was dismissed as a probable contaminant. Electron microscopy on both of the patients' cultures confirmed the presence of numerous herpesvirus-like particles.

Many additional blood and occasional CSF samples were found to induce a foamy vacuolating CPE in viral cultures on both human fibroblasts and rhesus monkey kidney cells. Adjustments to the culturing techniques, especially the use of serum free medium, frequent replacement of the medium, and freeze-thawing of the cells prior to culturing, led to the more rapid development and greater intensity of the CPE. The basic feature of positive cultures was the transformation of the normally thin, spindle-shaped fibroblasts into swollen, rounded, vacuolated, foamy (fat-filled) cells, with a tendency to form clusters. The extent of cell swelling, size of cell clusters, formation of syncytia from cells fusing with one another, amount of fat accumulation both in the cells and in the medium, and the extent and appearance of accompanying accumulations of fine to coarse pigments, all varied more or less independently between the cultures obtained from different individuals. Consistent positive/negative readings, as well as more detailed descriptions of the positive cultures were obtained by impartial observers and the distinction between patient and control populations confirmed in several double-blinded studies. The positive patients had a wide variety of neurological, neuropsychiatric and auto-immune diagnoses.[7-10] Families were identified with various diagnoses among the different family members, but with very similar cytopathic changes seen in cultures.

Animal Studies

Cells from the virus culture of the initial culture positive CFS patient were inoculated into cats. The cats experienced profound behavioral changes.[11] Within a week they transformed from friendly, frisky happy-go-lucky favorites of the University animal facility, to frightened, reclusive animals shying away from the light. Several of the cats developed bald areas of skin upon their heads and necks from rubbing against the cage. Some had exudates from their nose from scratching. Gentle handling of the animals revealed painful muscles and enlarged lymph nodes. The acute illness peaked at 1-2 weeks with significant clinical recovery seen in the animals maintained from 5-16 weeks post inoculation. Necropsy showed widespread non-inflammatory vacuolating cellular changes. Heat killed virus did not induce disease, and in fact, provided protection against subsequent inoculations of the same virus. The animal became ill when inoculated with a virus isolated from a different patient who had been diagnosed with systemic lupus erythematosus. Animals of patients were also reported to show various symptoms, and yielded positive cultures when blood samples were tested. Blood samples from ill cats were inoculated into healthy cats that subsequently became ill and virus culture positive.

Sequencing Studies

The primer sets that gave positive responses PCR reactions in the blood of the virus culture positive patient with CFS, also yielded strong reactions when tested on each of several repeat positive cultures from this patient. Two PCR products of approximately 1.5×10^3 nucleotides were isolated and sequenced.[6] One product (GenBank accession number U09212) contained a sequence that showed a statistically significant homology to a region of the genome of HCMV corresponding to the UL34 gene (UL refers to the Unique Long segment of the human CMV genome. The genome also contains numbered genes from the Unique Short, US, segment, with both segments flanked by a series of repeated sequences). The other product (GenBank accession number

U09213) did not match to the HCMV genome, or at the time to any other viral, cellular or bacterial sequence. A description of the initial culturing and growth characteristics of the virus, along with an analysis of the sequence data was published in 1994.[6]

DNA was isolated from the prototype culture using ultracentrifugation to isolate viral particles. In the first series of experiments, the DNA was cut with the enzyme EcoRI. In a second experiment agarose gel banded purified DNA was cut with the enzyme SacI. The cut DNA was cloned into plasmids and many of the clones were sequenced. As each sequence became available for analysis, it was compared with the known complete sequence of HCMV, and the available sequence data of all other viruses, including CMV of various animals.

The sequence of some of the clones corresponded to regions for which the sequence data were available on both rhesus cytomegalovirus (RhCMV) and African green monkey simian cytomegalovirus (SCMV). Sequence comparison led to the unequivocal conclusion that the virus had originated from a SCMV.[12] By comparing PCR results using primers sets directed against sequences of other parts of the stealth virus and other regions of SCMV, it was clear that the stealth virus had diverged significantly from SCMV. It lacked several genes corresponding to known regions of HCMV that would also be expected in SCMV. An update of these data that also includes comparison with baboon CMV (BaCMV) is provided in Table 1.

Much of the sequence data were included in a series of peer-reviewed publications,[13-18] and submitted to GenBank. Most noteworthy were the findings indicating: i) Apparent lack of genes coding the virus components known to be targeted by the majority of cytotoxic T lymphocytes reactive with CMV infected cells. ii) Genetic instability and fragmentation of the virus genome. iii) Incorporation of cellular sequences, including a potential cancer causing gene, into the stealth virus. iv) Increased gene-copy number, when compared to HCMV, of the US28 gene. This gene codes a cell receptor molecule that binds to a class of cell activators, termed chemokines, which belong to a group of cell signaling molecules called cytokines. More interestingly, the US28 gene product provides a cell entry molecule for HIV. v) Presence of unusual bacteria-derived sequences in the culture. Several of the sequences matched very closely (but not identically) to alpha-proteobacteria, especially Brucella. The homology extended to the spirochete Borellia (the cause of classic acute Lyme disease). Other sequences matched more closely to mycoplasma (implicated in CFS and Gulf was syndrome), Streptococcus (implicated in childhood obsessive compulsive disorders and PANDAS syndromes) and to other distinct types of bacteria. The finding of these sequences raised the distinct possibility that stealth virus infected patients could be mistakenly identified as being infected with these various types of bacteria. vi) The second PCR product isolated from the original stealth virus culture was shown to unequivocally correspond to the UL19 region of SCMV. vii) A closely related sequence was identified in a PCR product isolated from the patient with the bi-polar illness.[19]

Disease Outbreaks

The infectious nature of stealth-adapted viruses has been repeatedly suggested by individuals reporting their illness beginning shortly after a sexual and even non-sexual encounter with a symptomatic individual. Some patients have linked their illness to having received a blood transfusion or gamma globulin injection. Distinct community wide outbreaks have also come to the attention of various clinicians. One such outbreak occurred in the Mohave Valley region of Western Arizona in the spring of 1996.[20] Over 100 patients presented during a three month period with an acute gastrointestinal syndrome, followed by persisting fatigue, cognitive dysfunction and personality changes. Since then the community has experienced widespread chronic illnesses, including heightened allergies, chemical sensitivities especially to pesticides, higher than expected learning disorders among its children, and a high incidence of both depression and psychosis. Stealth virus cultures from these patients were consistently positive and distinguishable from control cultures in both routine and in double blinded independent studies. The clinician overseeing many of these cases encountered at least 10 fatalities among symptomatic, virus-culture positive, middle-aged individuals. He also referred a child with a slowly evolving behavioral and learning disorder.[21] An organic component was finally recognized by his parents, both of whom were physicians. In spite of an essentially normal clinical neurological examination, an MRI showed extensive sub-cortical T1 and T2 abnormalities. A brain biopsy showed the characteristic foamy vacuolating changes seen in other patients and in stealth virus inoculated cats. Mitochondria damage and intracellular inclusions were also evident. Both the mother and her son were culture positive. The child showed a significant response to ganciclovir (an anti-herpesvirus drug). The infecting virus could not be identified as a derivative of SCMV using a PCR based assay. This virus was also deposited with the American Type Culture Collection (Accession number VR-2568). Other community wide outbreaks of stealth-adapted viruses have been partially investigated and include Joelton, TN and

197

Peoria, IL. As noted above, illnesses occurring within families also attest to the potential infectious nature of stealth-adapted viruses. Positive cultures and PCR based assays have also frequently been obtained from cancer patients including patients with salivary gland tumors,[22] breast cancers and multiple myeloma.

Stealth Virus Inhibitor

Experience gained with the culturing of stealth-adapted viruses indicated the need to frequently replace the tissue culture medium. It was presumed that some factor or factors were accumulating in the cultures that were preventing the progression of the CPE. Insight into the inhibitory substances has come from recent studies on the aggregated pigmented intracellular materials seen in the virus cultures[23] and in the vacuolated cells present in various brain biopsies.[24] These materials are potentially providing an alternative energy source for the infected cells. The mineral containing pigments appear capable of converting various forms of physical energies to chemical energy.[23] They have been designated alternative cellular energy pigments (ACE)-pigments.

Summary and Conclusions

The following conclusions can be drawn from the work. i) Atypically structured, non-inflammation inducing (stealth-adapted) cytopathic viruses definitely exist. ii) Some of these viruses were derived from SCMV and have presumably entered the human population from SCMV contaminated polio vaccines. iii) Stealth-adapted viruses can be regularly cultured from patients with complex multi-system illnesses, including various cancers. iv) Stealth-adaptation is considered to be a generic process that can involve many types of cytopathic viruses. It presumably occurs through the loss of genes coding for major antigens normally targeted by the cellular immune system. v) Tissue culture provides the best means to screen for stealth-adapted viruses. Viral cultures can also provide useful insights into pathology, including formation of lipids, and energy transducing pigments. vi) The production of lipids and pigmented material is viewed as a reparative process helping to maintain cell viability. This is based on the marked reduction in the intensity of the cytopathic effect if the culture medium is not frequently replaced. vii) Bacterial and cell derived genes are in the SCMV-derived stealth virus culture. This finding indicates the potential intermixing of cellular, viral and bacterial genes in the creation of new highly pathogenic microorganisms. viii) Stealth viruses are found in certain cancer patients, most of whom have symptoms of an underlying neuropsychiatric illness. The prospect of bacteria transmitting cancer causing viruses is a very serious and urgent public health concern. ix) Bacterial genes can help explain partial and inconsistent serological and PCR diagnostic findings for different bacteria in stealth virus infected individuals. x) The present studies provide support for an infectious process as a major contributing factor in the etiology of many chronic illnesses including cancer.

References

1. Lederle. Cytomegalovirus Contingency Plan. August 1972.

2. Sierra-Honigmann A, Krause P. Live oral poliovirus vaccines and simian cytomegalovirus. Biologicals 30:167-174, 2002.

3. Martin W.J. Detection of viral related sequences in CFS patients using the polymerase chain reaction. in "The Clinical and Scientific Basis of Myalgic Encephalomyelitis Chronic Fatigue Syndrome." Byron M. Hyde Editor. Nightingdale Research Foundation Press. Ottawa Canada pp 278283, 1992.

4. Martin W.J. Viral infection in CFS patients. in "The Clinical and Scientific Basis of Myalgic Encephalomyelitis Chronic Fatigue Syndrome." Byron M. Hyde Editor. Nightingdale Research Foundation Press. Ottawa Canada pp 325327, 1992.

5. Martin W.J. Chronic fatigue syndrome (letter). Science 255: 663, 1992.

6. Martin WJ, Zeng LC, Ahmed K, Roy M. Cytomegalovirusrelated sequences in an atypical cytopathic virus repeatedly isolated from a patient with the chronic fatigue syndrome. Am. J. Path. 145: 441452, 1994.

7. Martin W.J. Stealth viruses as neuropathogens. CAP Today 8 6770, 1994

8. Martin WJ. Stealth virus isolated from an autistic child. J. Aut. Dev. Dis. 25:223224,1995.

9. Martin WJ. Severe stealth virus encephalopathy following chronic fatigue syndromelike illness: Clinical and histopathological features. Pathobiology 64:18, 1996.

10. Martin WJ. Stealth viral encephalopathy: Report of a fatal case complicated by cerebral vasculitis. Pathobiology 64:5963, 1996.

11. Martin WJ, Glass RT. Acute encephalopathy induced in cats with a stealth virus isolated from a patient with chronic fatigue syndrome. Pathobiology 63: 115118, 1995.

12. Martin WJ, Ahmed KN, Zeng LC, Olsen JC, Seward JG, Seehrai JS. African green monkey origin of the atypical cytopathic 'stealth virus' isolated from a patient with chronic fatigue syndrome. Clin. Diag. Virol. 4: 93103, 1995.

13. Martin WJ. Genetic instability and fragmentation of a stealth viral genome. Pathobiology 64:917, 1996.

14. Martin WJ. Stealth adaptation of an African green monkey simian cytomegalovirus. Exp Mol Path. 66:3-7, 1999.

15. Martin WJ. Cellular sequences in stealth viruses. Patobiology 66:53-58, 1998.

16. Martin WJ. Bacteria related sequences in a simian cytomegalovirus-derived stealth virus culture. Exp Mol Path. 66: 8-14, 1999.

17. Martin WJ. Melanoma Growth stimulatory activity (MGSA/GRO-alpha chemokine

genes incorporated into an African green monkey simian cytomegalovirus (SCMV)-derived stealth virus. Exp Mol Path. 66: 15-18,1999.

18. Martin WJ. Chemokine receptor-related sequences in an African green monkey simian cytomegalovirus (SCMV)-derived stealth virus. Exp Mol Path. 69: 10-16, 2000.

19. Martin WJ. Simian cytomegalovirusrelated stealth virus isolated from the cerebrospinal fluid of a patient with bipolar psychosis and acute encephalopathy. Pathobiology 64:6466, 1996.

20. Martin WJ, Anderson D: Stealth virus epidemic in the Mohave Valley. Initial report of viral isolation. Pathobiology 65:51-56, 1997.

21. Martin WJ, Anderson D. Stealth Virus Epidemic in the Mohave Valley: Severe vacuolating encephalopathy in a child presenting with a behavioral disorder. Exp Mol Pathol. 66:19-30 1999.

22. Gollard RP, Mayr A, Rice DA, Martin WJ. Herpesvirusrelated sequences in salivary gland tumors. J. Exp. Clin. Can. Res. 15: 14, 1996.

23. Martin WJ. Stealth virus culture pigments: A potential source of cellular energy. Exp Mol Path (in press).

24. Martin WJ. Complex intracellular inclusions in the brain of a child with a stealth virus encephalopathy. Exp Mol Path (in press).

Table 1. SCMV Related Genes in the Cultured Stealth-Adapted Virus

CMV-Related Contiguous Sequences			Matching of Stealth Virus Sequences to			
			SCMV	BaCMV	RhCMV	HCMV
			nucleotide matching (Expect value)			
Genes	Length	Match				
UL 14	1,458					
UL 19-48	38,262					
UL 48-54	8,407	UL50	571/598 (0.0)	422/465 (e-180)	384/453 (e-95)	159/185 (e-38)
UL 56	767					
UL 57 + ori	2,748	UL57	1226/1246 (0.0)	499/577 (0.0)	1011/1265 (0.0)	370/492 (e-62)
UL 61-69	4,459					
UL 70	1,729					
UL 71-76	6,328					
UL 77-78	1,884					
UL 84-104	25,023	UL93	630/656 (0.0)	575/647 (0.0)	368/406 (e-137)	266/317 (e-57)
UL 104-105	1,464					
UL 111-112	1,955	UL111	760/807 (0.0)	709/901 (e-171)	74/89 (e-7)	75/89 (e-15)
UL 115-132	8,628					
UL 141-144	5,820					
US 20-29*	16,011					
US 30-32	3,978					

Analysis based on >300 clones of DNA of stealth virus culture. There is good evidence that the virus has a fragmented, unstable genome. Note apparent absence of UL 83 and UL 55. The UL 123 showed significant mutations. The products of these three genes would ordinarily provide the major antigens recognized by anti-CMV cytotoxic T cells. The nucleotide homology data establish that this particular stealth-adapted virus was derived from SCMV.

- 5 copies of US28 related gene: a chemokine (and also HIV) receptor.

TOPIC 3: VACCINES

The rationale for vaccines and potential inadvertent consequences including autism, AIDS and other epidemics

W. John Martin

Center for Complex Infectious Diseases A Component of S3Support Rosemead, California
91770 Email: S3support@email.com
Published in Medical Veritas, 2004: Volume 1: 78-82.

Abstract

Humans and animals can be protected from epidemic infectious diseases by prior intentional stimulation of the immune system. This process is called immunization and has been hailed as the all time greatest contribution of science to human health. Such enthusiastic endorsements, together with compulsory legislation, have helped ensure widespread public acceptance and compliance with immunization programs. Dissenting or cautionary views on potential risks of certain vaccines have been largely ignored. The vaccine industry now has annual sales in excess of $6 billion with significant liability should adverse effects be proven. Society is facing alarming increases in various types of brain damaging and other illnesses consistent with an infectious process. A role for vaccine-derived "stealth-adapted" viruses in these illnesses, as well as in the emergence of the AIDS virus has been proposed. Such issues should be addressed by full disclosure and open participation of the public and independent researchers.

Introduction

A primary function of the immune system is to provide long-term protection against many types of infectious agents. Typically, the immune system responds to an initial exposure to a particular type of virus, bacterium or fungus, with a heightened capacity to subsequently respond to any further exposures to the same microorganism. With some types of microorganisms, the initial exposure will manifest as a time limited primary disease. This is seen, for example, with childhood exposure to common viral illnesses including measles and chickenpox. Once an individual has contracted measles, he or she is essentially immune from any further episodes of measles.

Epidemics of infectious illnesses have historically taken an enormous toll on mankind. Common examples include plague, smallpox and syphilis during the middle ages; polio, influenza and tuberculosis during the early 20th century; and currently AIDS and hepatitis A, B and C. Explorations of remote areas of the world were severely hindered by diseases such as yellow fever and malaria.

Man has learned to cope with some of these illnesses by harnessing the power of the immune response. Although the mechanism was not understood at the time, Dr. Edward Jenner was able to prevent primary disfiguring infection with smallpox virus by intentional infection with a related virus, termed vaccinia that was infecting cows (vacca in Latin). The process was referred to as vaccination, and from 1796 began to replace the widespread custom of trying to limit the severity of primary infection using minimal exposure to pus collected from a smallpox skin blister.

The concept of germs causing infectious illnesses was jointly developed by Dr. Robert Koch and by Louis Pasteur in the 1880's. While Koch failed in his attempts to develop a tuberculosis vaccine, Pasteur was successful with the development of a rabies vaccine. Essentially, he was able to grow the rabies virus in rabbits and to inject dried spinal cord material into patients exposed to the bite of a rabid animal.

Drs. Landsteiner, Shope, and Walter Reed discovered infectious agents responsible for such serious viral illnesses as polio, influenza and yellow fever in the early 20th century respectively. Bacteria responsible for diphtheria, tetanus, pneumonia and whooping cough (pertussis) were also identified. Unlike viruses, bacteria could be readily cultured away from living cells. Viruses had to be transmitted between living animals. Fortunately, fertile chicken eggs could be used to propagate influenza and yellow fever viruses. In 1948, Dr. John Enders developed the first animal-free tissue culture method to grow poliovirus. With each successfully grown microbe, efforts were made to develop material that could be used to immunize individuals (or animals) against challenge with the same disease-causing microbe.

The immune system was also becoming better understood. The issue of whether protection was being provided by soluble factors (antibodies) or cells was addressed in the early 1900's without a clear consensus. Foreign material, referred to as an antigen, was initially thought to "instruct" certain cells to make antibodies that selectively bound to and neutralized the antigen. In 1957 MacFarlane Burnet postulated that the body was pre-equipped with cells that collectively could recognize all foreign antigens but that individual immune cells, identified as comprising lymphocytes and plasma cells, were responsive to only one particular antigen. Selective outgrowth of antigen-specific responding cells explained the heightened antibody and cellular reactivity to subsequent exposure to the same antigen. The levels of antibody reactivity correlated with the levels of resistance to many infectious diseases providing a ready assay to determine efficacy of various immunization protocols.

Killed bacteria and bacteria-derived products provided the mainstay for bacteria immunization programs. Infusion of serum antibodies collected from immunized animals could also be used to provide passive immune protection of patients in the early stage of bacterial illnesses. Modified toxins (toxoids) produced by diphtheria and tetanus bacteria were relatively easy targets for vaccination. Crude extracts and more purified subcomponent vaccines are available for pertussis, cholera, typhoid, meningococcus, pneumococcus, hemophilus influnza and a tuberculosis related bacteria (BCG).

Formalin killed egg-grown influenza viruses were successfully developed into clinical vaccines. The transfer of the yellow fever virus from monkeys to mice was shown by Dr. Max Theiler to significantly reduce its capacity to induce disease in humans. This led to the production in fertile eggs of a live yellow fever vaccine that is still in use today. Experience with influenza and yellow fever vaccines provided contrasting models of how to best develop a polio vaccine once it was successfully cultured. Dr. Jonas Salk used formalin to inactivate disease causing poliovirus. Drs. Albert Sabin and Hilary Koprowski independently tried to reduce the virulence of polio viruses by extensive tissue culturing. Dr. Sabin was more successful in isolating weakened strains that were still able to induce a protective antibody response. His vaccine replaced that of Dr. Salk in the early 1960's, although in the United States, the use of inactivated polio vaccine was again mandated in 2000.

The introduction of polio immunization was followed by successful efforts to develop live vaccines against measles, mumps and rubella (MMR) viruses and more recently varicella zoster virus. Inactivated and more recently genetically synthesized hepatitis A and B antigenic materials have become available for vaccines. Experimental programs are underway to produce vaccines against many other viruses including herpes simplex viruses, cytomegalovirus, Epstein-Barr virus, human papillomavirus, rotavirus, Japanese B encephalitis virus and human immunodeficiency virus (HIV).

Numerous infectious agents have also been rendered as vaccines for animal use. Prominent examples include Newcastle disease virus in poultry, canine distemper virus in dogs, feline leukemia virus in cats and brucella bacteria in cattle.

Efficacy of Vaccines

The global eradication of smallpox has been attributed to vaccination and has served as a model for other illnesses, including polio. Common childhood infections with measles, mumps and rubella are less frequent in developed countries compared to the developing world. While some of this reduction can be traced to vaccine use, improved sanitation and nutrition were probably more important variables. Influenza mortality among the elderly and infirm is reduced in immunized populations. Because influenza virus can undergo antigenic changes, it is necessary to provide a vaccine that contains the virus responsible for an ongoing outbreak. The detection of a new influenza virus triggers a rapid response for vaccine production in time to provide protection to those not yet exposed to the current strain of influenza. Diphtheria and tetanus are rarely seen today and essentially only in individuals who have not been immunized. Mortality for meningococcus and pneumonia is reduced for those strains for which vaccines have been produced. Similar success stories apply to vaccinated livestock and domestic pets.

Lifelong protection against many infectious diseases is clearly achievable by vaccination. Moreover, the world remains at risk for newly emerging infectious agents, including common viruses with drastically altered antigens. Advances in biotechnology are likely to streamline vaccine manufacturing. Specifically, recombinant DNA technology is allowing the production in bacteria of structurally well defined antigens of viral, bacterial, fungal and parasitic microorganisms. A greater understanding of the immune system should also enable more directed approaches at eliciting the type of immunity that is most appropriate for a given type of infection. Effective vaccines are not yet available for several major illnesses, including tuberculosis, malaria and AIDS. A potential difficulty in the development and use of such vaccines is the growing reluctance of the public to accept the Government's blanket assurance that vaccines are safe and effective.

Adverse Effects of Vaccines

The use of vaccines has been justified as an important Public Health measure to stem the occurrence of epidemic illnesses. To be effective, it has commonly been argued that universal compliance with immunization programs is necessary. Frivolous concerns such as sprouting cow horns from taking vaccinia virus were aggressively countered by common sense. More serious concerns have periodically arisen and afforded less than stellar attention by Public Health authorities. The reluctance is explained in part by a protective reaction of those responsible for apparent oversights and by the considerable exposure of Industry to potential litigation. Historical examples include the probable transmission of syphilis and tetanus as inadvertent contaminants of vaccinia vaccine lots; the transmission of bovine leukemia virus to cattle herds because of contaminated experimental babesia vaccines; and an outbreak of Venezuela equine infectious virus in horses that was due to a contaminated vaccine.

Field testing of vaccines with live viral challenge can potentially explain the out-of-season cases of polio that occurred in the early 1950's in the United States. Actual polio cases developed among some of those receiving initial lots of Dr. Salk's vaccine because of inadequately assessed inactivation protocols. Simian virus 40 (SV-40) was a common contaminant of both live and killed polio vaccines produced in the freshly grown cells from the kidneys of rhesus monkeys. A switch was made to African green monkeys for further production of polio vaccine without the recall of known contaminated vaccine lots. Concerns about using fresh tissues from African green monkeys arose during the 1960's but simple suggestions such as using serum antibodies from the monkeys to test for possible contaminating viruses were disregarded.

In 1972 a joint Government-Industry study showed that kidney cell cultures from all eleven African green monkeys tested were contaminated with simian cytomegalovirus. Only 4 of the 11 isolates were detectable using the then mandated screening test. The Industry's contingency plan essentially concluded that the Bureau of Biologics would be unwilling to take the current product off the market in favor of a competing vaccine being produced in England from an established human cell line. African green monkeys continued to be used even after the Director of the Bureau of Biologics was informed in 1977 that licensed polio vaccines contained foreign DNA that was not of monkey cell origin. Of 8 vaccines lots from around this period that were recently tested in-house by the FDA Office of Vaccine Safety, 3 have DNA of simian cytomegalovirus. In a related study, British authorities reported that 32 of 34 polio vaccine lots from one manufacturer alone were contaminated with monkey cytomegalovirus DNA. FDA and British officials state they are unable to culture replicating cytomegalovirus from these vaccines. FDA was unwilling to provide samples of the vaccines for independent testing, ostensibly because of proprietary restrictions imposed by industry. This issue is important for at least two reasons: First, I have reported the definitive isolation of simian cytomegalovirus-derived cell damaging viruses from two patients with brain damaging illnesses, and as yet uncharacterized viruses from numerous additional patients with illnesses ranging from autism and learning disorders in children, chronic fatigue syndrome and fibromyalgia in adults, and various cancers and neurodegenerative illnesses in the elderly. The viruses were termed stealth because they were essentially not being recognized by the cellular immune system. Based on available DNA sequence data, it appears that the lack of effective immune recognition is due to the loss of the few major critical antigens that are targeted by the majority of virus reactive lymphocytes. Parents of stealth virus positive children have occasionally reported exacerbation of symptoms following vaccination. Vaccine viruses can promote the growth of certain stealth viruses in cultures. Furthermore, non-specific stimulation of the immune response could potentially trigger an anti-viral response directed at a few minor antigens retained by the stealth-adapted virus. Arguably, potential vaccine recipients should be screened for stealth virus infection prior to receiving the vaccine.

Cytomegalovirus, whether from African green or rhesus monkey, has multiple copies of the genes that promote cell entry of HIV and its precursor, the simian immunodeficiency virus (SIV) of chimpanzee. Cytomegalovirus contaminated experimental polio vaccines were used in chimpanzees in Central Africa. It is quite reasonable, therefore, that the use of experimental polio vaccine in Africa led to the conversion of SIV to HIV. Chimpanzees from Africa were also used to experiment with hepatitis B vaccine, again suggesting a possible link of vaccines with the spread of HIV in the United States. Requests to CDC to test stored human sera collected from polio vaccine immunized African children or hepatitis B immunized United States citizens have been ignored.

Political and Economic Considerations

These and other politically sensitive issues are seemingly not being addressed by our Public Health agencies. Undoubtedly, there is a reluctance of those in control to challenge the past performance of those entrusted with ensuring the Nation's health. The Pharmaceutical Industry also maintains a privileged position within our society. Not only does its financial strength curry support from Government, but it is likely to be heavily relied upon in the case

of biological warfare. Unfortunately, the primary motivation of this industry appears to have shifted from global Public Health concerns to simple profit motivation. An enormous price differential exists between charges for pediatric vaccines in Westernized countries compared to the developing world. Part of this differential is attributed to refinements in vaccine production, for example use of more purified bacteria products, or the use of inactivated versus live but weakened polio virus. Still the differences are staggering, for example US$0.07 versus US$10.65 for diphtheria-tetanus-pertussis (DTP) vaccine and US$0.10 versus US$8.25 for polio vaccine. Far more money is to be made vaccinating children from affluent countries, as well as international travelers from these countries, than addressing the world's health needs. The multi-national Pharmaceutical Industry has essentially withdrawn from servicing the developing world leaving this responsibility and low profit margin to a Developing Country Vaccine Manufacturers Network with facilities in countries such as India , Iran and Thailand .

The public is justifiably skeptical of the willingness of Government officials to request a full accounting of past and present vaccine manufacturing practices. Compulsory polio vaccination was legislated in the late 1950's to help reduce stockpiles of relatively ineffective lots of inactivated polio vaccines. Collusion between Government and vaccine producers may have occurred in the development and testing of agents of biological warfare. Intentional feeding of mentally retarded children with hepatitis B virus contaminated feces was justified as being necessary to protect other children. Possible responsibility for diseases such as AIDS, autism, sudden infant death syndrome, chronic fatigue syndrome and mental illnesses is vehemently denied and those making such suggestions attacked. To a large measure, vaccine and vaccine-related research have become money-driven endeavors with emphasis on perception rather than reality. This unfortunate trend needs to be addressed with forthright discussions that involve both the public and independent researchers.

Selected References:

History of Vaccines

Bazin H. A brief history of the prevention of infectious diseases by immunisations. Comp Immunol Microbiol Infect Dis. 2003; 26:293–308.

Hilleman MR. Vaccines in historic evolution and perspective: a narrative of vaccine discoveries. J Hum Virol, 2000;3(2):63–76.

Horaud F. Albert B. Sabin and the development of oral poliovaccine. Biologicals. 1963; 21:311–6.

Jenner E. An inquiry into the causes and effects of the variolae, a disease discovered in some of the counties of England , particularly Gloucestershire and known by the name of smallpox. London Samson Low. 1796.

Madsen T. Whooping cough: Its bacteriology, diagnosis, prevention and treatment. Bost Med Surg J, 1925; 192:50–60.

Pasteur L, Chamberland C-E, Roux E. [On the vaccination of sheep]. CR Acad Sci Paris; 1881; 92:1378–83.

Parish HJ. A History of Immunization. E & S Livingston, London , 1965.

Plotkin SA, Mortimer EA. Vaccines. WB Saunders, Philadelphia , 1988.

Salmon DE, Smith T. On a new method of producing immunity from contagious diseases. Am Vet Rev, 1886; 10:63–69.

The Public Health Service and the control of biologics. Public Health Rep, 1995; 110:774–5.

Specific Vaccines

Artenstein MS, Gold R, Zimmerly JG, et al. Prevention of meningococcal disease by group C polysaccharide vaccine. N Engl J Med, 1970; 282:417–20.

Austrian R, Douglas RM, Schiffman C, et al. Prevention of pneumococcal pneumonia by vaccination. Trans Assoc Am Phys, 1976; 89:184–92.

Clenny AT, Hopkins BE. Diphtheria toxoid as an immunizing agent. Br J Exp Path, 1923; 4:283–8.

Enders JF, Weller TH, Robbins FC. Cultivation of the Lansing strain of poliomyelitis in cultures of various human embryonic tissues. Science, 1949; 109:85–87.

Horsfall FL Jr., Lannette EH, Rickard ER, Hirst GK. Studies on the efficacy of a complex vaccine against influenza A. Public Health Rep, 1941; 56:1863–75.

Katz SL, Kempe CH, Blaid FL, Lepow ML, Krugman S, Haggerty J, Enders JF. Studies on an attenuated measles vaccine VIII. General summary and evaluation of results of vaccine. N Engl J Med, 1960; 263: 180–4.

Koprowski H, Jervis GA , Norton TW. Immune response in human volunteers upon oral administration of a rodent-adapted strain of poliomyelitis. Am J Hyg, 1952; 55:108–26.

Krugman S. The newly licensed hepatitis B vaccine. Characteristics and indication for use. JAMA, 1982; 247:2012–5.

McAleer WJ, Buynak EB, Margetter RZ, et al. Human hepatitis B vaccine from recombinant yeast. Nature, 1984; 307:178–80.

Provost PJ, Hughes JV, Miller WJ, et al. An inactivated hepatitis A vaccine of cell culture origin. J Med Virol, 1996; 19:23–31.

Sabin AB , Hennessen WA Winsser J. Studies on variants of poliomyelitis virus. I Experimental segregation and properties of virulent variants of three immunological types. J Exp Med, 1954; 99:551–76.

Salk JE, Krech V, Youngner JS, Bennett BL, Lewis LJ, Bazeley PL. Formaldehyde treatment and safety testing of experimental poliomyelitis vaccines. Am J Public Health, 1954; 44:563–70.

Smith W, Andrews CH, Laidlaw PP. A virus obtained from influenza patients. Lancet, 1933; 2:66–68.

Takahashi M, Otsuka T, Okuno Y, et al. Live vaccine used to prevent the spread of varicella in children in hospitals. Lancet, 1974; 2: 1288–90.

Theiler M, Smith HH. The use of yellow fever virus by in vitro cultivation for human immunization. J Exp Med, 1937; 65:787–800.

Immunology

Ada , GL. Vaccines in Fundamental Immunology . 3rd Edition. Ed. W.E. Paul. Raven Press, New York , 1993:1309–52.

Burnet FM. A modification of Jerne's theory of antibody production using the concept of clonal selection. Aust J Sci, 1957; 20:67–9.

Polio Vaccine Contamination

Baylis SA, Shah N, Jenkins A, Berry NJ, Minor PD. Simian cytomegalovirus and contamination of oral poliovirus vaccines. Biologicals, 2003; 31:63–73.

Kops SP. Oral polio vaccine and human cancer: a reassessment of SV40 as a contaminant based upon legal documents. Anticancer Res,2000; 20:4745–9.

Poinar H, Kuch M, Paabo S. Molecular analyses of oral polio vaccine samples. Science, 2001; 292:743–4.

Sierra-Honigmann AM, Krause PR. Live oral poliovirus vaccines and simian cytomegalovirus. Biologicals, 2002; 30:167–74.

Rizzo P, Di Resta I, Powers A, Ratner H, Carbone M. Unique strains of SV40 in commercial poliovaccines from 1955 not readily identifiable with current testing for SV40 infection. Cancer Res. 1999;596103–8.

Stealth-Adapted Viruses

Martin WJ, Zeng LC, Ahmed K, Roy M. Cytomegalovirus-related sequence in an atypical cytopathic virus repeatedly isolated from a patient with chronic fatigue syndrome. Am J Pathol,1994; 145:440–51.

Martin WJ, Ahmed KN, Zeng LC, Olsen J-C, Seward JG, Seehrai JS. African green monkey origin of the atypical cytopathic 'stealth virus' isolated from a patient with chronic fatigue syndrome. Clin. Diag. Virol, 1995; 4:93–103.

Martin WJ, Glass RT. Acute encephalopathy induced in cats with a stealth virus isolated from a patient with chronic fatigue syndrome. Pathobiology, 1995; 63:115–8.

Martin WJ. Stealth virus isolated from an autistic child. J Autism Dev Disord, 1995; 25:223–4.

Martin WJ. Simian cytomegalovirus-related stealth virus isolated from the cerebrospinal fluid of a patient with bipolar psychosis and acute encephalopathy. Pathobiology, 1996; 64:64–6.

Martin WJ. Genetic instability and fragmentation of a stealth viral genome. Pathobiology, 1996; 64:9–17.

Martin WJ. Severe stealth virus encephalopathy following chronic fatigue syndrome-like illness: Clinical and histopathological features. Pathobiology, 1996; 64:1–8.

Martin WJ. Stealth viral encephalopathy: Report of a fatal case complicated by cerebral vasculitis. Pathobiology, 1996; 64:59–63.

Martin WJ, Anderson D. Stealth virus epidemic in the Mohave Valley . Initial report of viral isolation. Pathobiology, 1997; 65:51–6.

Martin WJ. Cellular sequences in stealth viruses. Patobiology, 1998; 66:53–8.

Martin WJ. Bacteria related sequences in a simian cytomegalovirus-derived stealth virus culture. Exp Mol Path, 1999; 66:8–14.

Martin WJ. Melanoma Growth stimulatory activity (MGSA/GRO-alpha) chemokine genes incorporated into an African green monkey simian cytomegalovirus (SCMV)-derived stealth virus. Exp Mol Path, 1999; 66:15–8.

Martin WJ. Stealth adaptation of an African green monkey simian cytomegalovirus. Exp Mol Path, 1999; 66:3–7.

Martin WJ. Chemokine receptor-related genetic sequences in an African green monkey simian cytomegalovirus-derived stealth virus. Exp Mol Path, 2000; 69:10–6.

Martin WJ, Anderson D. Stealth Virus Epidemic in the Mohave Valley : Severe vacuolating encephalopathy in a child presenting with a behavioral disorder. Exp Mol Path, 1999; 66:19–30.

Martin WJ. Complex intracellular inclusions in the brain of a child with a stealth virus encephalopathy. Exp Mol Path, 2003; 74:179–209.

Martin WJ. Stealth Virus Culture Pigments: A Potential Source of Cellular Energy. Exp. Mol. Path, 2003; 74:210–23.

Polio Vaccines and the Possible Origin of AIDS

Elswood BF, Stricker RB. Polio vaccines and the origin of AIDS. Med Hypotheses, 1994; 42:347–54.

Gisselquist D. Emergence of the HIV type 1 epidemic in the twentieth century: comparing hypotheses to evidence. AIDS Res Hum Retroviruses, 2003; 19:1071–8.

Hooper E. Experimental oral polio vaccines and acquired immune deficiency syndrome. Philos Trans R Soc Lond B Biol Sci., 2001; 356:803–14.

Hooper E. The River: A Journey to the Source of HIV and AIDS . Little Brown, Boston 1999.

Lena P, Luciw P. Simian immunodeficiency virus in kidney cell cultures from highly infected rhesus macaques (Macaca mulatta). Philos Trans R Soc Lond B Biol Sci., 2001; 356:845–7.

Reinhardt V, Roberts A. The African polio vaccine-acquired immune deficiency syndrome connection. Med Hypotheses, 1997; 48:367–74.

Contamination of Other Vaccines

Bagust TJ, Grimes TM, Dennett DP. Infection studies on a reticuloendotheliosis virus contaminant of a commercial Marek's disease vaccine. Aust Vet J, 1979; 55:153–7.

Barkema HW, Bartels CJ, van Wuijckhuise L, Hesselink JW, Holzhauer M, Weber MF, Franken P, Kock PA, Bruschke CJ, Zimmer GM. [Outbreak of bovine virus diarrhea on Dutch dairy farms induced by a bovine herpesvirus 1 marker vaccine contaminated with bovine virus diarrhea virus type 2.] Tijdschr Diergeneeskd, 2001; 126:158–65.

Böni J, Stalder J, Reigel F, Schüpbach J. Detection of reverse transcriptase activity in live attenuated vaccines. Clin. Diagn. Virol, 1996; 5: 43–55.

Caramelli M, Ru G, Casalone C, Bozzetta E, Acutis PL, Calella A, Forloni G. Evidence for the transmission of scrapie to sheep and goats from a vaccine against Mycoplasma agalactiae. Vet Rec, 2001; 148:531–6.

Falcone E, Cordioli P, Tarantino M, Muscillo M, Sala G, La Rosa G, Archetti IL, Marianelli C, Lombardi G, Tollis M. Experimental infection of calves with bovine viral diarrhoea virus type-2 (BVDV-2) isolated from a contaminated vaccine. Vet Res Commun, 2003; 27:577–89.

Johnson JA, Heneine W. Characterization of endogenous avian leukosis viruses in chicken embryonic fibroblast substrates used in production of measles and mumps vaccines. J Virol, 2001; 75:3605–12.

Kappeler A, Lutz-Wallace C, Sapp T, Sidhu M. Detection of bovine polyomavirus contamination in fetal bovine sera and modified live viral vaccines using polymerase chain reaction. Biologicals, 1996; 24:131–5.

Levings RL, Wilbur LA, Evermann JF, Stoll IR, Starling DE, Spillers CA, Gustafson GA, McKeiman AJ, Rhyan JC, Halverson DH, Rosenbusch RF. Abortion and death in pregnant bitches associated with a canine vaccine contaminated with bluetongue virus. Dev Biol Stand, 1996; 88:219–20.

Rogers RJ, Dimmock CK, de Vos AJ, Rodwell BJ. Bovine leucosis virus contamination of a vaccine produced in vivo against bovine babesiosis and anaplasmosis. Aust Vet J, 1988; 65:285–7.

Senda M, Parrish CR, Harasawa R, Gamoh K, Muramatsu M, Hirayama N, Itoh O. Detection by PCR of wild-type canine parvovirus which contaminates dog vaccines. J Clin Microbiol, 1995; 33:110–3.

Thornton DH. A survey of mycoplasma detection in veterinary vaccines. Vaccine, 1986; 4:237–40.

Weaver SC, Pfeffer M, Marriott K, Kang W, Kinney RM. Genetic evidence for the origins of Venezuelan equine encephalitis virus subtype IAB outbreaks. Am J Trop Med Hyg, 1999; 60:441–8.

Failure of Vaccine Virus Inactivation

Nathanson N, Langmuir AD. The Cutter incident. Poliomyelitis following formaldehyde inactivated poliovirus vaccination in the United States during the spring of 1955. II. Relationship of poliomyelitis to Cutter vaccine. Am J Hyg, 1963; 78:29–60.

Brown F. Review of accidents caused by incomplete inactivation of viruses. Dev Biol Stand, 1993; 81:103–7.

Person-to-Person Transmission of Infection

Watanabe M. [The Tuberculosis outbreak due to whooping cough inoculation at Iwagasaki-machi in 1949] Nippon Ishigaku Zasshi, 2003; 49:479–92.

Montella M, Crispo A, Grimaldi M, Tridente V, Fusco M. Assessment of iatrogenic transmission of HCV in Southern Italy : was the cause the Salk polio vaccination? J Med Virol, 2003; 70:49–50.

Chernin E. Richard Pearson Strong and the iatrogenic plague disaster in Bilibid Prison, Manila , 1906. Rev Infect Dis, 1989; 11:996–1004.

Gisselquist DP. Estimating HIV-1 transmission efficiency through unsafe medical injections. Int J STD AIDS, 2002; 13:152–9.

Marx PA, Alcabes PG, Drucker E. Serial human passage of simian immunodeficiency virus by unsterile injections and the emergence of epidemic human immunodeficiency virus in Africa. Philos Trans R Soc Lond B Biol Sci, 2001; 356:911–20.

Other Complications of Vaccination

Geier MR, Geier DA, Zahalsky AC. Influenza vaccination and Guillain Barre syndrome small star, filled. Clin Immunol, 2003;. 107:116–21.

Communicating Vaccine Risks

Ball LK , Evans G , Bostrom. A . Risky Business: Challenges in Vaccine Risk Communication . Pediatrics, 1998; 101:453–8.

Coulter HL, Fisher BL. A Shot in the Dark . Avery Publishing Group, New York , 1991.

Kilch EW. The vaccine dilemma . Issues Sci Tech, 1986; 2:108.

Zhou W, Pool V, Iskander JK, English-Bullard R, Ball R, Wise RP, Haber P, Pless RP, Mootrey G, Ellenberg SS, Braun MM, Chen RT. Surveillance for safety after immunization: Vaccine Adverse Event Reporting System (VAERS)–United States, 1991–2001 . MMWR Surveill Summ, 2003; 52:1–24.

Lanctot, L. The Medical Mafia . Here's the Key Inc., Miami , 1995.

Martin B. The burden of proof and the origin of acquired immune deficiency syndrome . Philos Trans R Soc Lond B Biol Sci, 2001; 356:939–43.

Scheibner V. Vaccination . Published by Australian Print Group, Maryborough, 1993.

Protection Against Infectious Diseases, Including Stealth Adapted Virus Infections, Mediated by The Alternative Cellular Energy (ACE) Pathway*

W. John Martin, MD, PhD.
Institute of Progressive Medicine
South Pasadena, CA 91030

I will briefly review how the Germ Theory led to the development of vaccines; describe some of the major vaccine efficacy and safety issues and finally discuss the Alternative Cellular Energy or ACE Pathway as a potential replacement of vaccines, at least in certain individuals.

The Germ Theory, proposed by Louis Pasteur in the late 1800's, attributed infectious diseases to the mere existence of specific microbes. His concept prevailed over the opposing viewpoint, held by Béchamp and others, that microbes would not ordinarily invade healthy tissues. Rather, Béchamp had argued that infections were primarily an outcome of underlying damage to the tissue or the terrain as he called it, which allowed for the microbial growth.

Pasteur did, however, support the idea that the terrain could indeed be strengthened in its resistance to specific infections. He proved that exposure to a weakened microbe, either before infection as in the case of anthrax; or before clinical disease developed as in rabies, was protective. He further established that the protection was achieved by stimulation of an immune response against specific antigens of each type of microbe; thereby essentially equating terrain resistance with immunity. Pasteur's work led to the development of numerous additional vaccines throughout the 20th century and continues today.

What are some of the vaccine efficacy issues?

Two major issues of vaccine efficacy became apparent. The first was that the immune system comprises both an antibody mediated component and a, somewhat more difficult to induce, cell mediated component. For certain diseases, an antibody response by itself was inadequate in providing effective resistance. The second issue was that more purified antigens were rather weak in triggering an immune response. A concomitant generalized priming or boosting of the immune system was necessary and this could only be achieved by using an immunological adjuvant.

What are some of the vaccine safety issues?

As with all medical interventions, there has been an underlying concern of possible adverse effects of vaccinations. Actual mishaps have occurred and each has typically resulted in a strengthening and restructuring of the Government regulatory oversight of vaccine manufacturing. Three examples pertain to the development and use of polio vaccines. These include: i) Incomplete inactivation of some of initial batches of Salk vaccine; ii) Simian Virus 40 contamination of many batches of polio vaccines produced in kidney cell cultures of rhesus monkeys; iii) cytomegalovirus contamination of kidney cell cultures of both rhesus and African green monkeys. This latter issue has received scant attention by either industry or the FDA.

Why is there restrained oversight on vaccines?

Knowing that vaccines may provide an essential lifeline in response to the anticipated emergence of new infections, may explain why vaccine oversight still appears rather limited. A pressing issue is whether some individuals are unusually susceptible to adverse effects resulting from the non-specific adjuvant boosting of their immune system, especially when multiple vaccines are used. Increasingly, the FDA is being asked to require vaccine safety studies, not only in healthy individuals, but also in those with preexisting illnesses. FDA is also obligated to acknowledge data confirming the presence of derivative viruses of monkey cytomegalovirus origin in humans with a range of chronic neurological and psychiatric illnesses.

Why are these human isolates of particular concern?

It is because they have lost or mutated the genes coding for the relatively few antigens normally targeted by the cellular immune system. Accordingly, they have been termed "stealth adapted" viruses. This immune evasion mechanism can potentially occur with all viruses and poses an important challenge to public health.

What are some of these challenges?

One is the possibility that adjuvant induced generalized immune activation may help trigger low level cellular immune reactivity even against persisting stealth adapted viruses, thereby exacerbating the extent of brain damage. Another possibility is direct stimulatory interaction between live vaccine virus and stealth adapted viruses.

On a much more promising note, the immune system is not alone as the body's defense mechanism against viruses. Indeed, a broader definition of terrain includes the "alternative cellular energy (ACE) pathway." This pathway was demonstrated in the striking cellular repair process reversing the cytopathic effect (CPE) in many stealth adapted virus cultures. It also explains the recovery seen in animals inoculated with these viruses, in the absence of any inflammatory reaction. Enhancing the ACE pathway has also been shown to help suppress conventional infectious diseases of both virus and bacterial origin. Employing simple methods to enhance the ACE Pathway may well be a safer and more promising approach to providing disease resistance in some individuals than using a vaccine.

* Presented at the "Vaccine and Vaccination" conference organized by the Omics Group and held at Las Vegas on July 29-31, 2013.

Vaccine Provocation of Stealth Adapted Virus Induced Encephalopathy (VP-SAVIE)

MARCH 1, 2011 BY ADMIN LEAVE A COMMENT (Posted on Sanevax.org)

By W. John Martin, MD, PhD., Institute of Progressive Medicine

Vaccination is intended to stimulate the body's immune response against specific microorganisms or their toxic products. Vaccination is typically the administration via injection of non-living, or otherwise attenuated (weakened), microorganisms or antigenic products, along with an adjuvant intended to enhance the recipient's immune response. With the exception of attenuated viruses (which typically replicate over a period of a few days after inoculation), repeat doses of each vaccine are usually given to ensure an adequate immune response. By doing so, vaccination can ordinarily reduce the level of illness which would otherwise occur upon encountering the living microorganism by environmental exposure.

While generally considered to be safe, immune responses generated by vaccination can actually contribute to the cellular damage that can accompany an infection. In a small percentage of individuals, prior vaccination can lead to a more severe illness when exposed to the vaccine-relevant microorganisms than if the person had not been previously immunized. Another concern is that occasionally, the immune stimulation elicited by vaccination can trigger or augment an autoimmune response directed inwardly at the recipient's own tissues.

Another potential hazard to those being vaccinated could come from a pre-existing infection with stealth adapted viruses. These viruses lack the relatively few, major critical antigens that are normally targeted in the corresponding conventional virus from which they are derived*. Consequently, little or no cellular immune response normally accompanies stealth adapted viral infections, which may persist over many years. The heightened immune response induced by vaccination can potentially trigger tissue damaging reactions against remaining minor antigens expressed by stealth adapted virus infected cells.

Because of its regional separation of functions, the brain is particularly prone to clinical illnesses caused by stealth adapted viruses. This unique susceptibility to localized cellular damage probably accounts for the predominance of neurological and psychiatric symptoms in the majority of individuals experiencing prolonged severe adverse vaccine reactions. A suitable descriptive term for this condition is "vaccine provocation of stealth adapted virus induced encephalopathy," or VP-SAVIE.

Autism following vaccination in infants is possibly a prime example of VP-SAVIE. Undue focus on the vaccine, rather than on a potentially pregnancy acquired pre-existing stealth adapted viral infection, has unfortunately distracted many within the autism community.

Similarly, the tragic deterioration in the health of some adolescent recipients of human papillomavirus (HPV) vaccines is likely to be attributed to the provocation of a pre-existing stealth adapted viral infection.

When stealth adapted viruses were first brought to the attention of Public Health authorities, they actively resisted open discussions because one of the earliest stealth adapted viruses discovered arose from the African green monkey simian cytomegalovirus (SCMV). Monkeys infected with SCMV were knowingly used in the production of live poliovirus vaccines, a sad commentary on the diligence of the Food and Drug Administration (FDA).

An important insight gained from the study of stealth adapted viruses is that the body can resist viral infections without the participation of the cellular immune system. This non-immunological defense and healing mechanism involves the alternative cellular energy (ACE) pathway and occurs in the absence of inflammation or scarring.

As published over five years ago, the ACE pathway can be effectively activated in the sustained suppression of conventional herpes simplex virus (HSV) and herpes zoster virus (HZV) infections. These experimental protocols were further developed for testing in children with autism.

Simple screening methods are also being developed to potentially monitor the levels of activity of the ACE pathway. A reasonable suggestion is that vaccines should not be administered to anyone in whom the ACE pathway is not fully functional. This issue needs to be urgently addressed by the Public Health system. They hold the key to allow further clinical testing to proceed. Answers need to be uncovered.

Patient support groups and advocates could help by promoting internet based dialogue among themselves and Public Health officials on the issues of stealth adapted viruses and the ACE pathway.

The author would be pleased to participate in any such dialogues.

Email the author at s3support@email.com

*[note: human cytomegalovirus is composed of nearly 200 genes, yet only a single protein, termed UL83, leads to activation of approximately 60% of the entire cytotoxic T-cell response, with two additional proteins accounting for another 30% of the cellular immune response.]

Stealth Adapted Viruses: A Bridge Between Molecular Virology and Clinical Psychiatry

Abstract

Cytopathic "stealth-adapted" viruses bypass the cellular immune defense mechanisms because of molecular deletion or mutation of critical antigen coding genes. Stealth viruses establish persistent, systemic virus infections, which commonly involve the brain. The term "stealth adapted virus infection of the brain" or (SAVI-B) is suggested for infected patients with prominent neuropsychiatric symptoms. Additional symptoms commonly occur in stealth adapted virus infected patients and can be mainly explained by the direct effects of the brain on other organ systems of the body. Symptoms can also result from: i) induced autoimmunity, ii) antibody formation against virus antigens, iii) virus-induced cellular damage to non-brain tissues and iv) induced heightened overall immune reactivity, such that normally unrecognized components of the virus begin to become targeted by the cellular immune system. Certain stealth adapted viruses may acquire cancer causing (oncogenic) capacity. Appropriate medical care for stealth adapted virus infected patients clearly extends beyond regular psychiatry. Furthermore, disregarding the infectious component in these illnesses can place family members and others at risk for becoming infected.

Overview of Traditional Neurology and Psychiatry

Dysfunctional brain syndromes are erroneously viewed as comprising two distinct groups of illnesses: neurological and psychiatric. Neurologists mainly address diseases that can be attributed to discrete anatomic lesions, with readily elicited physical signs pertaining to the affected region of the brain. Although the therapeutic options are usually limited, the causes of these illnesses have a rational basis in terms of defined neuro-anatomical lesions. As opposed to neurologists, psychiatrists and other mental health personnel, mostly address diseases lacking precise anatomic localization or biologic explanation. These diseases are expressed in terms of varying degrees of altered emotions, behaviors and cognitive processes; functions that are viewed as

expressions of the "mind" rather than of the "organic brain." The availability of mind-altering drugs has helped shift the therapeutic emphasis for these diseases from simply trying to coerce the patient to change his or her ways (psychotherapy) to the somewhat more successful (if still empirical) psychopharmacological approach. The use of therapeutic drugs in psychiatry is predicated on the assumption that patients have an underlying disturbance in neural metabolism that can be at least partially restored pharmacologically. The etiology of the "chemical imbalance" is rarely addressed and often assumed to be a result of an inappropriate behavioral adaptation to life stressors. This paper provides an alternative explanation for many psychiatric illnesses; one that is based on altered brain function resulting from infection with stealth adapted viruses.

Spatial Distribution of Normal Brain Function

The brain is unique among the body's organs in the spatial distribution of its many functions. Unlike other organs, damage to one area of the brain cannot be readily compensated for by heightened activities in other brain areas. Moreover, individual components of the brain participate in complex neural networks, which can sub-serve a variety of integrated functions. Even minimal damage to neurological tissue has the potential for profound symptomatic effects; compared to the effects of limited cellular damage occurring in extra-neural tissues. Not only is the brain tissue spatially complex, it is hampered by the inability of mature neuronal cells to replicate and to replace neurons damaged by either illness or normal senescence.

Assessment of Brain Function

The brain is responsible for motor, sensory, autonomic and cognitive functions. It also determines personality, mood, self-perception and social interactions. Assessment of gross deficits of sensory and motor functions is readily achieved in routine neurological examinations. Psychiatrists rarely, if ever, employ testing for more subtle sensory and motor changes, and for possible derangements of the autonomic nervous system. Many neurologists have also disregarded such tests, as only providing inconsequential "soft signs." Complex assays, such as tilt-table testing for orthostatic hypotension can provide a quantitative measure of a specific autonomic function, but are unsuitable for everyday clinical practice. Neuroimaging techniques, such as computerized EEG, PET scans and functional MRI can also provide measures of brain activity but are also unsuitable for routine psychiatric practice. Furthermore, the etiological foundations for the minor imaging changes that have been seen in psychiatric patients are not yet established. Neuropsychiatric testing for minor personality disorders and for mild cognitive impairments requires an in-depth knowledge of the individual's pre-illness performance; information which is not generally available. One-time testing will, therefore, usually not reveal the early changes in personality or cognitive abilities that patients themselves or their friends may perceive.

Diagnostic Labeling of Psychiatric Illnesses

In spite of the shortcomings in assessments of many brain functions, psychiatrists have managed to categorize psychiatric illnesses into distinct clinical entities by grouping symptoms into a variety of syndromes (1). These groupings obscure the fact that many symptoms are common to various disease categories. Moreover, the naming of an illness tends to overlook the considerable variability in symptoms, and especially their relative severity between patients and even in a single patient over time. The lack of true diagnostic precision in reflected in such terms as "co-morbidity" and "borderline condition." The assumption that different syndromes have different underlying etiologies has also hampered efforts to find common causes of mental illnesses.

Etiology of Psychiatric Illnesses

In a similar way that diagnostic labels have tended to artificially sub-divide a spectrum of neuropsychiatric illnesses, the proponents of various etiologic theories have also tended to be exclusive rather than inclusive. The notion that organic brain illness is genetic, infectious, autoimmune or toxic; precludes the known interactions between all of these components. The aging process itself can slowly erode the limited functional reserves that may have survived an earlier insult. This can lead to a delay in the clinical expression of an illness years after the initiating event has occurred. Of the four etiologic categories listed above, an infectious cause has the promise of being the most readily targeted for therapy, as well as having the added concern of being potentially transmissible between individuals. Viral infections of the brain can present in many different ways depending simply on its localization to different regions of the brain and on the preexisting functional capacity levels prior to infection. Infections can also render individuals susceptible to normally tolerated environmental factors and to other stressors of brain function (2).

Viruses and Psychiatric Illnesses

The digression of psychiatry from basic molecular biology is seen in the minimal attention currently given to the potential role of viral infections in psychiatric illnesses. Historically, such conditions as encephalitis lethargica, subacute sclerosing panencephalitis, multifocal leukoencephalopathy and general paresis of the insane were belatedly accepted as infectious. The reality of AIDS dementia is also now unquestioned (4). On the other hand early attempts to detect viruses in patients with schizophrenia (schizoviruses) and other major psychiatric illnesses, failed to provide convincing and readily reproducible findings (5). In spite of the availability of more sensitive technologies, such as the polymerase chain reaction (PCR), few psychiatrists are intellectually poised to consider viral infections as a likely cause of their patients' illnesses.

The prevailing model of a viral brain infection is that of herpes simplex virus (HSV) encephalitis.

Typically the patient will present with an acute onset (<2 weeks from the initial symptoms to severe illness); have progressively diminishing level of consciousness; show localizing signs, often to the temporal lobes, on clinical, radiologic and EEG examinations; and have numerous lymphoid cells in the cerebrospinal fluid (CSF) with increased protein levels (6). Relatively mild meningitis/encephalitis-like illnesses are also commonly encountered in General Practice. If pursued vigorously, serological assays, changes in CSF, and virus cultures of feces will occasionally indicate an enteroviral infection (7). The illnesses are considered to be short lasting without sequela. The notion of a persisting, sub-acute, non-inflammatory viral encephalopathy is rarely considered clinically or tested for using either viral cultures or molecular probe based assays.

Virus Classification

One reason for the lack of consideration of viruses in psychiatric illnesses is the plethora of different viruses that in many ways seem unconnected to each other (8). Emphasizing differences rather than similarities has impeded a clear overview of molecular virology in much the same way that the over categorization of dysfunctional brain syndromes has confounded, rather than simplified psychiatry. A working model to help bypass the complex viral classifications schemes that are currently in place can be arrived at through the simple principle that viruses must replicate. While larger, more complex DNA viruses code their own DNA dependent DNA polymerase, smaller DNA viruses need to make use of either one of the cell's own DNA polymerases, or alternatively the polymerase of a larger DNA virus. Most of the very small DNA viruses, such as parvoviruses, have no polymerase coding gene and will only replicate within either already dividing cells or cells co-infected with a DNA polymerase coding virus. Somewhat larger DNA viruses (e.g. papovaviruses) can replicate in more cell types since they possess an additional protein capable of inducing proliferation in normally resting cells. With the exception of hepatitis D virus (an RNA virus that the cell misreads as DNA), conventional RNA viruses generally need to provide their own polymerase. Negative (non-reading) single-stranded (ss) RNA viruses encode their own RNA dependent RNA polymerase, which is packaged along with the viral genome. Positive ssRNA viruses and most double stranded (ds)RNA viruses, synthesized their polymerase soon after infection occurs. Retroviruses are positive strand RNA viruses, which encode a reverse transcriptase. This RNA dependent DNA polymerase replicates virus ssRNA into ssDNA and then to dsDNA. Integration of the dsDNA into the cellular genome, allows for reformation of ssRNA through the action of cellular DNA dependent RNA polymerase. As part of evolution, the human genome contains numerous copies of previously acquired retroviruses, some with genes coding for RNA dependent DNA polymerases. Activation of these endogenous retroviruses, as occurs in several neurological diseases, can similarly potentially lead to the conversion of virus ssRNA sequences into DNA copies.

Viral survival also depends on the ability to spread to other cells. Simple viruses can rely merely on passive uptake or "hitching a ride" within other viruses. The survival of viruses between cells is generally achieved by enclosing the viral genome within an insoluble protein structure formed by the self-assembly (aggregation) of proteins, giving rise to the viral capsid or coat. Many plant viruses establish tubular structures formed by the ordered assembly of

so called "movement proteins." These tubes penetrate between cells and act as conduits for the intercellular transfer of viral nucleic acids. Complex animal viruses may incorporate sequences coding for molecules, which become embedded into a cellular membrane surrounding the viral capsid. These molecules can engage with receptors expressed on the membranes of cells susceptible to becoming virally infected. The actual genes encoding the capsid, polymerase and envelope proteins can exist on a single nucleic acid molecule or on different molecules packaged into a single viral particle (segmented viruses). Bipartate and tripartite viruses requiring infection with 2 or 3 different viruses have also been described, especially among viruses affecting both plants and insects. Complex viruses may possess additional protein components within the virus particle. These can facilitate the metabolic interactions between the virus and the host cell. Such interactions reflect the close relatedness of certain viral genes to their cellular counterparts. In many cases, the virus genes have been captured from host cells at an earlier phase of their evolution. The requirement of multiple viral-host gene interactions can help explain much of the restricted host species, and even cell type specificity, exhibited by many viruses. Specific gene interactions can also account for the distinctive cytopathic effects (CPE) manifested by certain viruses, beyond those simply attributed to metabolic competition.

Viral Pathogenesis

All viruses have the potential to mediate cellular changes by altering the normal metabolic balance within the cell through over utilization of the cell's energy resources (9). While this can eventually lead to cell death, an earlier cost can be the failure of the cell to perform all of its specialized functions (10).

Continued metabolic drain on the cell can lead to a loss of essential components such as adenosine triphosphate (ATP). A tipping point is reached when there is insufficient ATP to import magnesium into the cell; a required co-factor for ATP activity. Energy starved cells can show foamy vacuolization, swelling and intercellular fusion. Some viruses trigger a more active form of cellular death, called apoptosis, characterized by shrinkage and condensation of cellular components. Herpesviruses, especially HSV and human cytomegalovirus (hCMV), are especially cytotoxic when cultured with normal cells. So too are adenoviruses, influenza, polio and many enteroviruses. Certain human viruses, however, are essentially incapable of inducing a readily discernable CPE in viral cultures on human cells. For example, rubella; hepatitis A, B, C and D; HTLV; Borna and Hartaan viruses are non-cytopathic. Moreover, primary clinical isolates of German measles and mumps viruses induce rather minimal CPE when directly cultured on human cells. For many of these non- or minimally- cytopathic viruses, the in vivo tissue damage is a consequence of immune activation and lymphocyte killing of the infected cells (11).

Viral Immunity

The immune system can also both reduce and enhance the extent of viral damage. Antibodies can provide an effective blockade preventing viruses from gaining access to normally permissive cells. In particular, antiviral antibodies can help prevent viruses passing through the blood to the brain. Cellular immunity can reduce viral load by destroying infected cells prior to the release of infectious viral particles. On the other hand, cellular immunity against viral antigens or against modified or inappropriately expressed cellular antigens can lead to immune damage of cells beyond that achieved by the virus itself.

Viruses have evolved various mechanisms to help evade the immune system. One such mechanism is the deletion of the genes coding for the major antigens recognized by the cellular immune system (12-13). This mechanism of bypassing the cellular immune defenses has been referred to as a "stealth adaptation."

Stealth Adapted Viruses

A corollary of the Clonal Selection Theory of Immunology (14) is that to be effectively recognized, a viral infected cell must present multiple copies of the antigen that is targeted by the responding antigen specific lymphocyte. This requirement restricts the number of different viral antigens, which can be presented to the cellular immune system. Even with large complex viruses, relatively few viral components are targeted for cellular immune defenses (15). For certain viruses, e.g. hCMV, experimental studies suggested that the complete deletion (or mutation) of the three genes coding for the major viral components recognized by the cellular immune system, would likely yield a defective, non-replicating, non-cytopathic viral sequence. The remaining sequences could, however, provide potential building blocks towards the evolution of a cytopathic non-immunogenic "stealth adapted virus." Potentially, the downsized gene-depleted virus could, for example, form a synergy with a replicating non-cytopathic virus and/or incorporate certain cellular genes by recombination, to yield an atypically structured cytopathic virus. These concepts are embodied in the following definition of stealth adapted viruses:

"Molecularly heterogeneous grouping of atypically structured, cytopathic viruses, that cause persistent systemic infection, often with neuropsychiatric manifestations, in the absence of significant antiviral inflammation. Stealth adaptation is a generic, derivative process in which conventional viruses have lost or mutated the relatively few genes encoding the major antigens normally targeted by the cellular immune system. Stealth adapted viruses typically induce a vacuolating cytopathic effect (CPE) in a range of human and animal cells. The formation, progression, and/or host range of the CPE distinguish stealth adapted viruses from the CPE caused by conventional human cytopathic viruses, including herpesviruses, enteroviruses, and adenoviruses. Additional distinctions can be made on the basis of electron microscopy, serology, and molecularbased studies."

Origins and Replication of Stealth Adapted Viruses

Certain stealth viruses contain genetic sequences that are nearly identical to sequences found in African green monkey simian cytomegalovirus (SCMV) (13). Other regions of these viruses are clearly different from SCMV. Genetic sequences related to other human and animal associated viruses, and additional sequences more closely related to various cellular genes, have been detected in the DNA and/or RNA fractions of stealth adapted virus infected cultures. Electron microscopy has also revealed differences between stealth viral cultures in terms of the types and relative abundance of distinctive accumulations of viral-like materials within cells showing the characteristic vacuolated CPE. An interesting observation is the apparent genetic instability and fragmentation of a stealth viral DNA genome (16). A potential mechanism of stealth viral DNA replication is through the bridging of viral fragments with long RNA molecules. If so, this scaffolding effect could potentially be inhibited in the presence of short RNA molecules competing with the longer RNA molecule for binding to one of the fragments.

Relevance of a Vaccine Origins of Certain Stealth Adapted Viruses

While stealth adapted viruses have presumably existed for eons, the increasing incidence of many current disease entities is consistent with the introduction of additional stealth adapted viruses through vaccines. Public health authorities have largely ignored the issue of SCMV as a possible contaminant of polio vaccines. This disregard occurred even though in 1972, all 11 monkey kidney culture tested using sensitive indicator cell lines showed the presence of SCMV. Only 4 of these isolates would have been detected using the standard detection procedures, which remained in place despite of the above finding.

A diverse array of animal cell lines has been used for the many animal vaccines that have been developed. It is not unreasonable to suggest some of these vaccines have been sources of stealth adapted viruses. It is quite conceivable that vaccine viruses can contribute genetic elements to contaminating herpes and other viruses and that this could facilitate the emergence of replicating, non-immunogenic (stealth-adapted) viruses. Once within the human population, stealth adapted viruses can be passed via direct human-to-human contact as well as potentially by interspecies transmission.

Detection of Stealth Adapted Viruses

The most reliable method for detecting the diversity of stealth viruses is to co-culture the patient' blood with a variety of indicator cell types and observe the cultures for the induction of a transmissible CPE (17). Typically, rhesus monkey kidney cells and a human fibroblast cell line such as MRC-5 cells are inoculated with the patient's mononuclear cells and observed for 2-4 weeks. Frequent re-feeding of the cultures can help promote the development of the CPE. It is quite unusual (<10%) to observe a rapidly developing CPE in blood samples from randomly selected hospital outpatients. Conversely, it is unusual not to observe a strong positive CPE in cultures from patients with otherwise unexplained neurological or behavioral disorders.

The stealth virus CPE is best characterized by the formation of foci of enlarged, rounded cells, often with cell fusion (syncytia). Proliferation foci of affected cells can occasionally be seen. The actual appearance of the CPE differs between cultures and is best followed by repeated examination of individual cultures by the same observer. The CPE can be transferred to fresh cultures. Positive cultures can be further examined by staining cell smears or sectioned cell pellets using the patient's and other sera. Electron microscopic studies can also be performed. Cell derived DNA and RNA can also be used for molecular characterization. A series of PCR primer sets based on previously characterized stealth viruses can be used to screen for virus-derived DNA and RNA sequences. The primers can also be used to test for DNA and RNA dependent polymerases. Finally, the viral cultures can be used to test the effects of various anti-viral therapies.

As noted above, stealth viral infections are not necessarily confined to the brain and indeed blood samples are routinely used for stealth viral cultures. Other serological signs of viral infections can include unusually high levels of anti-herpesvirus antibodies. This may reflect the presence of the stealth virus or the two-way cross stimulation that can be seen between stealth adapted viruses and conventional herpesviruses. Broadly reactive herpesvirus primers can also be used in low stringency PCR based assays on DNA and RNA directly isolated from the patient's blood (17). Other primer sets have been shown to cross react with several stealth virus isolates in low stringency PCR assays. Cloning and sequencing of the PCR products can be used to design more specific primer sets. The possible role of stealth adapted herpesviruses in secondary activation of parvo- and papovaviruses, including monkey-derived SV40, can also be assessed using serological and molecular probe based assays for these agents.

Stealth Adapted Viral Infection

Stealth viruses have been isolated from blood and CSF of patients with a spectrum of illnesses with neurological and neuropsychiatric manifestations (17-22). The clinical diagnoses have included autism and attention deficit learning disorders in children, CFS, fibromyalgia, Gulf war syndrome and depression in adults, and dementia/ Alzheimer's disease in the elderly. Severe acute encephalopathy and major psychotic reactions have also been associated with positive stealth adapted viral infections. The clinical diversity seen in stealth viral infected patients may relate to the predominant areas of the brain that are infected as well as the timing and intensity of the infection. Stealth viruses from humans have induced acute neurobehavioral diseases in experimental animals (23) accompanied by similar histological and electron microscopic changes as seen in brain biopsies of infected humans.

Hisopathology

The predominant histological characteristic in both humans and in the animal model is the presence of occasional cells with distinctly vacuolated, lipid-rich cytoplasm and distorted abnormal nuclei (18, 22-23). The affected cells may show varying granules positive with periodic acid Schiff (PAS) stain. Deposited material can also accumulate around small blood vessels, possibly impairing gas exchange and nutrient delivery. The marked vacuolization seen in some biopsies is certainly suggestive of Creutzfeld Jacob prion disease (22).

Animal studies confirmed that the cellular changes were not confined to the brain but that signs of infection could be found in various organs throughout the body (23). Nonetheless, the predominant clinical manifestations in the animals were neurobehavioral, consistent with the unique susceptibility of the brain to limited damage.

Clinical Manifestations of Illnesses Seen in Certain Stealth Adapted Virus Infected Patients

Stealth viral culture positive patients, whether presenting with a psychiatric or neurological illness, will not uncommonly report symptoms attributed to illnesses occurring elsewhere in the body. In some cases, such as in low back pain, the essence of the illness is a lowered pain threshold, more than the severity of the musculoskeletal changes. Similarly, pelvic pain can have a strong central nervous system component. Clinicians can err in over treating the localized area of pain without due regard to the underlying hypersensitivity of the patient's nervous system. The treated disease will likely recur or be replaced by an equally disabling painful condition occurring elsewhere in the body.

Various cardiovascular diseases, affecting the heart and/or peripheral circulation, can be ascribed to dysregulation by an impaired autonomic nervous system. Postural hypotension occurring in CFS patients is well described. An interesting illness encountered in some patients is erythromelalgia, in which inappropriate shunting of blood via opened arteriovenous connections compromises the capillary circulation with resulting pain (24).

The autonomic nervous system also controls aspects of gastrointestinal functioning and this can account for dysphagia and irritable bowel syndrome. It may also contribute to malabsorption, leaky gut syndrome and dysbiosis (abnormal gut flora). Bacteria sequences have been identified in some stealth adapted virus cultures (25). This is consistent with some of these viruses being able to pass within bacteria. Indeed, atypical bacteria have been seen in fecal cultures of some CFS patients. Stealth adapted virus infection of the gut bacteria could contribute to dysbiosis.

Another diagnosis experienced by stealth adapted virus infected patients is delusional parasitosis (26). Patients can produce pigmented particles with striking electrostatic properties, which the patients can easily mistake for living movements. When detected in the hair, the particles have also been misidentified as lice. The mother of a child reported severe mental deterioration requiring institutionalized care, after the anti-lice medication, Lindane, had been applied to her daughter's scalp and the house fogged with an insecticide.

A patient from whom strikingly positive cultures were repeatedly observed had numerous lipomas, which she said tended to come in episodes and slowly resolve. Her diagnosis was Dercum's disease, which is a condition largely unknown in conventional medicine (27). This disease can be likened to periodic outbreaks of shingles. The difference is that instead of vesicular lesions developing in the skin, localized areas of excessive subcutaneous lipids are being produced; sometimes distributed within a single dermatome. Virus induced overproduction of intracellular lipids can explain liver steatosis seen in some stealth adapted virus infected patients. Using the same reasoning, intracellular lipids can inhibit glucose transport into cells, rendering cells somewhat insulin resistant. Obesity may result from the unsightly but possibly necessary disposal of lipids overproduced by virus infected parenchymal organs.

Virus infections can also provoke autoimmune reactions, especially to cellular DNA, as in systemic lupus erythematosus. Autoantibody production against mitochondrial phospholipids can lead to hypercoagulation, while antibodies to clotting factors can lead to excessive bleeding. Hashimoto's disease and Graves' disease can occur with auto-antibodies against thyroid antigens. Autoimmunity can also be directed to the nervous system, as can be seen with the extension of an apparent CFS to multiple sclerosis. Cases of limbic encephalopathy (28) can also be explained as an autoimmune response to neurotransmitters or to their receptors. Antibodies may also form against the virus, since many more virus antigens evoke humoral immunity than the relatively few antigens targeted by the cellular immune system. These antibodies can explain the vasculitis occurring in some stealth adapted virus culture positive patients (19). The immune system itself can be directly damaged by virus infection leading to immune dysregulation.

An exception to the statement that limited localized damage is relatively less significant outside of the nervous system, is the occurrence of malignancy. Upon inquiry, breast cancer patients not uncommonly report on fatigue for years prior to the detection of their cancer and even after its removal. Multiple myeloma patients commonly have a prior history of neuropsychiatric illnesses. The consistent finding of positive cultures for stealth adapted viruses in multiple myeloma patients was confirmed in a double blind study (29). Direct evidence for stealth adapted viruses was also obtained by the polymerase chain reaction (PCR) in patients with salivary tumors (30). A patient with strikingly positive virus cultures had a glioblastoma, which is consistent with published evidence linking CMV to this tumor.

Virus infection of germ cells can potentially lead to genetic disorders in offspring. This has been noted in some patients with a genetic abnormality, not present as a somatic mutation in either of the parents. CMV is especially prone to infect the gonads and may induce genetic change in germ cell even prior to fertilization. This could account for the circumstances in which a de novo genetic disease occurs in children in whom one or more parent was stealth adapted virus infected.

Transmission of Stealth Adapted Virus Infections

The occurrence of family illnesses of presumptive infectious origin has been noted on several occasions. To cite one of these families: The mother openly declared she had CFS. She believed her husband also had the illness but was in denial since he was still required to work. The woman's mother was diagnosed with Parkinson's disease, while her son was diagnosed as schizophrenic. Within the family, they all recognized a basic similarity of their illnesses. Several other families with differing illnesses but all of presumptive infectious origin have been seen.

Community wide transmission of stealth adapted virus induced illnesses can easily explain many of the reported outbreaks of CFS-like illnesses. One such epidemic occurred in 1996 in the Mohave Valley area of Arizona and adjoining town of Needles, California (21). The nation's blood and blood products supply is also an expected mode of transmission of stealth adapted viruses.

Occupational exposure is also a potential risk factor, especially in individuals likely to come into close contact with others (31). This group includes healthcare providers, schoolteachers, prison guards, etc. Avoiding becoming stealth adapted virus infected is of special concern to women anticipating pregnancy. A difficult issue, which also applied to conventional CMV infection, is the risk of daycare facilities and of pediatricians' medical offices in spreading infection to other children. Freshly infected children can expose their mother to the virus prior to or during her next pregnancy.

Summary

The specialization of medicine has focused attention on disorders that are essentially restricted to a single organ system. Multi-system diseases tend to fall outside the purview of most physicians and a balanced, comprehensive approach to their assessment is often lacking. The clinical evaluation of patients with a stealth viral induced encephalopathy should not be confined to disorders of brain function. Rather, the clinical evaluation should include seeking evidence for viral involvement of additional organs, as would be expected for a systemic virus infection. Efforts should also be undertaken to restrict the likely transmission of stealth adapted virus infections, both within families, certain occupations and whole communities.

References

1. American Psychiatric Association (2013) Diagnostic and Statistical Manual of Mental Disorders, Fifth Edition (DSM-5). Amer Psychiatric Pub Inc., Arlington, VA.

2. Noshpitz JD Coddington RD. Editors (1990). Stressors and the adjustment disorders John Wiley and Sons, New York, NY

3. Tselis A, Booss J (2003) Behavioral consequences of infections of the central nervous system: with emphasis on viral infections. J Am Acad Psychiatry Law 31: 289-98.

4. Ances BM, Ellis RJ (2007) Dementia and neurocognitive disorders due to HIV-1 infection. Semin Neurol. 27: 86-92.

5. Torrey EF (1988) Stalking the schizovirus. Schizophr Bull. 14: 223-9.

6. Tyler KL (2004) Herpes simplex virus infections of the central nervous system: encephalitis and meningitis, including Mollaret's. Herpes 11 Suppl 2: 57A-64A.

7. Rhoades RE1, Tabor-Godwin JM, Tsueng G, Feuer R (2011) Enterovirus infections of the central nervous system. Virology 411: 288-305.

8. White DO, Fenner FJ. (1994). Medical Virology 4th edition Academic Press San Diego, CA

9. Carrasco L. Editor (1987). Mechanisms of Viral Toxicity in Animal Cells. CRC Press Boca Raton Florida

10. de la Torre JC1, Borrow P, Oldstone MB (1991) Viral persistence and disease: cytopathology in the absence of cytolysis. Br Med Bull. 47: 838-51.

11. Zinkernagel RM (1997) Immunology and immunity studied with viruses. Ciba Found Symp. 204: 105-25.

12. Martin WJ. (1994) Stealth viruses as neuropathogens. College of American Pathologist's publication "CAP Today" 8: 6770.

13. Martin WJ. (1999) Stealth adaptation of an African green monkey simian cytomegalovirus. Exp Mol Path. 66: 3-7.

14. Burnet, FM (1959) The Clonal Selection Theory of Acquired Immunity. Vanderbilt University Press, Nashville TN

15. Wills MR1, Carmichael AJ, Mynard K, Jin X, Weekes MP, et al (1996) The human cytotoxic T-lymphocyte (CTL) response to cytomegalovirus is dominated by structural protein pp65: frequency, specificity, and T-cell receptor usage of pp65-specific CTL. J Virol. 70: 7569-79.

16. Martin WJ (1996) Genetic instability and fragmentation of a stealth viral genome. Pathobiology 64: 917.

17. Martin WJ, Zeng LC, Ahmed K, Roy M (1994) Cytomegalovirusrelated sequences in an atypical cytopathic virus repeatedly isolated from a patient with the chronic fatigue syndrome. Am J Path. 145: 441452.

18. Martin WJ (1996) Severe stealth virus encephalopathy following chronic fatigue syndromelike illness: Clinical and histopathological features. Pathobiology 64:18.

19. Martin WJ (1996) Stealth viral encephalopathy: Report of a fatal case complicated by cerebral vasculitis. Pathobiology 64:5963.

20. Martin WJ. (1996) Simian cytomegalovirusrelated stealth virus isolated from the cerebrospinal fluid of a patient with bipolar psychosis and acute encephalopathy. Pathobiology 64:6466.

21. Martin WJ, Anderson D (1997) Stealth virus epidemic in the Mohave Valley. Initial report of viral isolation. Pathobiology 65: 51-56.

22. Martin WJ, Anderson D (1999) Stealth Virus Epidemic in the Mohave Valley: Severe vacuolating encephalopathy in a child presenting with a behavioral disorder. Exp Mol Pathol. 66:19-30.

23. Martin WJ, Glass RT (1995) Acute encephalopathy induced in cats with a stealth virus isolated from a patient with chronic fatigue syndrome. Pathobiology 63: 115118.

24. Ljubojević S1, Lipozencić J, Pustisek N (2004) Erythromelalgia. Acta Dermatovenerol Croat. 12: 99-105.

25. Martin WJ (1999) Bacteria related sequences in a simian cytomegalovirus-derived stealth virus culture. Exp Mol Path. 66: 8-14.

26. Martin WJ (2005) Alternative cellular energy pigments mistaken for parasitic skin infestations. Exp. Mol. Path 78: 212-214.

27. Hansson E, Svensson H, Brorson H (2012) Review of Dercum's disease and proposal of diagnostic criteria, diagnostic methods, classification and management. Orphanet J Rare Dis. 7: 23.

28. Ramanathan S, Mohammad SS, Brilot F, Dale RC (2013) Autoimmune encephalitis: Recent updates and emerging challenges. J Clin Neurosci. pii: S0967-5868.

29. Durie BG, Collins RA, Martin WJ (2000). Positive stealth virus cultures in myeloma patients: A possible explanation for neuropsychiatric co-morbidity. Blood 96: (suppl 1) abstr 1553.

30. Gollard RP, Mayr A, Rice DA, Martin WJ (1996) Herpesvirusrelated sequences in salivary gland tumors. J Exp Clin Can Res. 15: 14.

31. Joseph SA, Béliveau C, Gyorkos TW (2006). Cytomegalovirus as an occupational risk in daycare educators. Paediatr Child Health 11: 401-407.

Complex Multi-system Illnesses Occurring Within a Family: Presumptive Evidence for an Infectious Disease Process

W. John Martin, M.D., Ph.D. Institute of Progressive Medicine, South Pasadena CA 91030

Author Telephone number: (626) 616-2868   e-mail: wjohnmartin@hotmail.com

Address: 1634 Spruce Street, South Pasadena CA 91030

Abstract:

Society is witnessing an increasing incidence of illnesses with neuropsychiatric manifestations. Prominent examples include autism and learning disorders in children, depression and chronic fatigue syndrome in adults and neurodegenerative diseases in the elderly. Although clinically diverse, all of these illnesses could be contributed to by infectious agents. If so, one might expect to trace the occurrence and progression of various brain damaging illnesses among various family members. An example of this occurring in three generations of a family is presented in this paper.

Keywords: Stealth adapted virus, Epidemic, family, chronic granulomatous disease, amyotrophic lateral sclerosis, chronic fatigue syndrome, Parkinson's disease, delusional parasitosis, Morgellons disease, ACE pigments.

Family History

A Caucasian family enjoyed upper middle class living in the United States. The wife had a walking impairment, similar to that of her father. It had been diagnosed as a mild case of Charcot-Marie-Tooth disease and had not caused any major disability outside of restricting sports activities. Her father was head of data processing for a bank and lived happily with his wife of 30 years, enjoying strong community and social support. His life changed dramatically when, four years ago, at 65 years of age he experienced a discrete but rather non-specific flu-like illness with fatigue and muscle aching. Rather than resolving, the daughter recalls a gradual but progressive deterioration in his demeanor and

personality. He began to express anger, complain of impaired memory and slept excessively during the day. He would sit idly with a blank stare. The illness forced his retirement. Severe headaches soon began and hydrocephalus developed requiring placement of a shunt. His voice became mono-tonal and his muscles began to atrophy with notable rigidity. Clinical diagnoses have included Alzheimer's disease for which he was prescribed donepezil hydrochloride (Aricept), and Parkinson's-like illness. At least 3 outbreaks of shingles-like eruptions have occurred over the last 4 years. Persisting small ulcerating skin lesions and recurrent mouth ulcerations have also been noted.

The man's wife was forced to deal with family matters. Within several months, however, this task became more difficult. She too began to complain of tingling and weakness in her arms, tinnitus, shortness of breath with occasional panic attacks, and irregular bouts of diarrhea. She had worked as a school teacher with excellent social and communicative skills. These functions changed dramatically as she progressively became more distant with paranoia. She was diagnosed as being hypothyroid and hypertensive, for which medications was prescribed. Radiologically identified lesions in her lung were considered granulomatous without any definitive diagnosis. Thrombocytopenia was also noted. The woman became obese and neglectful of her own and her husband's care.

Because of the deterioration of her parents' health, her daughter arranged for them to move into her home. The move occurred approximately a year after the onset of her father's illness. Plans were made for renovating the house to accommodate her parents. Not long thereafter both the daughter and her husband became ill and the renovations have remained unfinished. The daughter recalls feelings of excessive fatigue, irregular bouts of diarrhea, dull persisting headaches and insomnia. Small ulcerating, acne-like pruritic skin lesions were noted from which she could extract pigmented, irregularly shaped structures. Draining lymph nodes were slightly enlarged and painful. Her hair became more brittle and her face acquired a drained look with dark circles beneath her eyes. A biopsy of a skin lesion was classified as prurigo nodularis (1-2) from neurotic excoriation. Her description of the structures from the lesions was interpreted as delusional parasitosis, also called Ekbom's disease (3-4).

The wife's husband, who is also now aged 38, was a successful marketing manager for a major telephone company. Within several months of his in-laws moving into the house, he experienced an episode of diarrhea with occasional vomiting accompanied by bouts of coughing. As these symptoms resolved, he experienced overwhelming fatigue, insomnia, headaches and unprovoked perspiration. His personality changed dramatically from that of a caring husband and father to an indifferent, neglectful and occasionally angry individual. His wife noted significant dementia aggravated by dysarthria. Neurological examination revealed wasting of small muscles of his hands and generalized hyper-reflexia and increased muscle tone. He exhibited bilateral extensor plantar responses, a positive Hoffmann's sign and mild fasciculations. He was diagnosed as having amyotrophic lateral sclerosis and prescribed Riluzole. He was also placed on Baclofen and Celexa for depression, Accupril for hypertension and Prevacid for indigestion. His wife has noted a peculiar body odor and acne-like lesions. He now walks with difficulty and is unable to rotate door handles. Both shoulders are frozen from lack of movement. His wife's major concern is the apparent indifference he shows to his own care, or to the deteriorating health of their four children.

His eldest son is now 20 and has noted a marked loss of short term memory and diminishing muscle strength with frequent tingling. A daughter of 14 had a distinct mono-like illness with sore throat and fatigue that has not fully resolved. Her school and sporting performance changed from being a gifted student active on the softball team to barely being able to cope with her studies and relegated to a back-up cheerleading squad. Her mother withdrew her from school to try to provide home schooling to make up for the shortcomings in her learning capacity. The daughter is on Prozac for depression. She experiences frequent migraines and has become somewhat obese. Daughters now 11 and 5 years are also experiencing short term memory loss and are unable to attend regular school because of an attention deficit disorder. The 11 year old developed a severe thrombocytopenia, with platelet levels as low as 10,000/ml. Platelets have been maintained between 50-100,000/ml with repeated gamma globulin injections. A 1 cm cyst was identified in the pineal gland and is being managed conservatively. The child has slight eosinophilia. The children's hair has been thinning with a defined bald spot on the child with thrombocytopenia. All three children regularly develop small ulcerating mouth lesions that persist for 1-2 weeks. They express little joy in life, have lost most of their friends and obtain little or no emotional support from their father.

The children's mother has also had to deal with recent onset illnesses among her siblings, several of whom have tried to help with the care of their parents. A now 30-year-old brother moved in with the parents in their own home soon after his father became ill. From being a successful artist he has become socially withdrawn, depressed with outbursts of anger and suicidal thoughts. He has been unable to hold a steady job and now works as a part time bar tender. A sister, now 36, has developed a lupus-like illness with a positive ANA, recurrent miscarriages, aching joints and muscles, shortness of breadth, chronic fatigue, depression and marked weight gain. Her husband has sinusitis and what have called fungal skin lesions. Another brother, now aged 44, and his wife, have had two outbreaks of shingles in the last year. Four years ago, their two children stayed with the grandparents. A baby girl at the time and now aged 4 has been diagnosed as having chronic granulomatous disease, which is extremely rare in girls. The diagnosis was offered to explain skin lesions infected with Chromobacterium violaceum.

A pet dog belonging to the grandparents was euthanized because of protruding skin lesions, unsteady gait and uncontrolled in-house urination. A dog belonging to the daughter became unusually skittish and aggressive shortly after the parents moved in the house and was also euthanized.

Discussion

The above description underscores several of the problems in the current medical system. First, none of the attending physicians has bothered to focus beyond an individual family member. Second, the labeling of the grandfather's illness as idiopathic hydrocephalus and her husband's illness as ALS has seemingly obviated the need to delve further into possible causes. Third, in spite of the daughter's pleas, no one has lent credence to the possibility of an infectious process progressively taking its toll on her husband, herself and now her children. She has repeatedly been put down as being neurotic, with delusional parasitosis. Her task is made all the harder by the "la belle indifference" shown by her husband and by not wanting to emotionally burden her children.

The woman made contact with the Institute of Progressive Medicine because of published findings on stealth-adapted viruses. These viruses are not effectively recognized by the cellular immune system and, therefore, do not provoke an inflammatory response typical of most infectious diseases (5-13). They can be detected by culturing with fibroblasts causing the normal spindle shaped cells to form clusters of foamy vacuolated cells with syncytia (5,13). The clusters commonly acquire pigmented inclusions, similar to complex structures seen in tissue biopsies of stealth virus infected patients (12-13). These mineral containing, electron donating materials accumulate in the tissue culture supernatants of stealth virus cultures and seemingly provide the striking repair process that occurs in infrequently re-fed stealth virus cultures. Cell survival in spite of marked mitochondria disruption, as well as various energy-based studies, supports the concept that these pigmented materials provide an alternative (non-mitochondria) source of cellular energy (12-13). They have accordingly been termed alternative cellular energy pigments (ACE pigments).

Several stealth virus infected patients and animals have been noted to excrete ACE-pigment-like materials in perspiration. As in the stealth virus cultures, conglomerates of fine materials can form visible structures that can assume irregular ribbon and thread shapes and can display varying colors. Patients have reported seeing such structures on bed sheets and in bath water. They can also be seen and felt attached to hair strands and sometimes misidentified as lice. Patients are sometimes led to believe they are parasites because of the tendency to show marked electrostatic attraction and repulsion. The particles are typically auto-fluorescent and contain various minerals when examined using energy dispersive X-ray (EDX) analysis. They have electron donating and electron accepting properties. A patient support group has encouraged the use of the term Morgellon skin disease to refer to similar unusual skin lesions (http://morgellons.org). Belated apologies are probably due to many patients accused of being delusional because of sensing and observing ACE-pigment-like particles within their skin.

Electrostatically active, auto-fluorescent fine particles can occasionally be observed in patients' blood samples. Similar structures were likely misidentified by earlier investigators, such as Antoine Bechamp, Virginia Livingston and Gaston Naessens, as pleomorphic microorganisms, especially when seen in association with ribbon and thread-like materials. ACE-pigment-like threads can also be collected from environmental sources adding to the confusion of their origins and underlying nature.

The performance of diagnostic stealth virus cultures was prohibited by the Federal Government in late 2002 on questionable grounds (a copy of correspondence relating to the inspection process is available at www.s3support.com). Efforts to develop clinical assays for ACE-pigments and for stealth virus infected bacteria were also stymied by Federal Regulations. These actions have delayed progress in addressing an important Public Health issue.

The Nation is facing a growing incidence of chronic illnesses with characteristic neuropsychiatric symptoms. Among these illnesses, are nearly 20,000 cases of encephalitis diagnosed annually within United States hospitals (14). Even when subjected to detailed diagnostic procedures, no etiological cause can be identified in over 60% of patients with encephalitis (15). Autism rates are increasing nationwide and childhood learning and behavioral problems are overwhelming the educational facilities. The concept of stealth-adaptation is not particularly difficult to grasp, and the published data, including findings in animals inoculated with these viruses are compelling (16). The conclusive finding that several of these viruses were derived from African green monkey simian cytomegalovirus (SCMV) has implicated their origin from live polio vaccines (17-18). These vaccines were grown on kidney cells from SCMV contaminated monkeys and virus DNA was present in licensed polio vaccines (19-20). The argument that FDA investigators are unable to culture virus from contaminated vaccines is unconvincing especially given the argument that they cannot release any of the vaccines for independent testing because of proprietary restrictions. FDA has also argued in legal proceedings against a mother wishing

231

to have the polio vaccine tested that her child had received prior to developing a severe and eventually fatal neurological illness. Efforts to test for stealth-adapted viruses in illnesses such as autism and learning disorders in children, psychiatric illnesses in adults, and neurodegenerative diseases in the elderly has been steadfastly avoided by Public Health officials. Arguably, members of the family described in this paper, as well as other individuals with brain damaging illnesses, etc, ought to bring legal suits to compel Public Health testing for infectious agents. Such action may help protect other members of society and will surely lead to the development of more targeted therapy, including the use of natural products with ACE-pigment-like activity.

References

1. Linhardt PW, Walling AD. (1993) Prurigo nodularis. J Fam Pract. 37:495-8.

2. Accioly-Filho LW, Nogueira A, Ramos-e-Silva M.(2000) Prurigo nodularis of Hyde: an update. J Eur Acad Dermatol Venereol. 14:75-82.

3. Wykoff RF. (1987) Delusions of parasitosis: a review. Rev Infect Dis. 9:433-7.

4. Ait-Ameur A, Bern P, Firoloni MP, Menecier P. (2000) Delusional parasitosis or Ekbom's syndrome. Rev Med Interne. 21:182-6.

5. Martin WJ, Zeng LC, Ahmed K, Roy M. (1994) Cytomegalovirus-related sequences in an atypical cytopathic virus repeatedly isolated from a patient with the chronic fatigue syndrome. Am. J. Path. 145: 441-452.

6. Martin WJ. (1995) Stealth virus isolated from an autistic child. J. Aut. Dev. Dis. 25:223-224.

7. Martin WJ.(1996) Simian cytomegalovirus-related stealth virus isolated from the cerebrospinal fluid of a patient with bipolar psychosis and acute encephalopathy. Pathobiology 64:64-66.

8. Martin WJ. (1996) Severe stealth virus encephalopathy following chronic fatigue syndrome-like illness: Clinical and histopathological features. Pathobiology 64:1-8.

9. Martin WJ. (1996) Stealth viral encephalopathy: Report of a fatal case complicated by cerebral vasculitis. Pathobiology 64:59-63.

10. Martin WJ, Anderson D. (1997) Stealth virus epidemic in the Mohave Valley . Initial report of viral isolation. Pathobiology 65:51-56.

11. Martin WJ, Anderson D. (1999) Stealth Virus Epidemic in the Mohave Valley : Severe vacuolating encephalopathy in a child presenting with a behavioral disorder. Exp Mol Pathol. 66:19-30.

12. Martin WJ. (2003) Complex intracellular inclusions in the brain of a child with a stealth virus encephalopathy. Exp Mol Path 74: 179-209.

13. Martin WJ. (2003) Stealth virus culture pigments: A potential source of cellular energy. Exp. Mol. Path. 74: 210-223, 2003.

14. Khetsuriani N, Holman RC, Anderson LJ. (2002) Burden of encephalitis-associated hospitalizations in the United States, 1988-1997. Clin Infect Dis. 35: 175-82.

15. Glaser CA, Gilliam S, Schnurr D, Forghani B, Honarmand S, Khetsuriani N, Fischer M, Cossen CK, Anderson LJ. (2003) California Encephalitis Project, 1998-2000. In search of encephalitis etiologies: diagnostic challenges in the California Encephalitis Project, 1998-2000. Clin Infect Dis. 36:731-42.

16. Martin WJ, Glass RT. (1995) Acute encephalopathy induced in cats with a stealth virus isolated from a patient with chronic fatigue syndrome. Pathobiology 63: 115-118.

17. Martin WJ, Ahmed KN, Zeng LC, Olsen J-C, Seward JG, Seehrai JS. (1995) African green monkey origin of the atypical cytopathic 'stealth virus' isolated from a patient with chronic fatigue syndrome. Clin. Diag. Virol. 4: 93-103.

18. Martin WJ. (1999) Stealth adaptation of an African green monkey simian cytomegalovirus. Exp Mol Path. 66:3-7.

19. Sierra-Honigmann AM, Krause PR. (2002) Live oral poliovirus vaccines and simian cytomegalovirus. Biologicals. 30:167-74.

20. Baylis SA, Shah N, Jenkins A, Berry NJ, Minor PD. (2003) Simian cytomegalovirus and contamination of oral poliovirus vaccines. Biologicals. 31:63-73.

Summary of Research provided in Press Release (9-24-2012)

W. John Martin, MD, PhD. Summary of Research on Stealth Adapted Viruses in Chronic Fatigue Syndrome (CFS) Patients and the Potential for Therapy of These Viruses via Activation of the ACE Pathway

As widely anticipated, an NIH sponsored multicenter study failed to identify XMRV in CFS patients. It is fitting; therefore, that I restate the politically sensitive, yet compelling evidence for stealth adapted viruses as the primary cause of CFS.

SOUTH PASADENA, CA, September 24, 2012 /24-7PressRelease/ -- I began using the polymerase chain reaction (PCR) on blood samples from CFS patients in the late 1980's and reported low-level positive results in approximately a third of the tested samples. A 1990 brain biopsy from a patient, whose illness began as CFS, was also PCR positive. Yet microscopic examination of her brain tissue showed no inflammation; the accepted hallmark of an active virus infection. This disparity implied that the brain-infecting virus, which was presumably responsible for the positive PCR, was not capable of activating the cellular immune defense mechanism. Cellular immunity typically targets relatively few of the components coded for by the virus genome. It is possible, therefore, for a virus to lose or mutate small portions of its entire genome as a means of evading effective immune recognition. I termed this immune evasion mechanism "stealth adaptation."

Renewed efforts at culturing cell-damaging (cytopathic) viruses from CFS patients and from patients with more severe psychiatric and neurological illnesses, yielded consistent and unmistakable positive results. Cell damage was primarily seen as the formation of foamy vacuolated cells, which tended to fuse into small clusters. The cultures were shown to other CFS researchers, including Dr. Paul Cheney, who was being supported by the CFIDS Association of America. The Centers for Disease Control and Prevention (CDC) was also informed of the results. In 1991, an aliquot of a clearly positive culture derived from the cerebrospinal fluid (CSF) of a comatose patient with a history of bi-polar psychosis, was provided to the Los Angeles County Department of Health, which subsequently sent the culture to the California State Health Department.

The early microscopic appearances of the cultures were somewhat suggestive of foamy viruses. These are retroviruses, which have incorporated an additional gene, which by itself, is responsible for the foamy cell appearance of infected cells. Actual sequence data obtained on two of the early isolates, however, indicated a more probable origin from cytomegalovirus (CMV), a type of herpes virus.

Definitive DNA sequence data were obtained in 1995 on a virus, repeatedly cultured from the blood and CSF of a CFS patient. The data showed an unequivocal origin from African green monkey simian cytomegalovirus (SCMV). So too did the 1991 virus isolate provided to the County Department of Health. Some other cultured viruses appeared to be related to SCMV, while others were more likely derived from human herpes and human adenoviruses. Stealth adaptation was, therefore, viewed as a generic process, which could potentially occur with all viruses.

The SCMV origin of several stealth adapted viruses clearly implicated probable contamination from African green monkeys, which were still routinely being used to produce live poliovirus vaccines. In June 1995, I conveyed this important information to the Food and Drug Administration (FDA), CDC, Los Angeles County Health Department, the polio vaccine manufacturer and officials at the University of Southern California, where I was working as a tenured professor of pathology.

In 2002, I proceeded with an approved study with the Blood Bank at the University of California, Irvine. Consistent with other surveys, approximately 10% of the donated blood samples tested positive by culture. This left the CDC with the dilemma of having to accept the results or fabricate deficiency with the testing methodology. Personal from the California Department of Health acting as inspectors for the Centers for Medicare and Medicaid (CMS) and, by their own admission, working in conjunction with CDC, deemed that testing for stealth adapted viruses had placed the Nation's health in "Immediate Jeopardy" and that all further clinical testing or even use of stored blood samples was prohibited. Detailed copies of all procedures were taken by the State Health Department along with my request that they please undertake their own testing.

A disgruntled patient and vindictive CFS patient support group played into the hands of the CDC by making false and malicious assertions regarding stealth adapted viruses. Unfortunately, the assertions are easily viewed on the Internet and have discouraged many from either supporting or learning from my research.

With limited resources, the research focus moved to understanding an alternative cellular energy (ACE) pathway. This pathway provides a non-immunological healing mechanism, which is able to suppress and even reverse the cellular damage caused by stealth adapted and other viruses. The ACE pathway can be easily self-monitored using a fluorescence screening method and can be enhanced with various simple interventions, including lifestyle changes. Community based implementation and evaluation of methods of enhancing the ACE pathway in the prevention and suppression of illnesses caused by stealth adapted and other viruses should become major public health goals.

Federal health authorities have been willing to spend millions of dollars in pursuit of obviously flawed prior studies on the possible role of mouse retroviruses in CFS patients. They have been far less willing to publicly acknowledge research on the existence of stealth adapted viruses. This reluctance to do so is explained in part by the unequivocal origin of some of these viruses from the monkeys used to provide live poliovirus vaccines. Therapy for stealth adapted viruses, based on activation of an alternative cellular energy (ACE) pathway, appears promising and warrants further evaluation.

Cause, Prevention and Treatment of Autism and Related Disorders

By W John Martin MD, PhD.
Institute of Progressive Medicine

Definition: Autism is an illness defined by symptoms recognizable in infants prior to three years of age, which indicate severe difficulties with social interactions. Autism defining symptoms comprise impaired verbal and non-verbal communications, accompanied by socially inappropriate mannerisms. Affected children commonly display other abnormalities, including epileptic seizures; repetitive movements, such as hand flapping; emotional intolerance to unexpected changes in routine activities; sleeping problems; allergies; and gastrointestinal disorders. There is a wide spectrum of disease severity and children also differ in the age at which symptoms are initially recognized. Significant improvements occur over time in some children, such that their earlier diagnosis of autism ceases to be warranted.

Primary Cause: Brain damage due to stealth adapted virus infection acquired during pregnancy. Stealth adapted viruses are not readily recognized by the immune system and, therefore, do not evoke an inflammatory reaction. The best studied of these viruses arose from African green monkey simian cytomegalovirus (SCMV), a type of herpes virus.

Potentially, any virus can become stealth adapted by losing or mutating the relatively few genes of most viruses, which actually code for antigens normally targeted by the cellular immune system. Infections with stealth adapted viruses also explain many mental illnesses and such common conditions as the chronic fatigue syndrome (CFS).

Dynamics: Various factors contribute to the wide variability seen among children diagnosed as having an autism spectrum disorder. These include: i) Severity and timing of infection of the developing embryo during pregnancy, as will be reflected by the extent of brain damage present at birth. ii) Whether the virus infection progresses or regresses during infancy. iii) The amount of social training the child receives in preparation for the challenge of initiating and maintaining socialization. These variables can account for why some infants never learn to engage in personal interactions, while other infants regress from previously gained social achievements. Marked regressive autism has been observed in some infants following vaccination and this may be attributed to activation of the underlying stealth adapted virus infection or to the triggering of a low-level anti-stealth adapted virus inflammatory response. Even if autism is avoided, the persistently infected child remains at risk for other stealth adapted virus associated illnesses, including learning and behavioral disorders, impaired social attachments and acute psychotic episodes. Life long infection may not become clinically manifest until adult life, with illnesses such as CFS, depression, anxiety disorders and neurodegenerative diseases.

Perspective of the Child with Autism: It is undoubtedly very difficult and confusing for an autistic child to understand his or her predicament. The child likely suffers from an impaired sense of personal identity. Difficulties seemingly exist in recognizing and responding to others as comparable, yet distinguishable individuals. In spite of struggling efforts, the child is unable to readily express thoughts and feelings, especially by using spoken language. There is very limited recall of emotionally driven learning experiences, which would be easily remembered by normal children. This limitation leads to apparent confusion and social errors (e.g. laughing in response to another child's crying). Autistic children can seemingly experience some gratification from repetitive actions and from following predictable routines. The affected children are also likely to have many related symptoms, including headaches, muscle pains, seizures, delusions, non-restorative sleep, gut-related problems, hypersensitivity to sensory stimulations, impaired autonomic neural responses, etc.

Prevention of Autism: A social obligation is to inform women of childbearing age with overt signs and symptoms indicative of an active stealth adapted virus infection that they are at an increased risk for having an autistic child. Prevention of autism can, thereby, occur by the decision of such women to refrain from becoming pregnant. Treatment guidelines, discussed below, can be crafted for adults with illnesses, such as CFS, and pregnancy delayed till recovery is achieved. Therapeutic support is similarly indicated if pregnancy is already underway in symptomatic women, as it should probably be for all pregnant women. Following birth, the focus on prevention extends to the infant. The two major goals are: i) Suppress virus activity and ii) expand capacity for interpersonal relationships. The former relates to treatment guidelines designed to enhance the alternative cellular energy (ACE) pathway, while the later entails determined efforts at developing the child's brain through effective mother-infant engagement.

Alternative Cellular Energy (ACE) Pathway: A Mechanism for Suppressing Stealth Adapted Viruses. Culturing of stealth adapted viruses led to identification of the ACE pathway. Essentially, the ACE pathway involves a capturing and utilization of physical energies to effectively reverse the cell damaging, energy-draining effects caused by viruses. The energy transfer involves mineral containing complex organic macromolecules termed ACE pigments, which can be likened to miniature batteries. ACE pigments can be sampled from saliva, urine and dried perspiration and their energy status assessed by testing for ultraviolet (UV) light inducible fluorescence in the presence and absence of neutral red dye. Fluorescence occurs when the dye interacts with uncharged ACE pigments, whereas partially charged pigments will even directly fluoresce with UV illumination. The preferred situation is when no fluorescence is seen and is taken as a presumptive indicator of an adequately charged ACE pathway. Guided by monitoring of the ACE pathway, parents can institute various simple approaches to ensure adequacy of their child's ACE pathway. These approaches include regular consuming of enerceutical™ foods; drinking ACE Water™; avoidance of toxic, energy-inhibiting chemicals; elimination of emotional stressors; and reinforcement of joyful playtime. This latter factor is based on growing evidence that an individual's joyful mindset may allow the body to become a direct receiver of ACE pathway enhancing environmental energies. Conversely, stress inducing fear and hostility can undermine this capacity. If needed, a variety of more direct methods can be undertaken to enhance the ACE pathway. An ACE Phototherapy method based on UV illumination of a solution of activated neutral red dye has the most supportive data, not only in autism, but also in the therapy of more conventional virus infections, such as herpes simplex virus (HSV) and human papillomavirus (HPV).

ACE Phototherapy: In this procedure, a plastic bag containing an energized solution plus neutral red dye is placed on the soles of the child and illuminated using a 13 watt UV light for 30 minutes. Successful activation of the ACE pathway is shown by the appearance of direct UV inducible fluorescence in various areas of the skin and/or either the fresh development of UV intraoral fluorescence or significant enhancement of preexisting intraoral fluorescence.

Examples: An autistic teenage girl received the above therapy by her mother. Soon, thereafter, the girl was noted to be looking intently at herself in a mirror. It seemed as if, she had previously not been fully aware of herself. The mother further observed that her daughter became better able to cope with situations, which would have ordinarily caused emotional distress. Upon inquiry the child said she had simply recalled her mother's prior advice and had no reason to be upset. More exciting for the mother was that her daughter now enjoyed joking with her, demonstrating the capacity for empathy and interpersonal communication. A similar neutral red dye light therapy protocol was also strikingly effective in suppressing epileptic seizures in a 4-year-old autistic child. From essentially one hospital admission for epilepsy per month over the prior 6 months, the child became seizure-free even with the subsequent discontinuation of anti-seizure medication. Many other autistic children improved using the protocol, but not when the solution being used lost its ability to effectively activate the neutral red dye.

Social Education in the Prevention and Therapy of Autism: Preschool educators are beginning to distinguish between a child's analytic capacity of recalling simple facts and the more comprehensive social mindfulness of being able to appreciate one's individuality and yet connectivity with others. Various approaches at further developing this latter talent have come to be realized in advanced training classes of so-called "gifted" children. As these teaching programs are becoming more refined, they offer important clues to better educating children either at risk for or already diagnosed with autism. Examples of beneficial training include: improvised playacting and role reversals, e.g., the child becoming the teacher; integrating music with language; self-drawing and photography; etc. As the symptoms of autism subside, major emphasis needs also to be placed on providing "catch-up" analytic educational input to cover the period during which the child was not learning. Unless unable to do so, the fulltime teaching role should fall to the parents with government provided direct reimbursement, comparable to the money currently being provided to more questionable governmental and commercial endeavors.

Barriers to Progress: A major barrier is the reluctance of public health officials to acknowledge the existence of stealth adapted viruses. This is partially due to lack of innovative thinking within the scientific community but can also be attributed to some officials actively antagonistic to disclosing prior vaccine errors or acknowledging susceptibility of autism-prone children to vaccination. Other barriers are from health practitioners specializing in autism, but with no special expertise in virology; and from pharmaceutical companies focused on identifiable targets for specific drug therapies. There is also a lack of enthusiasm for an infectious cause of autism among the leadership in autism support community. First, it attributes a role of the mother in transmitting an illness and possibly remaining somewhat impaired in her current capacities. It also raises the prospect that children with autism might be shunned as being potentially a source of infection to others. It is far more attractive to focus blame on current vaccine manufacturers with the prospect of large financial settlements. Another barrier is simply the logistics of conveying useful information and having it translated into Food and Drug Administration (FDA) approved clinical trials and subsequent publication and acceptance of positive findings.

Ten-Point Autism Prevention and Therapy Program: 1) Culture of stealth adapted viruses from children with autism and from infants presumptively at risk for becoming autistic. 2) Animal inoculations to better define the in vivo cellular pathology and modes of transmission of illness caused by stealth adapted viruses. 3) Characterize the production, composition and mode of action of ACE pigments generated in virus cultures, patients and inoculated animals. 4) Analyze the respective roles of the ACE pathway and the mitochondria oxidative metabolic pathway in various cellular functions, including biosynthesis, proliferation, differentiation and longevity. 5) Institute an available screening program for assessing the ACE pathway in humans, with a special emphasis on its use in pregnant women and infants. 6) Evaluate various corrective actions to be taken when a deficiency in the ACE pathway is identified; including assessing the benefits of regularly consuming enerceutical™ foods and ACE Water™, as enhancers of the ACE pathway and, if necessary, employing the phototherapy method as previously described. 7) Conversely, identify possible adverse effects of environmental and food toxins on the ACE pathway. 8) Explore the role of brain activity as a stimulus and as an inhibitor of the ACE pathway. 9) Create educational programs geared towards children with autism and related disorders; based on improving self-awareness and empathy and on achieving better integration between analytic and emotional knowledge. 10) Extend studies on the ACE pathway to other infectious diseases, wound healing, ageing and in the therapy of illnesses due to impaired metabolism resulting from deficiencies in the supply of oxygen and/or other metabolites to cells.

Public Support: A major responsibility of adults is to care for the health and welfare of the coming generations. Between 1-2% of children are autistic with 20% of all children having a diagnosable mental illness. The problem is severe and its correction is urgent. I can assure the readers that the preceding listed endeavors offer a real opportunity to see a major decline in the incidence of autism and related disorders. Financial support along with collaborative scientific efforts can greatly facilitate progress. Interested contributors and collaborators should contact the author at the Institute of Progressive Medicine. This is the major component of MI Hope Inc., a non-profit public charity founded in 1988. The author can be reached at 626-616-2868 or by email to wjohnmartin@hotmail.com

Autism, Vaccines, Stealth Adapted Viruses and Mitochondria Dysfunction

W. John Martin, MD, PhD. Institute of Progressive Medicine
Burbank CA 91502

Parents of autistic children are understandably seeking an explanation of their child's illness. Many have been encouraged to blame the government sponsored vaccination programs. Initially, the major focus was on the impurity of the pertussis (whooping cough) vaccine and its alum additive.[1] The next issue was live measles virus included in the measles-mumps-rubella (MMR) vaccine combination.[2] More recently, the emphasis has shifted to killed vaccines and especially to the widely used mercury containing thimerosal preservative.[3] In addition to concerns regarding the actual constituents in vaccines, an overriding issue has been the increasing number of vaccines being given to infants, some of which may be less essential than commonly argued.[4] It has not been difficult for various lay spokespersons to gather followings of supporters of the vaccine-induced-autism hypothesis and to preach about the dangers of vaccination. Attributing autism to vaccines also carries a potential financial benefit in terms of possible liability settlements within the government's Vaccine Injury Compensation Program (VICP).[5]

The initial legal testing of this hypothesis was disappointing to many observers given the poor quality of the data presented on behalf of a carefully selected family.[6] Recently, however, the government did not contest a claim that vaccination had precipitated autism in a child with genetically defective mitochondria.[7] Although the actual reasoning behind this decision has yet to be publicly released, the presumption appears to have been that vaccination induced fever placed an intolerable added energy requirement on the child, leading to potential progression of the underlying disorder. Government spokespersons have emphasized that a similar outcome would have likely occurred had the child developed one of the illnesses for which the vaccine was intended to prevent. Moreover, the spokespersons have stressed the relative infrequency of the type of inherited mitochondria disorder that had affected the child (estimated 1:4000) and that childhood vaccinations are saving several thousand lives annually.

Anti-vaccine proponents have been quick to question how frequently autistic children may have defective mitochondria. As with many other chronic diseases, it is relatively easy to document metabolic imbalances in autistic patients that are attributable, at least in part, to mitochondria dysfunction.[8-10] Some reports have placed this figure as high as 20%, especially if markers such as elevated blood lactate levels are equated with impaired mitochondria. Mitochondria defects are potentially genetically inherited from either parent or may be acquired. Again, the anti-vaccine proponents have argued for that vaccines are potentially damaging even to normal mitochondria because of chemical contaminants, including mercury. In the public relations arena, the government is clearly being outmaneuvered by the anti-vaccine advocates. The rules of engagements appear to include the legitimacy of using passionate celebrities, articulate lay authors and founders of the rapidly emerging non-profit groups in the anti-vaccine camp. These spokespersons can avoid scientific accountability for their statements, yet expect full disclosure by the government. It is unlikely that full disclosure will ever be provided because of the proprietary restrictions imposed on the government by the pharmaceutical industry. Instead, somewhat hapless pediatricians or representatives of major clinical organizations are being asked to defend vaccination programs with which they are not fully knowledgeable. The real tragedy of this divisive contest is that it directs efforts away from addressing the issue of what can be done now for autistic children. There is also an urgent need to prevent autism from occurring among the increasing number of susceptible infants. There is little excuse for the anti-vaccine proponents to dismiss critical data simply because they do not fit into a political or economic driven agenda. Similarly, the government owes their constituency a full accounting of largely suppressed information concerning vaccines. Toping this agenda is much more open discourse on African green monkey simian cytomegalovirus (SCMV) contamination of previously licensed polio vaccines (discussed below).

There is evidence that the seeds of autism are sown before birth. Dr. Leo Kanner reported in his sentinel publication of 1943 on the enlarged head size of some of his patients.[11] This observation has been substantiated in studies showing that although relatively smaller than normal at birth, the circumference of the head of most children who subsequently become autistic increases at an accelerated rate over the first year of life.[12-13] Neuropeptide levels have been measured in stored cord blood samples of children who were subsequently diagnosed as autistic. The levels differed significantly from levels in children who developed normally.[14-15] Serum abnormalities, including the presence of brain reactive auto-antibodies, are present in mothers of autistic children, as well as the affected children and some of their siblings.[16-19] Many of these mothers have prior and/or ongoing evidence of less than optimal health with fatigue being among the more prevalent symptoms.[20] It is difficult, therefore, to avoid the inference of mother to newborn transmission of a pervasive disease inducing factor that can predispose the developing child to autism.

In the face of an epidemic, the first issue ought to be to search for a potential infectious agent. A number of studies have shown that congenital infection with cytomegalovirus (CMV) can result in autism.[21-25] It is inexcusable that the Government dismissed early reports of the existence of atypical viruses that failed to provoke an inflammatory response.[20, 26-32] This decision was undoubtedly prompted by politically damaging data that the first two of such virus isolates were unequivocally derived from SCMV.[26-27] One isolate was cultured from a patient with chronic fatigue syndrome and the other from a patient with a bi-polar psychosis. Additional culture based evidence, along with polymerase chain reaction (PCR) supportive data, argued for a generic process whereby various cell damaging (cytopathic) viruses could avoid immune elimination by simply deleting or mutating the few critical viral genes normally involved in direct cellular immune recognition.[32] In the case of human CMV, only 3 of over 150 virus components are in aggregate targeted by more than 90% of the anti-CMV cytotoxic T cells. The generic process of avoiding cellular immune recognition by the deletion or mutation of critical virus antigenic components was coined stealth adaptation. In can potentially occur with all types of viruses and has definitely occurred with SCMV.[27,32] The African green monkeys whose kidneys were used to provide cells for producing polio virus vaccines were known to be infected with SCMV. Moreover, a 1972 joint

Food and Drug Administration (FDA) / Industry study showed SCMV contamination in all eleven kidney cultures derived from infected monkeys. SCMV DNA was detected in three of eight polio vaccine lots subsequently released in the mid 1970's.[33] FDA researchers have reported on their failure to culture virus from these contaminated vaccines but were unwilling to disclose details of culturing or to provide samples for independent analysis.

Numerous examples, some of which were published and otherwise reported provided culture evidence for the presence of stealth adapted viruses in blood and, when tested, cerebrospinal fluids, of autistic children.[34] Logical explanations have been provided as to how immune stimulation caused by vaccination could potentially trigger an immune response to some residual minor antigens in the embedded stealth adapted viruses, or how a live vaccine virus could directly stimulate the replication or cytopathic effect (CPE) of a stealth adapted virus. Similarly, environmental toxins, such as pesticides or even mercury could cause additive damage to that being caused by the stealth adapted viruses, allowing the underlying illness to become clinically manifest. Normal brain functioning may simply become overwhelmed by the natural demands of speech and interpersonal socializing during the child's second year. An underlying virus infection is consistent with the co-morbidity issues commonly present in children with autism, including prematurity,[35] epilepsy,[36] psychosis,[37-38] gastrointestinal dysfunction[2] and other illnesses.

A beneficial outcome of research on stealth adapted viruses has been the identification of a virus defense mechanism that is distinct from that provided by the cellular immune system. It is powered by what has been termed the alternative cellular energy (ACE) pathway.[39-42] This pathway became apparent during the culturing of stealth adapted viruses. Clearly positive cultures showing a marked CPE, would regularly undergo a repair process if the cells were simply maintained in their culture medium. Replacement of the culture medium with fresh medium led to the rapid reappearance of the CPE. The reactivation process was inhibited if particulate matter that accumulates in cultures undergoing repair was added to the fresh replacement medium. These particulate materials were electrostatic, autofluorescent, commonly pigmented and occasionally magnetic. They could act as electron donors and could also self assemble into long ribbon and thread-like structures.[39] Similar complex structures were seen within cells in brain biopsies of stealth adapted virus infected patients.[40] These cells commonly displayed marked disruption of their mitochondria. The proposal was posited that these structures represented a non-mitochondria source of cellular energy and were accordingly called alternative cellular energy (ACE) pigments.

Analysis of ACE pigments indicates a lattice-like structure in which the predominant organic components are simple, yet diverse, aromatic compounds and which contain rather restricted groupings of a wide range of mineral elements. Organically bound minerals are a mainstay of products such as chlorophyll and it is reasonable to suppose that ACE pigments are representative of a natural biological energy source. Indeed, using the repair of stealth adapted virus cultures as the testing method, several natural compounds were shown capable of providing cellular energy. Some of these, such as humic and fulvic acids and terpenes (essential oils) have known plant growth promoting activities. Others had found use in some clinical studies in humans. The term "enerceutical" was introduced to define cellular energy delivering products that could i) enhance the vitality of plants, animals and humans; ii) ameliorate a wide of illnesses through a non-disease specific process and even promote better overall performance in some individuals without an overt illness; and iii) work by creating an energy field rather than necessarily having to be actually present within the affected cells. Enerceuticals are, therefore, clearly distinguishable from pharmaceuticals.

Materials with ACE pigment (enerceutical) activity can be retrieved from the skin, urine and saliva of patients suspected of having stealth adapted virus infections, including patients with autism. They can also be identified within active lesions and in the surrounding skin areas of patients infected with herpes simplex virus (HSV) and herpes zoster virus (HZV) infections. In other studies, herpes virus skin lesions were shown to fluoresce under ultraviolet-A light in the presence of the dye neutral red.[43] Various enerceutical products were formulated and examined for inducible fluorescence. One such product was found to be particularly effective in the therapy of HSV skin lesions.

Following Institutional Review Board (IRB) approval, this product was dye activated and placed on the outer surface of impermeable surgical towels, which were laid onto skin areas of autistic patients and illuminated with ultraviolet-A lighting. Discernable skin fluorescence which progressively faded was seen during the 30-60 minutes exposure. In some patients, the procedure was repeated the next day. In all 14 patients so far treated, there have been rather remarkable clinical improvements that have persisted and have actually progressed since the therapy. Three of the subjects were inappropriately urinating during the day requiring change of clothing. These events have not occurred over the 1-3 months since therapy.

Two adolescents have begun employment for the first time in their lives. Reading skills have improved as well as obvious demonstrations of emotional joy at being able to demonstrate this capacity to the evaluating clinical investigator. Childcare is easier for mothers with one mentioning how nice it now is to have her previously autistic daughter help with household chores. Another mother commented upon how her child no longer hesitated before responding to a question or comment. Mothers also spoke of their children being less restless, able to ride calmly in car seats or to watch and follow attentively long television programs. Plans are underway to more fully document these findings and to further optimize the procedure. Once confirmed, widespread application of this simple therapy should become the unifying and highest priority of autism support groups worldwide. For more information on the current investigational trial please contact the Institute of Progressive Medicine at s3support@email.com

References

1. Coulter HL, Fisher BL. "A Shot in the Dark: Why the P in the DPT Vaccination May be Hazardous to your Child's Health." Avery 1991. ISBN:089529463X First published 1985 by Harcourt Brace Jovanovich.

2. Wakefield AJ, Murch SH, Linnell J, Casson DM, Malik M, Berelowitz M, Dhillon AP, Thomson MA, Harvey P, Valentine A, Davies SE, Walker-Smith JA, Anthony A. Ileal-lymphoid-nodular hyperplasia, non-specific colitis, and pervasive developmental disorder in children. Lancet 1998; 351:637-41.

3. Kirby D. Evidence of Harm: Mercury in Vaccines And the Autism Epidemic: A Medical Controversy. Macmillan 2006 ISBN:0312326459.

4. Miller NZ. Vaccines, Autism and Childhood Disorders: Crucial Data That Could Save Your Child's Life. Midpoint Trade Group Inc. 2003 ISBN: 9781881217329.

5. http://www.hrsa.gov/vaccinecompensation/

6. http://rationalwiki.com/wiki/Autism_omnibus_trial

7. http://www.time.com/time/health/article/0,8599,1721109,00.html

8. Poling JS, Frye RE, Shoffner J, Zimmerman AW. Developmental Regression and Mitochondrial Dysfunction in a Child With Autism. J Child Neurol. 2006; 21: 170-172.

9. Oliveira G, Diogo L, Grazina M, Garcia P, Ataíde A, Marques C, Miguel T, Borges L, Vicente AM, Oliveira CR. Mitochondrial dysfunction in autism spectrum disorders: a population-based study. Dev Med Child Neurol. 2005; 47:185-9.

10. García-Peñas JJ. Autism, epilepsy and mitochondrial disease: points of contact. Rev Neurol. 2008; 46 Suppl 1:S79-85.

11. Kanner L. "Autistic disturbances of affective contact". Nerv Child 1943; 2: 217–50.

12. Courchesne E, Carper R, Akshoomoff N. Evidence of brain overgrowth in the first year of life in autism. JAMA. 2003; 290: 337-44.

13. Redcay E, Courchesne E. When is the brain enlarged in autism? A meta-analysis of all brain size reports. Biol Psychiatry. 2005 ;58:1-9.

14. Nelson PG, Kuddo T, Song EY, Dambrosia JM, Kohler S, Satyanarayana G, Vandunk C, Grether JK, Nelson KB. Selected neurotrophins, neuropeptides, and cytokines: developmental trajectory and concentrations in neonatal blood of children with autism or Down syndrome. Int J Dev Neurosci. 2006; 24:73-80.

15. Nelson KB, Grether JK, Croen LA, Dambrosia JM, Dickens BF, Jelliffe LL, Hansen RL, Phillips TM. Neuropeptides and neurotrophins in neonatal blood of children with autism or mental retardation. Ann Neurol. 2001;49: 597-606.

16. Singer HS, Morris CM, Gause CD, Gillin PK, Crawford S, Zimmerman AW. Antibodies against fetal brain in sera of mothers with autistic children J Neuroimmunol. 2008; 194:165-72.

17. Braunschweig D, Ashwood P, Krakowiak P, Hertz-Picciotto I, Hansen R, Croen LA, Pessah IN, Van de Water J. Autism: Maternally derived antibodies specific for fetal brain proteins. Neurotoxicology. 2008; 29: 226-231.

18. Zimmerman AW, Connors SL, Matteson KJ, Lee LC, Singer HS, Castaneda JA, Pearce DA. Maternal antibrain antibodies in autism Brain Behav Immun. 2007; 21: 351-7.

19. Singer HS, Morris CM, Williams PN, Yoon DY, Hong JJ, Zimmerman AW. Antibrain antibodies in children with autism and their unaffected siblings. J Neuroimmunol. 2006;178:149-55.

20. Martin WJ. Stealth viruses as neuropathogens. CAP Today. 1994; 10: 67-70. and personal observations.

21. Ivarsson, S.A., Bjerre, I,, Vegfors, P, Ahlfors K. Autism as one of several disabilities in two children with congenital cytomegalovirus infection. Neuropediatrics. 1990; 21:102-3.

22. Markowitz PI. Autism in a child with congenital cytomegalovirus infection. J Autis m Dev Disord. 1983; 13: 249-53.

23. Stubbs EG, Ash E, Williams CP. Autism and congenital cytomegalovirus. J Autism Dev Disord. 1984; 14: 183-9.

24. Sweeten TL, Posey DJ, McDougle, CJ. Brief report: autistic disorder in three children with cytomegalovirus infection. J Autism Dev Disord 2004; 34:583-6.

25. Yamashita, Y, Fujimoto C, Nakajima E, Isagai T, Matsuishi T. Possible association between congenital cytomegalovirus infection and autistic disorder. J Autism Dev Disord. 2003; 33:455-9.

26. Martin WJ, Zeng LC, Ahmed K, Roy M.. Cytomegalovirus-related sequence in an atypical cytopathic virus repeatedly isolated from a patient with chronic fatigue syndrome. Am J Pathol. 1994; 145:440-51.

27. Martin WJ, Ahmed KN, Zeng LC, Olsen JC, Seward JG, Seehrai JS African green monkey origin of the atypical cytopathic 'stealth virus' isolated from a patient with chronic fatigue syndrome. Clin Diagn Virol. 1995; 4: 93-103.

28. Martin, WJ Severe stealth virus encephalopathy following chronic-fatigue-syndrome-like illness: clinical and histopathological features. Pathobiology. 1996; 64:1-8.

29. Martin, WJ. Simian cytomegalovirus-related stealth virus isolated from the cerebrospinal fluid of a patient with bipolar psychosis and acute encephalopathy. Pathobiology. 1996; 64:64-6.

30. Martin, WJ. Glass RT. Acute encephalopathy induced in cats with a stealth virus isolated from a patient with chronic fatigue syndrome. Pathobiology. 1995; 63:115-8.

31. Martin WJ, Anderson D. Stealth virus epidemic in the Mohave Valley: severe vacuolating encephalopathy in a child presenting with a behavioral disorder. Exp Mol Pathol. 1999; 66:19-30.

32. Martin WJ. Stealth adaptation of an African green monkey simian cytomegalovirus. Exp. Mol. Pathol. 1999; 66, 3–7.

33. Sierra-Honigmann AM, Krause PR. Live oral poliovirus vaccines and simian cytomegalovirus. Biologicals. 2002; 30:167-74.

34. Martin WJ. Stealth virus isolated from an autistic child. J Autism Dev Disord. 1995; 25:223-4.

35. Limperopoulos C, Bassan H, Sullivan NR, Soul JS, Robertson RL Jr, Moore M, Ringer SA, Volpe JJ, du Plessis AJ. Positive screening for autism in ex-preterm infants: prevalence and risk factors. Pediatrics. 2008; 121:758-65.

36. Canitano R. Epilepsy in autism spectrum disorders Eur Child Adolesc Psychiatry. 2007; 16:61-6.

37. Matson JL, Nebel-Schwalm MS. Comorbid psychopathology with autism spectrum disorder in children: an overview. Res Dev Disabil. 2007; 28:341-52.

38. Leyfer OT, Folstein SE, Bacalman S, Davis NO, Dinh E, Morgan J, Tager-Flusberg H, Lainhart JE. Comorbid psychiatric disorders in children with autism: interview development and rates of disorders. J Autism Dev Disord. 2006; 36:849-61.

39. Martin, WJ. Stealth virus culture pigments: a potential source of cellular energy. Exp Mol Pathol. 2003; 74:210-23.

40. Martin WJ. Complex intracellular inclusions in the brain of a child with a stealth virus encephalopathy. Exp Mol Pathol. 2003; 74:197-209.

41. Martin WJ. Progressive medicine. . Exp Mol Pathol. 2005; 78:218-20.Martin, W.J. Etheric biology. Exp Mol Pathol. 2005; 78:221-7.

42. Martin WJ, Stoneburner J. Symptomatic relief of herpetic skin lesions utilizing an energy-based approach to healing. Exp Mol Pathol. 2005; 78:131-4.

A Reinterpretation of the Image of "Refrigerator Mothers" of Autistic Children

Descriptions of emotional difficulties in mothers of autistic children are consistent with mother-to-child transmission of stealth adapted viruses, some of which are vaccine-derived. An epidemic of these viruses can explain the increasing incidence of autism and learning disorders in children.

/24-7PressRelease/ - BURBANK, CA, March 15, 2007 - Dr. Leo Kanner, who first described autism as a distinctive childhood illness, commented on the apparent paucity of affection shown by parents towards their autistic child. While he clearly believed that autism was an "inborn disturbance of affective contact," there was an implication that the parents were "short of interest in other people (and) had little more feeling for their own children." He once represented the cold and indifferent attitude as if certain mothers were "just happening to defrost enough to produce a child." This "refrigerator mother" concept was further promoted by the Austrian born psychiatrist Bruno Bettelheim in his book entitled "Empty Fortress." He postulated that the child was so emotionally starved for parental affection that the mind took refuge in its isolation from humanity. Parents of autistic children were rightfully indignant at being accused of contributing to their child's illness.

The incidence of autism has been steadily increasing among all social groups. Current estimates are that up to 1 percent of boys and 1 in 150 of all preteen children are autistic. High functioning autistic children overlap with an even larger group of children with various learning and/or behavioral disorders. Over a quarter of high school seniors now fail to acquire basic skills in English and over a third lack basic skills in mathematics. These figures do not include high school dropouts, many of whom are essentially illiterate. The downward performance trend will surely continue unless an explanation is found for the impaired brain functioning of children and corrective measures are undertaken.

Rather than dismissing Kanner's observations, it would have been more insightful to suggest a possible common agent causing brain damage in both mothers and their autistic children. Close questioning of mothers of autistic children will frequently uncover episodes of an emotional disorder and/or of unexplained fatigue occurring well prior to the child's birth. When medical help was sought, diagnoses such as chronic fatigue syndrome (CFS), fibromyalgia and depression were commonly applied. While parental symptoms may persist, they are clearly overshadowed by the tragedy that has befallen the children.

Also relevant is the distressingly large numbers of parents who voluntarily relinquish custody of their emotionally impaired children to the State. Government figures released for 2001 indicate that in just 19 mid-sized States, 12,600 children were transferred to State custodial care by parents for either financial or emotional reasons. One wonders if this too is not an indication that the parents are involved in a pervasive illness that affects them as well as their children.

Traditionally, when epidemiologists are faced with an apparent epidemic, primary consideration is given to a potential infectious cause. Neonatal infections from mother to unborn child are known to occur and autism has been reported in children infected before birth with human cytomegalovirus and rubella virus.

For nearly two decades, I have proposed a viral cause of CFS as well as of more severe psychiatric illnesses, including autism. I have isolated and described atypical cell damaging (cytopathic) viruses that possess a stealth-like quality of not being effectively recognized by the cellular immune system. I have reported that these viruses are actively infecting the vast majority of autistic children tested. Yet this message has been ignored by Public Health officials. The Centers for Disease Control and Prevention (CDC) has not responded to petitions or personal requests to perform virus cultures in autistic children to confirm the presence of an active virus infection. Sadly, this agency appears to be fearful of indications that live polio vaccines have allowed the entry of some stealth viruses into the community with widespread devastating consequences. This conclusion stems from extensive DNA sequencing performed on the initial stealth virus isolated from a CFS patient, and substantiated by sequencing of a separate isolate from a psychiatric patient. The data show an unequivocal origin from African green monkey simian cytomegalovirus (SCMV). The Food and Drug Administration (FDA) has confirmed the presence of DNA of SCMV in several licensed batches of live polio virus vaccines. Moreover, studies as early as 1972 confirmed that virtually all of the monkeys being used for polio virus vaccine production were infected with SCMV. Stealth adaptation is not restricted to SCMV but presumably is a generic process applicable to many types of cytopathic viruses.

More recent research has provided a clearer understanding of a non-immunological virus defense mechanism that can operate against stealth adapted viruses. It is mediated by an alternative cellular energy (ACE) pathway of cellular repair and can be activated by naturally occurring compounds termed "enerceuticals." Clinical evaluation of these compounds will be facilitated once there is widespread acceptance of the damage being caused to children and others by stealth adapted viruses.

Parents of autistic children should contact their political representatives and insist that CDC test a sampling of autistic children for stealth adapted viruses. Faced with positive findings, it will be harder for the CDC to maintain its indifference to the nationwide epidemic of stealth adapted virus related illnesses.

For further information, please refer to the web site http://www.s3support.com or send an e-mail to s3support@email.com

W. John Martin, MD, PhD.
Institute of Progressive Medicine
Burbank CA 91502

As the epidemic of autism continues, more autistic patients are becoming of marriageable age and beginning to have children. The fate of these children needs to be addressed as a major public health concern. Support is needed for enrolling children of autistic parents in an ongoing clinical trial.

/24-7PressRelease/ - BURBANK, CA, March 24, 2008 - A new facet of the autism epidemic is emerging. It is the plight of children born to autistic parents. This point was driven home to me as autistic children were being recruited for an investigational study that is evaluating enerceutical activation of the alternative cellular energy (ACE) pathway in various disease conditions, including autism. A simple non-invasive procedure for activating the ACE pathway has been shown to be very effective in suppressing both active and latent herpes simplex virus (HSV) infections. It is also designed to suppress stealth adapted viruses which, unlike conventional herpes viruses, are not actively resisted by the body's immune system. Although still ignored by many in the autism community, there is solid scientific support for prenatal infection with stealth adapted viruses as the underlying cause of autism in the vast majority of affected children.

I have previously noted the high incidence of fatiguing illnesses, such as the chronic fatigue syndrome (CFS), among mothers of autistic children. This observation is consistent with the premise that while stealth adapted viruses can cause CFS in adults; these same viruses being passed from mother to child can damage the developing brain so as to render a child susceptible to autism.

I now wish to report that several of the parents of children currently being recruited into the ACE pathway activation study are, themselves, on the autism spectrum. This is not unexpected as an increasing number of autistic individuals are now entering their 20's and becoming parents. The clinical investigator on the ACE pathway activation project was struck, however, by the severity of neurological/behavioral illnesses among the children of autistic parents; and the enormous difficulties being faced by these parents in caring for their children. Unlike the more publicized internet savvy, politically outspoken, upper middle class parents of many autistic children, these autistic families are floundering under the pressures of both economic and emotional deprivation. The parents have limited skills in even accessing basic social disability and healthcare services. Reading and writing are commonly beyond their capabilities and, for many families, the responsibility for essential household management falls to grandparents or neighbors. It is to be hoped that from within the competing autism support groups, much needed resources will be urgently directed towards autism's second generation.

The best answer to this growing problem will be effective therapy for both the parents and their affected children. Tax deductible donations are urgently needed to help offset the costs of enrolling additional impoverished autistic patients in the current clinical trial. For more information about the clinical trail, the ACE pathway and stealth adapted viruses, please go to the medical commentary section on autism at http://www.s3support.com or send an e-mail to s3support@mail.com

Can Epilepsy be Controlled Via the ACE Pathway?

The experience gained from studies on three autistic children with epilepsy would certainly suggest that the answer is affirmative. Two of the three children were described in Chapter 4 of this book, a 4-year-old boy and a 14-year-old girl. The boy's severe autism was further complicated by generalized (Grand Mal) seizures, which were occurring 3-4 times a day, in spite of his receiving Trileptal (oxcarbazepine) plus Valproic acid. Actual sedation using Diazepam (Valium) to suppress seizure activity was often required. Some of the more severe seizures left the boy paralyzed from the waist down for several hours. He had nine emergency room visits over the preceding 4 months, six were for uncontrolled seizures and one was for extended post seizure paralysis. From the very first treatment, all seizures stopped and have never returned. Moreover, within 3 months, he was taken off all anti-seizure medications, based on normal EEG findings.

The second autistic child, also described in Chapter 4, was a 14-year-old girl on medication from age 3 for daily recurring petit mal (absence) seizures and frequent complex partial seizures. The seizures would completely cease for 1-2 weeks after treatment and then slowly return till her mother applied the next treatment. The third child had a similar outcome with the parents monitoring the frequent daily seizure occurrences. No entries were made in the log of seizure activity during the week following ACE-enhancing therapy.

Two interesting extrapolations can be derived from these extremely promising results. First, is that the heightened sensitivity to epileptic discharges is quite possibly the result of impaired ACE pathway. As discussed in Chapter 6, the ACE pathway is envisioned as a mechanism of helping to maintain charge separation. It may assist, therefore, in maintaining the normal electrical potential of nerve cells. If this is the mechanism, it could well explain impaired functioning of other neuronal cells. The second extrapolation is that some of the other non-conventional approaches to epilepsy control may also be achieving their benefit through the ACE pathway.

The term enerceutical™ was coined for various natural products with ACE pathway enhancing activity. Several years ago, I asked whether marijuana might not be an enerceutical™ for some individuals (www.s3support.com). Given the recent reports of anti-seizure activity of the marijuana compound cannabidiol[1], it will be worthwhile to test this product for its water activating capacity via KELEA. Similarly, there are anecdotal reports of diminished epileptic seizures associated with the consumption of other enerceuticals™, including humic/fulvic acids[2], zeolites[3], moringa oleifera[4], HB-101 and others. On the other hand, emotional stress is a major trigger for epileptic seizures. This is consistent with various other indicators that stress may diminish ACE pathway activity.

It is sometimes necessary to surgically remove the portion of the brain prone to recurrent seizures. Histological examination of the tissue from one such case confirmed the presence of foamy, vacuolated cells, as seen in brain biopsies of stealth adapted virus infected patients[5]. It will be of interest to examine brain biopsies from more patients with epilepsy to see if this is a consistent finding. In the interim clinical studies are warranted using enerceuticals™ and KELEA activated water in both autistic and non-autistic patients with epilepsy.

References

1. Porter BE, Jacobson C (2013) Report of a parent survey of cannabidiol-enriched cannabis use in pediatric treatment-resistant epilepsy. Epilepsy Behav. 29: 574-7.

2. ww.zeolitetechnology.com/testimonials.html

3. http://quantuminplus.webs.com/apps/blog/show/7539029

4. Popoola JO, Obembe OO. (2013) Local knowledge, use pattern and geographical distribution of Moringa oleifera Lam. (Moringaceae) in Nigeria. J Ethnopharmacol. 150: 682-91.

5. Martin WJ (1996) Severe stealth virus encephalopathy following chronic fatigue syndromelike illness: Clinical and histopathological features. Pathobiology 64:18, 1996.

Schizophrenia and MI Hope Inc.

MI Hope Inc. was established in 1988 by parents of children with schizophrenia. It had largely been abandoned by 2002, when I met with one of the parents. His son had been diagnosed 30 years earlier at age 18.

Without hesitation the father had reassured his son that he could always stay at home. I was impressed by the serenity of the father as his son acted out as nighttime fell. I saw the same calmness on my second visit a week later. Leaving that night, I tried to compliment the father for his enduring efforts. He wanted no praise and simply stated " we brought him into this world, he's our responsibility." He further mentioned that in World War II, his patrol had come under fire. Everyone but he sustained bullet wounds, with many of his close friends dying. "I will never have reason to ever feel ungrateful."

He then suggested that I might be able to do something with the essentially abandoned MI Hope Inc. The next day, I relayed the offer to an IRS official. He not only agreed to resurrect MI Hope Inc., but also suggested that I apply to elevate its status to a provisional public non-profit charity. I assumed the management of MI Hope Inc., and five years later it was declared a fully authenticated public charity. Addressing the tragedy of schizophrenia will always be one of its major missions.

I had actually been interested in schizophrenia since the time of the second stealth adapted virus isolate. It has been cultured from the cerebrospinal fluid of a comatose 22-year-old woman admitted to the Los Angeles County Hospital in 1991. Her initial diagnosis at age 18 was schizophrenia, but changed to bipolar psychosis largely because of a therapeutic response to lithium. I soon realized the looseness of diagnostic labeling of psychiatric illnesses.

Another memorable patient from whom I had repeated culture evidence of an ongoing stealth adapted virus infection had previously volunteered at age 17 to work one summer at a veterinary virology laboratory. She soon became sick with a "mono-like" illness followed by persistently impaired cognition. Within 3 years she was diagnosed as schizophrenic with marked aggressive tendencies.

Her aggression made her care difficult even during admissions to major hospitals, including Stanford University Hospital. Her mother found it necessary to yield custody to the State. Motivated by the lack of substantial clinical improvements, her mother fought hard to regain medical conservatorship. Further, the mother once noted that her daughter briefly displayed the clarity of thought and expression of her earlier life.

This occurred as she was recovering from anesthesia from an elective operation. This spark of normal behavior was quickly replaced by her flat affect as her prescribed medicines were reinstituted.

Encouraged by the brief return to sanity, the mother successfully pursued the use of various enerceuticals™, such as humic/fulvic acids and HB-101. My gratification came when speaking with the much-improved daughter; I heard her empathetically express real concern for her mother's wellbeing. This was appropriate since the mother was beginning to experience CFS. More serious, was the deep depression developing in the girl's father. He later went on to develop multiple myeloma.

With Institutional Review Board (IRB) approval, I would culture occasional blood samples from patients admitted to the psychiatric ward of the Los Angeles County Hospital. Observing a strikingly positive culture, I visited the ward and asked if I might meet with the patient. The patient was clearly troubled by her habit of using the wrong or gabled words. Even more troubling was her very poor recall of even ongoing events and conversations. To a lesser extent, these same cognitive deficits affect many CFS patients. I thought how sad it is that CFS patients choose to distance themselves from psychiatric patients, instead of being willing to champion their illness as being an extension of CFS.

I began asking more severely ill CFS patients if they ever heard voices? "Yes," I was told by one patient "but I wouldn't tell anyone other than you or they would think I'm mad."

It is important that patients with schizophrenia be included in clinical trials based on enhancing the ACE pathway. It is also potentially of value to employ techniques, which may facilitate memory, speech and self-awareness. A nursing aide in a psychiatric facility would encourage patients to draw parts of their body, leading up to drawing their own face. She really thought this exercise was far more effective than the prescribed mind-dulling medications.

Morgellon's Disease – ACE Pigment Dermatitis: Possibly a Healing Process Gone Awry?

W. John Martin, MD, PhD.
Institute of Progressive Medicine
South Pasadena CA 91030

Abstract

During the course of investigations into stealth adapted viruses, the alternative cellular energy (ACE) pathway was identified as a major non-immunological defense mechanism. This pathway is mediated by energy transducing (converting) materials, termed ACE pigments, which can take the form of particles and fibers. ACE pigments are regarded as miniature batteries, which may seemingly be uncharged, partially charged or fully charged. Uncharged and partially charged ACE pigments can be identified by their ultraviolet (UV) fluorescence with and without, respectively, the addition of a suitable triggering dye, such as neutral red. An overproduction of uncharged and partially charged ACE pigments in perspiration may explain a variety of skin lesions occurring in some patients with persisting infections caused by stealth adapted viruses. Concomitant infection of the brain by these viruses can also help explain the association in some patients between impaired brain function and ACE pigments containing skin lesions. This understanding may help resolve the uncertainties with an illness called Morgellons disease and provide a rational approach to therapy to this and related illnesses.

Key Words: Morgellons, Delusional parasitosis, Delusional infestation, Ekbom syndrome, Psychiatry; Chronic fatigue syndrome, CFS, Lyme disease, Stealth adapted viruses, Alternative cellular energy, ACE pathway, ACE pigments, Neutral red, Fluorescence, Ultraviolet light, Perspiration, Skin lesions

Running Title: ACE Pigment Dermatitis

Author: W. John Martin, MD, PhD. Institute of Progressive Medicine, 1634 Spruce Street South Pasadena CA 91030. Telephone 626-616-2868. E-mail wjohnmartin@hotmail.com

Introduction

An illness characterized by the formation of persisting localized skin lesions, which evoke an irritating sense of prickliness and from which fibers and other particles are seemingly emanating, is being referred to as Morgellons disease (1-11); especially on patient directed internet sites (12). Other suggested diagnostic terms include the neuro-cutaneous syndrome (13, 14) and psycho-cutaneous disorder (15), which acknowledge the accompanying cognitive and psychological impairments manifested by many of the affected patients. Indeed, for some patients the skin lesions are considered self-induced and the reported fibers as simply being derived from clothing, carpets or house dust. The terms delusional parasitosis (16-26), delusional infestation (27-30) and Ekbom syndrome (31-33) are also used in the context of seemingly irrational psychiatric patients who insist that the fibers are living organisms, which have invaded their body. To the patients, the fibers are unquestionably living because, as discussed below, the fibers commonly show unexplained movements and apparent transformations between differing forms. Some of the other more extreme patient belief systems are that the fibers are (i) toxic chemicals that have possibly been seeded from chemtrails; (ii) nanomaterials that have escaped from highly secretive experimental laboratories; (iii) silicon based living creatures or (iv) possibly of extraterrestrial origin. The vehemence with which such views are often expressed is symptomatic of an underlying cognitive and emotional disorder, fueled by frustration of not being really helped by the medical profession.

Stealth Adapted Viruses and the ACE Pathway

Stealth adaptation is a generic process whereby viruses can evade effective cellular immune recognition through the deletion/mutation of the relatively few viral proteins actually targeted by the cellular immune system. (34-48). The prototype stealth adapted virus originated from an African green monkey simian cytomegalovirus (SCMV) and was cultured from a chronic fatigue syndrome (CFS) patient (36, 38). Stealth adapted viruses can also be consistently cultured from many patients with a wide range of neurological and psychiatric illnesses. A striking characteristic of cultures of stealth adapted viruses is the tendency for self-repair of the initial virus cytopathic effect (CPE). The repair process coincides with the formation of structured intracellular materials that can coalesce and be extruded as extracellular particles and as material that can assemble into fibers, including long threads and ribbons (44). Intracellular and extruded extracellular particles are typically black, while the fibers can display a range of colors, including white, blue, green, yellow and red. Removal of the extracelluar materials by re-feeding the cultures can

result in a rapid reactivation of the virus CPE (44). The reactivation can be prevented if some of the isolated particulate material from other cultures is included in the re-feeding medium.

Morgellons Disease in Stealth Adapted Virus Infected Patients

Several CFS patients in whom cultures provided clear evidence for infection with stealth adapted viruses, have subsequently reported to now have numerous fibers emanating from irritable skin lesions and to have been told they have Morgellons disease. Additional Morgellons patients have provided detailed information regarding the clinical manifestations of their illness and have kindly shared intriguing observations on the fibers and on other strange occurrences within their household. These individuals have periodically provided various samples and photographs. Several Morgellons patients have also been directly examined in conjunction with clinical colleagues. The remainder of this article addresses some of the major issues on which these patients have sought my opinion. Although, still the subject of ongoing research, the present understanding may be helpful to clinicians confronted by understandably confused and anxious patients.

Questions Posed by Patients Regarding Their Fiber Producing Skin Disease

1. Are the fibers actually originating from my skin lesions or simply coming from outside of my body and attaching to my skin? Undoubtedly, the fibers are originating within the patients. Several patients have described pulling out long strands of fibers

from deep within skin lesions. Fibers can form in capped urine samples that are stored over several days. Fibers can also be observed in fresh blood samples using dark field live cell analysis. Not knowing about ACE pigments, many of those performing dark field analyses have mistakenly labeled the structures as borrelia (Lyme disease) spirochetes, especially if they observe them to be vigorously vibrating. There is an interesting observation reported by several patients of presumably previously shed fibers moving rapidly from an environmental surface back onto the patient's skin. There has also been the impression that some of these returning fibers actually re-penetrate into the skin. This possible phenomenon will be discussed later.

2. Are the fibers and particles living? The categorical answer is no. Yet they can show movements, which may be experienced as a sporadic crawling sensation within areas of their affected skin. Patients may also occasionally experience a sharp pin-prick or stinging sensation, again understandably equated by the patient with the possible presence of a biting insect. Even outside of the body, some of the fibers display spontaneous curling movements and may suddenly "jump" from one location to another. This feature, along with some questionable photography, has suggested the possibility that the hexapod, Collembola, also known as springtails, may be present (59 -61). Other patients have been told they are infested with either scabies or lice, especially when multiple black particles, presumed to be eggs, are

seen attached to hair. Other suggestions of possible microorganisms offered in internet postings include fungi, nematodes, algae and especially Lyme disease causing borrelia spirochete (11, 60). In reality, the fibers and particles are far too variable even within individuals to be consistent as a single living species. Internal non-symmetrical heterogeneity can, for example, be readily shown within patients' particles using fluorescent microscopy (Fig. 1).

An interesting observation recounted by an affected gentleman provided good support that the fibers are being self assembled from smaller units. Essentially, he placed a fiber in a water droplet and observed it under a microscope. He then added increasing amounts of a saturated sodium chloride (salt) solution. The fiber disintegrated into minute particles that moved vigorously within the droplet. He then added water and saw the particles almost instantaneously reassembled into a fiber. When he had two or more fibers in a droplet, the salt would lead to their disintegration, but upon diluting the salt, long fibers would no longer form. Other particles from patients can be dissociated using alcohol or acetone, further supporting their aggregate nature. Rapid disintegration of a fiber into microscopic particles may explain the sense of some patients of fibers re-entering into their skin.

Self-assembly also implies a polarity of the interacting components and a responsiveness to an orienting force field. Electrostatic force and, with some particles, magnetic and electromagnetic forces, can clearly be shown to induce particle and fiber movements. An example of a magnetic particle provided by a patient is shown in Fig. 2. When suspended with another of the patient's particle in water, the two particles readily attached. A small vapor bubble slowly formed over the next 30 minutes between the two particles (Fig. 3).

Acknowledging that the particles and fibers are self assembled complexes of chemicals, possibly in the form of polymers, dendrimers or clathrates, does not exclude the additional presence of atypical life forms other than an underlying stealth adapted virus infection. The process of stealth adaptation can greatly expand the range of infectivity of the viruses. For example, the prototype SCMV derived stealth adapted virus grows well in insect cells and has seemingly also acquired genetic sequences from bacteria (41). Arguably, the virus has acquired these sequences via passage through bacteria. Indeed, bacteria with very unusual typing characteristics were isolated from the initially identified stealth adapted virus infected CFS patient. ACE pigment like material has also developed in routine bacterial cultures isolated from the feces and from throat swabs of CFS patients (46), including some with Morgellons (Fig. 4). It would not be surprising, therefore, if some of the Morgellons disease patients were to harbor energy-enhanced bacteria, fungi, worms or insects. Similarly, it is possible that ACE pigments, accumulating within the environments of Morgellons disease patients, may provide an overall energy field, which could be utilized by other living creatures,

some of which have not been previously recognized.

3. If the fibers are non-living, how can you explain their movements and more importantly, how can they increase in number when placed into a sealed container? As noted above, physical forces can clearly cause the particles to move. One example is the attraction or repulsion to electrostatic forces formed by rubbing plastic or Styrofoam. Some other less well-defined forces may also be at play. For example, some of the fibers will clearly sway towards silicone rubber. Another patient reported how black particles tended to cling to white surfaces and vice versa. Using a sound generator with tunable frequencies, patient derived particles can sometimes be shown to vigorously resonance with a particular frequency. Different particles from the same patient will typically resonate at different frequencies. There are also reports of low level sound being detectable from fibers placed over the input microphone of computer based sound systems. As mentioned above, some patient derived particles are clearly magnetic (Fig. 2). The particle shown in this figure came from a patient who installed a magnetic mattress to help with his sleeping. Each morning, he would awake to black particles on his bed sheets, having being drawn to the magnetic mattress. The particles from this patient could also generate vapor bubbles when placed into water (Fig. 3). Similar magnetic and vapor-forming particles have been obtained from several patients. Interestingly, the magnetic and electrostatic properties of the particles are labile and can be lost even

during the course of a single experiment. Variable and only limited success has been achieved in restoring the particles' magnetic activity, for example by placing the particles into sunlight or illuminating with UV light.

A harder question is to fully explain the apparent abiotic formation of additional particles, fibers and other materials. One patient described how she used the sponge from a jar of cosmetics on only one occasion to cleanse her face. The sponge was put back into the jar and over time she observed her cosmetic becoming riddled with black particles. Another gentleman reported the repeated accumulation of a waxy material on the top of his bathroom faucet. Several patients have microscopically observed fibers slowly sprouting out of some of the particles isolated from their body. Transformations between particles and fibers is consistent with the complexities of reversible self-assembly processes being influenced by environmental forces. When freshly mined humic acid particles are observed microscopically in water, one can occasionally see some of the particles gradually being replaced by tangled fibers. Presumably, the alignments of carbon atoms comprising the humic acid particles can undergo rearrangements, resulting in more linear formations. Similar particles to fibers transformations can occur with silicon-based zeolites and is explainable by the variable coordinating valancies of the silicon atom (62).

Abundant production of lipid-like materials, including needles, crystals, membranes and even pyramids, occurs in long held stealth adapted virus culture tubes that contain ACE pigments; even in those cultures without any remaining cells. Lipid like fibers can also slowly form around patient derived particles (Fig. 5). Some patients have described tenacious gel-like material oozing from their skin lesions and/or becoming deposited on objects within their household environment. If as suggested above, ACE pigments are capturing energy, it is not inconceivable that the energy is being expended in the bonding of carbon, hydrogen and other atoms. Interestingly, even ethanol in the presence of certain minerals is thought capable of catalyzing carbon-carbon covalent bonding (63). It is not inconceivable, therefore, that certain particles may, themselves, catalyze the abiotic synthesis of complex chemicals in addition to lipids. In a sense, therefore, the particles may have the capacity for essentially replicating.

4. Could ACE pigments explain the electrosensitivity to cell phones and other energy radiating devices, which is seemingly affecting some patients with Morgellons disease as well as some patients without overt Morgellons disease. As noted earlier, whether derived from stealth adapted virus cultures or more directly from patients, ACE pigments respond to many forms of energy input. It is not surprising; therefore, that microwave and radio wave emitting devices could be an added input sources of energy. It could be converted (transduced) to other energy forms, including electrical, which could well be perceived as stinging or prickling. These sensations could occur

even if the skin ACE pigments were not directly capable of inducing skin ulceration. Furthermore, ACE pigments in other locations within the body could similarly be stimulated by external energy sources with resulting organ dysfunction. This reasoning underscores the concept that Morgellons disease is best considered simply as a subset of a wider range of illnesses associated with the overproduction of ACE pigments.

5. If the fibers are not living, am I nevertheless a source of disease to others with who I am in contact? Public health authorities have been reluctant to accept an infectious cause for many chronic illnesses. I am aware of several families in whom Morgellons was diagnosed in one member, with other family members having new onset severe neurological/psychiatric diseases (64). Even household pets can develop marked behavioral illnesses and begin to show skin and hair lesions comparable to those of the patient. In extreme cases, limited areas of animal hair appear to be replaced by tufts of fibers coming out of hair follicles.

The evidence of infectivity of stealth adapted viruses extends to community wide outbreaks (40) as well as to animal inoculation studies (37). Not all viruses are short lived when outside of the body. Indeed, certain stealth adapted viruses can resist desiccation and can remain infectious in a dried form (34). Appropriate precautions against virus exposure, along with efforts to enhance the ACE pathway in family members and in close social contacts are indicated. It is certainly possible that some stealth adapted

viruses may become attached to the fibers or other particles released from patients.

As noted above, some stealth adapted viruses can be cultured in insect cell lines. I provided this information to a gentleman who was convinced that flies, wasps and spiders, in and near his house, were showing peculiar patterns of behavior. For example, flies and wasps would be seen flying in an erratic fashion, only to fall suddenly to the ground and soon die. When illuminated with a UV light, fluorescing material would be seen oozing from their bodies. Unusual spider webs were also noted, with the possible coating of some of the web strands with additional fibrous-like material.

It is also interestingly that agrobacterial DNA sequences may be present in some of the fibers isolated from Morgellons disease patients (65). Sequences from this genus of bacteria were also described as being present in a stealth adapted virus culture (41). As mentioned above, bacteria may be susceptible to infection by some stealth adapted viruses and this may allow for the intermixing of viral and bacterial genes (viteria). At this stage, caution is indicated in handling fibers from Morgellons patients and indeed the handling of any bodily fluids from presumptively stealth adapted virus infected individuals.

6. Can the concept of stealth adapted viruses explain the fatigue and impaired brain function that is commonly seen among patients being diagnosed as having Morgellons disease? There is a broad clinical spectrum in stealth adapted virus infected

patients. It extends from relatively minor illnesses, such as mild CFS, to very devastating major psychiatric and neurological illnesses. A practicing psychiatrist in a major US city sees severely ill Morgellons-like patients in both his inpatient and outpatient practice. Yet some other Morgellons patients have been extremely understanding of their somewhat impaired cognitive and emotional functioning and have continued to remain rational and provide important insights about their illness. Unlike other organs, the brain is uniquely susceptible to functional disruption caused by even limited, localized cellular damage. This is because of the spatial specialization of the central nervous system, such that no overall compensatory mechanism exists to correct impaired activity within localized areas of the brain. Spatial heterogeneity of the many brain functions can also help explain the wide range of psychological and neurological symptoms observed among stealth adapted virus infected patients.

7. Another important question is being asked is, even apart from the electrosensitivity issue addressed earlier, are the ACE pigment fibers harmful to the body? My basic answer is that while the ACE pathway is inherently helpful to the body, excessive quantities of uncharged or only partially charged ACE pigments may, in some patients, be deleterious to the body. Possibly, the materials which later comprise fibers through a self-assembly mechanism, are moving onto the body external surfaces as part of an energy recharging mechanism. This would require passage of the materials through sweat ducts and hair follicles.

Until they become recharged, for example by external sunlight, they may actually be a drain on the energies of neighboring cells. ACE pigment particles obtained from stealth adapted virus cultures can show both electron donating (reducing) and electron stealing (oxidizing) activities. It is the later activity, which is likely to lead to cellular damage. This can in turn cause secondary inflammation manifesting as dermatitis. Oxidative damage to hair follicles could lead to hair loss (alopecia). Consistent with the postulated skin damaging electrical and oxidizing properties of patient derived materials is a patient's observation that shed particles from his body were causing pitting erosion on metal surfaces, including aluminum and silver.

It is clear, however, that other patients can shed fluorescing fibers and not develop discernable skin lesions. This difference may relate to the intensity of postulated energy draining activity of the materials on skin cells. The detection of fluorescing materials on individuals is a very simple presumptive screening test for a deficiency in the body's ACE pathway. Excreting direct or dye induced fluorescent materials is not restricted to Morgellons patients, but can be seen in ACE pathway deficiency patients without skin lesions.

8. If this is a correct understanding of what is occurring to my skin, what can I do to minimize the amounts and/or deleterious activity of uncharged and/or partially charged ACE pigments that I am excreting? Therapeutic methods are being developed

for systemically activating the ACE pathway using UV light illumination of various energized solutions placed onto the skin. The methods are based on the supposition that the energized solutions can transfer energy into the body and help recharge the body's ACE pathway. Supportive data for this supposition have been obtained in preliminary studies in patients infected with stealth adapted viruses and follow similar successful efforts using neutral red dye to directly treat herpes simplex virus (HSV) skin infections (66).

In the interim, patients should follow the basic tenets of minimizing stress, avoiding toxins and eating fresh wholesome foods. Patients begin to learn of specific environmental factors that can either alleviate or conversely worsen their symptoms. For some patients, previously shed fibers accumulating within their household appear to exert a deleterious effect, as if they are continuing to drain the patient's energy. Similarly, some stealth adapted virus infected patients become extremely sensitive to sources of low-level electromagnetic radiations, including cell phones and television. Conceivably, the energy delivering function of the ACE pigments in these patients may be adversely affected by some forms of electromagnetism.

9. As a final question, it could be asked whether any good is likely to come out of an understanding of the ACE pathway? The answer is affirmative for physicists, biologists, clinicians and patients. With the caveat that the fibers and particles may pose a risk of possibly having attached stealth adapted viruses, they should yield important new information regarding their physical structure and energy transducing activities. Biologists may come to appreciate the ACE pathway as the probable forerunner to photosynthesis. Clinicians as well as patients themselves will become better equipped to provide more effective prevention and therapy. Hopefully, the Public Health system may also learn to be less dismissive of important clues provided by patients regarding the emergence of new diseases.

Summary

An explanation is provided for the association of fibers containing atypical chronic skin lesions with neuropsychiatric illnesses. The fibers are viewed as being part of the ACE pathway, which is enhanced as a potential defense mechanism against stealth adapted viruses. These viruses have a propensity for causing impaired functioning of the brain. They are not effectively controlled by the cellular immune system, but rather are resisted by the ACE pathway. The fibers form by the self-assembly of energy transducing ACE pigments. It is proposed that fully charged ACE pigments can provide energy to virus infected cells to help suppress the virus induced cellular damage. Once ACE pigments become enervated by losing some or most of their energy donating activity, they may begin to extract cellular energy from otherwise normal cells. ACE pigments are seemingly transported to the skin via sweat glands and hair follicles. Skin lesions can erupt as a result of the energy draining property of uncharged ACE pigments being concentrated within the skin. The damage may result in secondary inflammation, leading to the diagnostic term "ACE fiber dermatitis" for such skin lesions. Once outside of the body, ACE pigments can display remarkable properties as energy transducing elements and may also be involved in abiotic transformations and synthesis of additional ACE pigment-like materials.

Acknowledgement

I am especially grateful to the graduate student whose studies were interrupted because of her disease and who took the initiative to make telephone contact. Her questions provided the impetus for me to follow up with some earlier patients and to clearly formulate my ideas in the form of this manuscript. I also want to thank the many other patients who have provided useful information and in particular the gentleman from Indiana.

References

1. Savely VR, Leitao MM, Stricker RB. (2006) The mystery of Morgellons disease: infection or delusion? Am J Clin Dermatol. 7: 1-5.

2. Waddell AG, Burke WA. (2006) Morgellons disease? J Am Acad Dermatol. 55:914-5.

3. Koblenzer CS. (2006) The challenge of Morgellons disease. J Am Acad Dermatol. 55: 920-2.

4. Robles DT, Romm S, Combs H, Olson J, Kirby P. (2008) Delusional disorders in dermatology: a brief review. Dermatol Online J. 14: 2.

5. Accordino RE, Engler D, Ginsburg IH, Koo J. (2008) Morgellons disease? Dermatol Ther. 21: 8-12.

6. Harvey WT, Bransfield RC, Mercer DE, Wright AJ, Ricchi RM, Leitao MM. (2009) Morgellons disease, illuminating an undefined illness: a case series. J Med Case Reports. 3: 8243.

7. Lustig A, Mackay S, Strauss J. (2009) Morgellons disease as internet meme. Psychosomatics. 50: 90. PMID: 19213978

8. Harth W, Hermes B, Freudenmann RW. (2010) Morgellons in dermatology. J Dtsch Dermatol Ges. 8: 234-42.

9. Dovigi AJ. (2010) Intraoral Morgellons Disease or Delusional Parasitosis: A First Case Report. Am J Dermatopathol. 32: 603-5.

10. Marianne J Middelveen1, Divya Burugu2, Akhila Poruri2, Jennie Burke3, Peter J Mayne1, Eva Sapi2, Douglas G Kahn4, Raphael B Stricker (2013) Association of spirochetal infection with Morgellons disease. F1000 Research 2:25

11. Mayne P, English JS, Kilbane EJ, Burke JM, Middelveen MJ, Stricker RB. (2013) Morgellons: a novel dermatological perspective as the multisystem infective disease borreliosis. F1000 Research 2:118

12. www.morgellons.org;www.morgellons-disease-research.com; www.morgellons-uk.net; www.morgellons-sanctum.org; www.morgellonsexposed.com; www.thenmo.org; www.morgellons.ca; www.morgellonsgroup.proboards.com; www.morgellonsresearchfoundation.com; http://www.youtube.com/user/bannanny www.chemtrailagenda.com/

morgellons.htm; www.staningerreport.
com; www.morgellonspictures.org;
www.morgellonsfocus.com; www.
morgellonssupport.com; www.
crawlingsensations.com; www.morgellons-
disease-research.com; www.parasitetesting.
com

13. Amin OM. (2001) Neuro-cutaneous syndrome (NCS); a new disorder. Explore 10: 55-56.

14. Amin OM. (2006) An overview of neuro-cutaneous syndrome (NCS) with a special reference to symptomology. Explore 15: 41-49.

15. Ehsani AH, Toosi S, Mirshams Shahshahani M, Arbabi M, Noormohammadpour P. (2009) Psycho-cutaneous disorders: an epidemiologic study. J Eur Acad Dermatol Venereol. 23: 945-7.

16. Wilson FC, Uslan DZ. (2004) Delusional parasitosis. Mayo Clin Proc. 2004; 79: 1470.

17. Aw DC, Thong JY, Chan HL. (2004) Delusional parasitosis: case series of 8 patients and review of the literature. Ann Acad Med Singapore. 33: 89-94.

18. Vila-Rodriguez F, Macewan BG. (2008) Delusional parasitosis facilitated by web-based dissemination. Am J Psychiatry. 165: 1612.

19. Freudenmann RW, Lepping P. (2009) Delusional infestation. Clin Microbiol Rev. 22: 690-732.

20. Hanihara T, Takahashi T, Washizuka S, Ogihara T, Kobayashi M. (2009) Delusion of oral parasitosis and thalamic pain syndrome. Psychosomatics. 50: 534-7.

21. Boggild AK, Nicks BA, Yen L, Van Voorhis W, McMullen R, Buckner FS, Liles WC. (2010) Delusional parasitosis: six-year experience with 23 consecutive cases at an academic medical center. Int J Infect Dis. 14: e317-21.

22. Lee CS. (2008) Delusions of parasitosis. Dermatol Ther. 21:2-7. PMID: 18318879

23. Wong JW, Koo JY. (2013) Delusions of parasitosis. Indian J Dermatol. 58: 49-52.

24. Reichenberg JS, Magid M, Jesser CA, Hall CS (2013) Patients labeled with delusions of parasitosis compose a heterogenous group: a retrospective study from a referral center. J Am Acad Dermatol. 68 :41-6

25. Freudenmann RW, Lepping P. (2009) Delusional infestation. Clin Microbiol Rev. 22: 690-732.

26. Foster AA, Hylwa SA, Bury JE, Davis MD, Pittelkow MR, Bostwick JM. 2012 Delusional infestation: clinical presentation in 147 patients seen at Mayo Clinic. J Am Acad Dermatol. 67: 673.

27. Freudenmann RW, Lepping P, Huber M, Dieckmann S, Bauer-Dubau K, Ignatius R, Misery L, Schollhammer M, Harth W, Taylor RE, Bewley AP. (2012) Delusional infestation and the specimen sign: a European multicentre study in 148 consecutive cases. Br J Dermatol. 167: 247-51.

28. Heller MM, Wong JW, Lee ES, Ladizinski B, Grau M, Howard JL, Berger TG, Koo JY,

Murase JE. (2013) Delusional infestations: clinical presentation, diagnosis and treatment. Int J Dermatol. 52: 775-83

29. Bhatia MS, Jhanjee A, Srivastava S. (2013) Delusional infestation: a clinical profile. Asian J Psychiatr. 6:124-7.

30. Wolf RC, Huber M, Depping MS, Thomann PA, Karner M, Lepping P, Freudenmann RW. (2013) Abnormal gray and white matter volume in delusional infestation. Prog Neuropsychopharmacol Biol Psychiatry. 46C:19-24

31. Bourée P, Benattar B, Périvier S. (2007) Ekbom syndrome or delusional parasitosis Rev Prat. 57: 585-9.

32. Hinkle NC. (2010) Ekbom syndrome: the challenge of "invisible bug" infestations. Annu Rev Entomol. 55: 77-94.

33. Simonetti V, Strippoli D, Pinciara B, Spreafico A, Motolese A. (2008) Ekbom syndrome: a disease between dermatology and psychiatry. G Ital Dermatol Venereol. 143: 415-9.

34. Martin WJ, Zeng LC, Ahmed K, Roy M. (1994) Cytomegalovirus-related sequence in an atypical cytopathic virus repeatedly isolated from a patient with chronic fatigue syndrome. Am J Pathol. 145: 440-51.

35. Martin WJ. (1994) Stealth viruses as neuropathogens. CAP Today. 8: 67-70.

36. Martin WJ, Ahmed KN, Zeng LC, Olsen JC, Seward JG, Seehrai JS. (1995) African green monkey origin of the atypical cytopathic 'stealth virus' isolated from a patient with chronic fatigue syndrome. Clin Diagn Virol. 4: 93-103.

37. Martin WJ, Glass RT. (1995) Acute encephalopathy induced in cats with a stealth virus isolated from a patient with chronic fatigue syndrome. Pathobiology. 63:115-8

38. Martin WJ. (1996) Simian cytomegalovirus-related stealth virus isolated from the cerebrospinal fluid of a patient with bipolar psychosis and acute encephalopathy. Pathobiology. 64: 64-6.

39. Martin WJ. (1996) Severe stealth virus encephalopathy following chronic-fatigue-syndrome-like illness: clinical and histopathological features. Pathobiology. 64:1-8.

40. Martin WJ, Anderson D. (1999) Stealth virus epidemic in the Mohave Valley: severe vacuolating encephalopathy in a child presenting with a behavioral disorder. Exp Mol Pathol. 66:19-30.

41. Martin WJ. (1999) Bacteria-related sequences in a simian cytomegalovirus-derived stealth virus culture. Exp Mol Pathol. 66: 8-14.

42. Martin WJ. (1999) Stealth adaptation of an African green monkey simian cytomegalovirus. Exp Mol Pathol. 66: 3-7.

43. Martin WJ. (2003) Complex intracellular inclusions in the brain of a child with a stealth virus encephalopathy. Exp Mol Pathol. 74: 197-209.

44. Martin WJ. (2003) Stealth virus culture pigments: a potential source of cellular energy. Exp Mol Pathol. 74: 210-23.

45. Martin WJ. (2005) Alternative cellular energy pigments from bacteria of stealth virus infected individuals. Exp Mol Pathol. 78: 215-7.

46. Martin WJ. (2005) Alternative cellular energy pigments mistaken for parasitic skin infestations. Exp Mol Pathol. 78: 212-4.

47. Martin WJ. (2005) Progressive medicine. Exp Mol Pathol. 78: 218-20.

48. Martin WJ. (2005) Etheric biology. Exp Mol Pathol. 78: 221-7.

49. Lovley DR, Fraga JL, Coates JD, Blunt-Harris EL. (1999) Humics as an electron donor for anaerobic respiration. Environ Microbiol. 1: 89-98.

50. Shailaja J, Kaanumalle LS, Sivasubramanian K, Natarajan A, Ponchot KJ, Pradhan A, Ramamurthy V. (2006) Asymmetric induction during electron transfer mediated photoreduction of carbonyl compounds: role of zeolites. Org Biomol Chem. 4:1561-71.

51. Zagal'skaia EO. (1995) The magnetic susceptibility of the melanin in the eyes of representatives of different vertebrate classes. Zh Evol Biokhim Fiziol. 31:416-22.

52. Menter JM, Willis I. (1997) Electron transfer and photoprotective properties of melanins in solution. Pigment Cell Res. 10: 214-7.

53. Zhernovoĭ MV, Lebedev SV, Grigoriuk AA, Krasnikov IuA, Kharitonskiĭ PV. (1998) Reaction of melanocytes in vertebrate retina to the exposure to constant magnetic field. Tsitologiia. 40: 676-81.

54. Nicolaus BJ. (2005) A critical review of the function of neuromelanin and an attempt to provide a unified theory. Med Hypotheses. 65:791-6.

55. Dadachova E, Bryan RA, Huang X, Moadel T, Schweitzer AD, Aisen P, Nosanchuk JD, Casadevall A. (2007) Ionizing radiation changes the electronic properties of melanin and enhances the growth of melanized fungi. PLoS One. 5: e457.

56. Goodman G, Bercovich D. (2008) Melanin directly converts light for vertebrate metabolic use: heuristic thoughts on birds, Icarus and dark human skin. Med Hypotheses. 71:190-202.

57. Balaban TS, Fromme P, Holzwarth AR, Krauss N, Prokhorenko VI. (2002) Relevance of the diastereotopic ligation of magnesium atoms of chlorophylls in Photosystem I. Biochim Biophys Acta. 1556: 197-207.

58. Sheftel A, Stehling O, Lill R. (2010) Iron-sulfur proteins in health and disease. Trends Endocrinol Metab. 21: 302-14.

59. Altschuler DZ, Crutcher M, Dulceanu N, Cervantes BA, Terinte C, and Sorkin LN. (2004) Collembola (Springtails) (Arthropoda: Hexapoda: Entognatha) found in scrapings from individuals diagnosed with delusory parasitosis. J New York Entomol Soc. 112: 87-95.

60. http://morgellonspgpr.wordpress.com/2009/07/07/morgellons-springtails-collembola-and-fungal-gnats/

61. Shelomi M. (2013) Evidence of photo manipulation in a delusional parasitosis paper. J Parasitol. 99: 583-5.

62. Pierrefixe SC, Fonseca Guerra C, Bickelhaupt FM. (2008) Hypervalent silicon versus carbon: ball-in-a-box model. Chemistry. 14: 819-28. PMID: 18058957

63. Watson AJ, Williams JM. (2010) Chemistry. The give and take of alcohol activation. Science. 329: 635-6.

64. Martin WJ. Family illness of presumptive infectious origin. www.s3support.com

65. Stricker RB, Savely VR, Zaltsman A, Citovsky V. Contribution of agrobacterium to Morgellons disease. www.morgellons.org/suny.htm

66. Martin WJ, Stoneburner J. (2005) Symptomatic relief of herpetic skin lesions utilizing an energy-based approach to healing. Exp Mol Pathol. 78: 131-4.

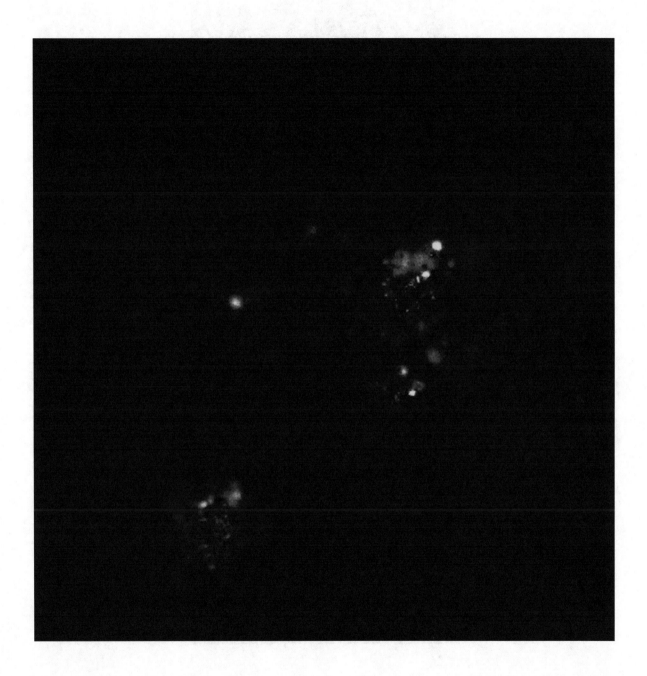

Fig. 1. Discrete areas of fine red, green and yellow fluorescence in a Morgellons patient skin-derived particle.

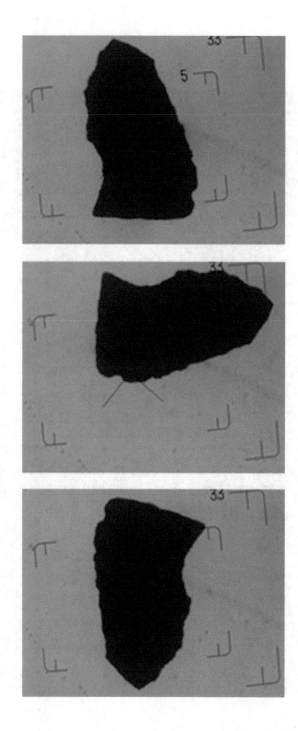

Fig. 2. Magnetic rotation of a skin-derived particle suspended in distilled water. The particle was viewed with an inverted microscope using a 10× objective. The small hand held magnet of 80 gauss was moved to within 3–5 in. of the water droplet containing the particle. The particle showed magnetic polarization with the top and the base showing opposing attraction or repulsion depending upon which pole of the magnet was pointing towards the particle. The particle was rotated by simply moving the magnet.

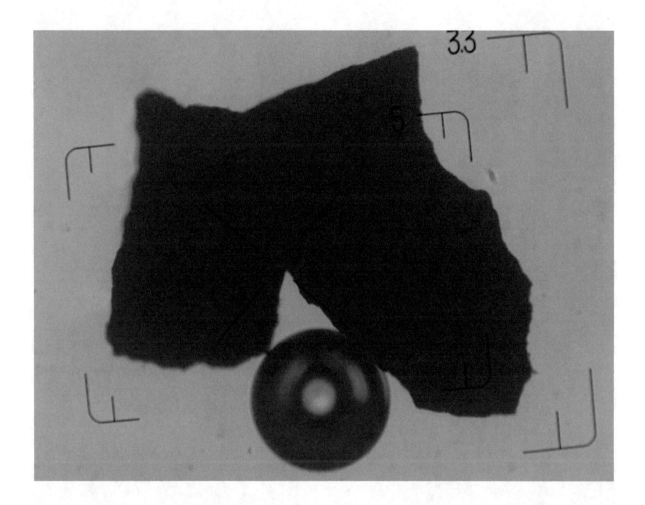

Fig. 3. Formation of a vapor bubble between two magnetically attached particles. The particle shown in Fig. 38, magnetically attached itself to another magnetic particle from the same patient. A vapor bubble began to appear between the particles. The vapor bubble that slowly increased in size over the period of approximately 1 h. Vapor bubbles formation by solitary particles has also been commonly observed.

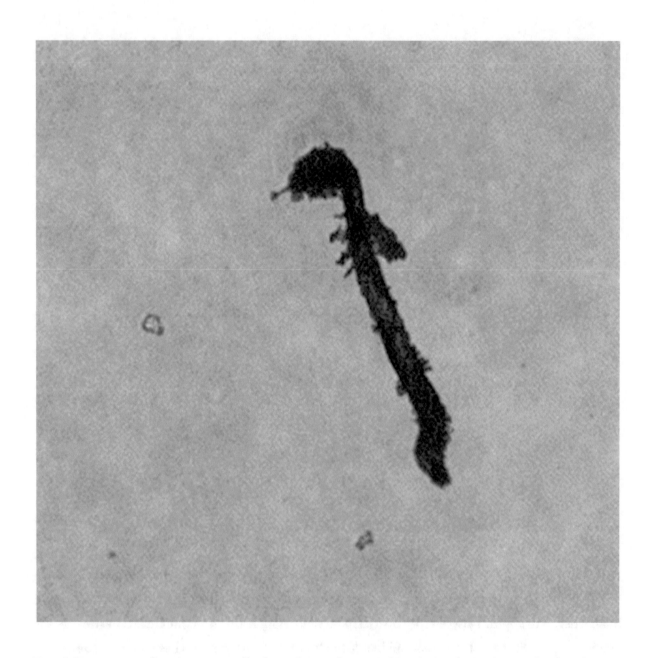

Fig. 4. Formation of numerous needle shaped crystals around a patient-derived particle placed on an agar plate. The needles formed over several days and are similar to those seen in cultures of stealth-adapted viruses.

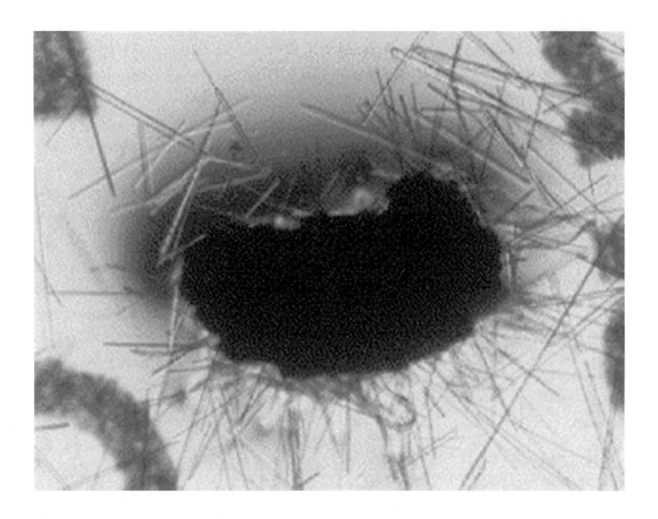

Fig. 5. An example of a complex pigmented structure that formed in a Petri dish that was growing bacterial colonies derived from a throat culture of a stealth-adapted virus positive patient. The structure measured approximately 1 in. and displayed various colors when viewed under either dark field or fluorescent microscopy.

Irritating Skin Fibers and the Epidemic of Stealth-Adapted Virus Infections

The epidemic of stealth-adapted virus infections explains the skin fibers forming in patients diagnosed as having delusional parasitosis or Morgellon's disease. The fibers comprise a source of alternative cellular energy (ACE). Clinical trials are underway using ACE generating Enerceuticals.

/24-7PressRelease/ - BURBANK, CA, February 05, 2007 - Stealth-adapted viruses are not effectively recognized by the cellular immune system and consequently do not provoke an inflammatory response, the accepted hallmark of an infectious disease. They can form through the loss or mutation of the relatively few viral components normally targeted by the cellular immune system. The current analogy is that of terrorists lacking a recognizable military insignia to avoid identification by Homeland Security.

The best characterized stealth-adapted viruses were derived from the African green monkey simian cytomegalovirus (SCMV), a likely contaminant of various batches of live polio virus vaccines. DNA of SCMV has been confirmed in licensed polio vaccines.

Although the immune system is ineffective in suppressing stealth-adapted viruses, the body can respond through an alternative cellular energy (ACE) pathway mediated by structures termed ACE pigments. These mineral containing organic structures readily develop in cultures of stealth-adapted viruses. Similar materials can also be seen in brain and skin biopsies of stealth-adapted virus infected humans and animals. Through a process

of self-assembly, these materials can take the form of fibers, threads, ribbons and discrete particles. They can be transparent or variously colored; will commonly fluoresce when illuminated with ultraviolet light; can display marked electrostatic and electron donating properties; and can occasionally be magnetic. Many of the organic components are aromatic terpenes and phenolics. Polymerization seemingly leads to clathrate formation with modified chelated minerals acting as transducers of physical energy to biological life support. ACE pigments can also participate in biosynthetic reactions, leading especially to the formation of their constituent components. ACE pigments are presumably a natural source of biological energy that may well represent a more fundamental process than photosynthesis.

Clearly, additional research is urgently required to more fully understand the dynamics of stealth-adapted virus infections and the ACE pathway of cellular repair. Extensive culturing for stealth-adapted viruses indicates that many of the common maladies becoming increasing prevalent in society are likely due to stealth-adapted virus infections. These include autism, learning disorders, allergies and even obesity in children; depression, chronic fatigue and neuropsychiatric disorders in young adults and neurodegenerative diseases in the elderly. Infections are commonly pervasive within individual families and are also seemingly readily transmitted through occupational and social exposures. Political considerations, partially relating to the safety of vaccines, have hindered the willingness of public health officials to seriously consider the concept of stealth-adapted viruses.

Some stealth-adapted virus infected patients have reported on the presence of strange fibers and particles that simply reflect the formation of ACE pigments. They can be seen on the skin, attached to hairs, left on bed sheets and forming in water used for bathing. They are occasionally erroneously identified as lice leading to the sometimes disastrous use of neurotoxic compounds such as Lindane. In some patients, the fibers cause severe irritation leading to scratching and ulceration of the skin. Electrostatic movements of the fibers can suggest a life form to some patients. This opinion may be rebuffed by physicians preferring instead to offer a psychiatric label of delusional psychosis. The term Morgellon's disease has been used to try to more narrowly define the phenomenon of skin ulceration with an emphasis on showing living properties of the fibers. Such patients are desperately seeking relief from their skin irritation and from the commonly accompanying emotional distress of not being understood by their physicians. The skin lesions are also of particular concern as potential sites of entry of toxic bacteria, such as community acquired methicillin resistant Staphylococcus aureus (MRSA).

Until efforts are undertaken to prevent transmission of stealth-adapted viruses, patients are wise to follow the lead of nature and to do whatever is possible to enhance the ACE pathway. One approach that has proven very effective in expediting the healing of both herpes virus and papillomavirus infected skin lesions is through a dye assisted ultraviolet light activation of ACE pigments. Another approach is to use products termed Enerceuticals with ACE generating activity. Ideally, these products will be evaluated in formal placebo-controlled clinial trials.

Patients who wish to be considered for such trials should contact the Institute of Progressive Medicine via an e-mail to [email=s3support@mail.com]

Morgellon's Disease: Not Really a Mystery!

The Institute of Progressive Medicine is inviting participants for a clinical trial on Morgellon's disease that is based on activating an alternative cellular energy (ACE) pathway. The studies should also help in the control of other illnesses caused by stealth adapted viruses.

/24-7PressRelease/ - BURBANK, CA, January 24, 2008 - Morgellon's disease is a somewhat inappropriate name given to a clinical condition characterized by ulcerating skin lesions accompanied by the production of particles and fibers. The fibers are irritating and can provoke scratching, which is the probable ultimate cause of the ulcerating skin lesions. The particles and fibers have characteristics of alternative cellular energy (ACE) pigments. They are commonly fluorescent and the intensity of fluorescence can be enhanced using various dyes.

ACE pigments were initially identified in patients infected with viruses that escape recognition by the cellular immune system. Termed stealth adapted viruses, they simply lack the few critical components viral normally targeted by cytotoxic lymphocyte. ACE pigments are also produced in response to conventional viruses, such as herpes simplex virus (HSV) and herpes zoster virus (HZV), the cause of shingles. Activation of the ACE pathway is proving a very effective method of controlling HSV and HZV infections. It is time to include some Morgellon's patients in these ongoing studies.

The Centers for Disease Control and Prevention (CDC) and patient supported organizations such as the Morgellon's Foundation have been very slow in understanding or accepting the connection between stealth adapted viruses and Morgellon's disease. Hopefully, they will be more responsive once marked clinical improvements can be demonstrated in a few well documented cases.

The issue of stealth adapted viruses will eventually prove to be a major embarrassment to the CDC. Similar to the belated response to the epidemic of methicillin resistant Staphylococcus aureus (MRSA) infections, CDC personnel have essentially ignored the many requests for open discussions on the issue of viruses causing the increasing prevalence of such illnesses as autism in children and the chronic fatigue syndrome in adults. Part of this reluctance is the unequivocal data indicating that some stealth adapted viruses clearly arose from simian cytomegalovirus (SCMV) and that the DNA of this virus is readily detectable in several lots of previously licensed polio virus vaccines. The story goes deeper since it can be reasonably argued that the testing of monkey cytomegalovirus contaminated polio vaccines in chimpanzees in Africa during the 1950's led to the formation of the AIDS virus.

Organizations representing Morgellon's patients may wish to assist in recruiting a few representative patients for participation in a research study aimed at activating the ACE pathway. The organizations should certainly convey to their members some of the available information concerning the postulated role of stealth adapted viruses in Morgellon's disease and related conditions. Some of this information is available within the Medical Commentary section at http://www.s3support.com For additional information, patients can contact the Institute via e-mail at s3support@mail.com or call 626-616-2868.

Stealth Adapted Virus Infections Possibly Mistaken for Bacterial Infections: Viteria and the Potential Over Diagnosis of "Chronic Lyme Disease," PANDAS and Mycoplasma Infections

DNA cloning and sequencing studies were performed on virus cultures obtained from a patient with chronic fatigue syndrome (CFS). In addition to virus sequences matching closely to those of African green monkey simian cytomegalovirus (SCMV), DNA sequences of bacterial origin were present. No evidence existed of actual bacterial contamination of the cultures. Moreover, many of the identified bacterial sequences were significantly different from common human bacterial pathogens. The data strongly suggests that bacterial DNA sequences were incorporated into the virus replication mechanism. Similar to the fragmented and genetically unstable virus genome, the bacterial sequences had also undergone complex mutational and recombination changes. The data also imply the probable replication of eukaryotic viruses through bacteria. Although not yet extensively studied, atypical bacteria with discordant typing profiles are being increasingly identified in clinical microbiology laboratories. Positive virus cultures were obtained on an extract from an atypical bacterial colony cultured from the feces of a CFS patient.

Based on rather imprecise serological and molecular assays, many CFS patients are diagnosed as having chronic Lyme disease since they have antibodies reactive with *Borrelia burgdorferi*, the spirochete that causes acute Lyme disease. Children with CFS-like illnesses who test positive for anti-DNase B and/or anti-streptolysin O antigens are being diagnosed as having PANDAS (Pediatric Autoimmune Neuropsychiatric Disorders Associated with Streptococcal infections). In both cases, actual culture confirmation of the presence of the respective microbe is rarely obtained. Yet cultures for stealth adapted viruses have near uniformly been positive in blood samples provided from patients diagnosed as having chronic Lyme disease and from a few children tested with PANDAS in their differential diagnosis.

A similar situation applies to the interpretation of molecular assays for mycoplasma in patients with CFS and related illnesses. A positive assay based on the detection of a small segment of the bacterial genome, does not necessarily indicate the entire microbe is present and causing the disease. Moreover, the molecular assays used can be rather non-specific in their reactivity and actual sequencing of the detected DNA sequence is rarely performed.

More convincing data attribute CFS to infection with stealth adapted viruses, which have either lost or mutated the relatively few genes coding for antigens targeted by the cellular immune system. It is proposed that these stealth adapted viruses have gained a propensity to replicate within and potentially be transmitted by bacteria. The term viteria is proposed for viruses containing bacterial sequences.

As noted above, it is not inconceivable that stealth adapted viruses may actually be propagated through bacteria. Attempts to actually infect bacteria with stealth adapted viruses were not performed because of safety considerations. It is possible that stealth adapted viruses play a role in altering the bacteria flora within the body of infected patients. Moreover, it is extremely difficult to prevent bacteria transmission within communities. The prospect of bacteria transmitted virus illnesses clearly has important public health consequences and deserves immediate study.

Are Stealth Virus Infections Misdiagnosed As Chronic Lyme Disease?

Abstract

This article reviews the basis for believing that stealth virus infected patients are being inadvertently diagnosed as having chronic Lyme disease. With only a few exceptions, blood samples from patients being treated for chronic Lyme disease have tested positive in an assay designed to detect stealth-adapted cytopathic viruses. It is known that a stealth virus has assimilated bacterial sequences, some of which are related to genes of *Borrelia burgdorferi*, the causative agent of classical acute Lyme disease. The presence of bacteria-derived sequences, rather than actual Borrelia bacteria, may account for the positive serological and molecular based assays that are often, but not always, seen in patients labeled as having chronic Lyme disease. The partial clinical response seen in response to antibiotic therapy may reflect the known capacity of certain antibiotics to suppress the production of virus-growth-enhancing chemokines. If the hypothesis of this paper is correct, then many patients currently being classified as having chronic Lyme disease, may respond to therapies being developed for stealth virus infections.

Lyme Disease Diagnosis

Laboratory support for a clinical diagnosis of chronic Lyme disease is currently provided by positive results in various antigen, antibody and/or molecular based assays for *Borrelia burgdorferi*. Inter-laboratory variability in the performance, reading and interpretation of Lyme disease testing has thrown into question the reliability of such assays, even to the extent that the clinical diagnosis is not infrequently sustained even in the face of negative or inconsistent laboratory findings. Conversely, over time, many patients already diagnosed as having Lyme disease on the basis of positive serological tests are now revealing additional positive assays to such diverse infectious agents as Babesia, Ehrlichiosis, Mycoplasma, Chlamydia, human herpesvirus-6, parvo virus, etc.

Chronic Fatigue Syndrome (CFS)

The clinical manifestations of patients currently labeled as having chronic Lyme disease are essentially indistinguishable from those exhibited by patients categorized as having chronic fatigue syndrome (CFS). In turn, a CFS diagnosis merges with vaguely defined conditions such as fibromyalgia, Gulf War syndrome and depression. Relevant to the present discussion, several CFS patients from non-Lyme disease endemic regions, such as Los Angeles, are showing positive serological test for Borrelia. Centers for treating Lyme disease are springing up with no basis beyond a positive Borrelia antigen and/or antibody test.

Stealth Viruses

The Center for Complex Infectious Diseases has focused on atypically structured viruses that evade the immune system because they no longer possess the major antigenic targets required to evoke effective anti-viral cellular immunity. Their capacity to evade immune defenses led to the use of the term "stealth" to describe such viruses. In spite of the loss of certain viral antigens, stealth viruses are still able to replicate in, and to cause damage to, cells. The mechanisms that allow for stealth viruses to retain and/or regain their cytopathic (cell damaging) activity are not yet understood, but are likely to be related to their capacity to capture additional genes from infected cells (discussed below). Although stealth viruses describe a diverse group of structurally distinct viruses, they share a general property of inducing a foamy vacuolating cytopathic effect (CPE) in cells of multiple species. They are, therefore, most readily detectable using in vitro culture techniques that are based on observing this characteristic CPE.

Stealth viruses were initially detected in patients with CFS. They have since been positively correlated with a wide range of neurological, psychiatric, auto-immune and malignant diseases. An obvious question is whether stealth virus infections could be misdiagnosed as chronic Lyme disease. If so, what is the explanation for the variably positive antigen, antibody and/or molecular probe based assays that use reagents that react with Borrelia bacteria?

In approaching this problem, CCID has been successful in demonstrating positive stealth virus cultures in blood samples from over 90% of patients referred with a diagnosis of chronic Lyme disease. By comparison, the incidence of a positive stealth virus culture among healthy blood donors and healthy non-medical personnel is less than 15%. The high prevalence of stealth virus infections in patients with chronic Lyme diseases matches well with similar high prevalence rates in CFS patients and patients with other forms of stealth virus associated illnesses (discussed below).

Viteria

An explanation for the positive Borrelia bacteria based assays may lie in the additional capacity of stealth viruses to assimilate bacterial genes. Normally, viruses that are infectious for human or animal cells (eukaryotic cells) will not infect bacteria (prokaryotic cells). Stealth viruses appear to have overcome this phylogenetic barrier. The term "viteria" has been coined to define eukaryotic viruses that have acquired bacteria-derived genetic sequences. The sources of the bacteria sequences include both eubacteria and archaebacteria and both cell wall containing and cell wall deficient bacteria.

Where do stealth viruses capture bacterial genes?

There is increasing evidence for the direct infection of various bacteria by stealth viruses, including the culturing of bacteria with unusual biochemical profiles from stealth virus infected patients. The assimilation could also occur within eukaryotic cells that are dually infected with a stealth-adapted virus and intracellular bacteria. Among the more notable intracellular bacteria are those that lack rigid cell walls. These include spirochetes, of which *Borrelia burgdorferi* is a prime example; *Mycoplasma fermentans* and *incognitus;* and *Rickettsiae.* Since these bacteria do not have sterol synthesizing capacities, they are entirely dependent on the infected cell to provide their lipid cell walls. As noted above, lipid ladden cells are a hallmark of the CPE induced by stealth viruses. By simply co-culturing stealth viruses and *Borrelia burgdorferi*, one can observe a marked enhancement in the intracellular growth of the Borrelia (personal observation). To one extent, this could suggest that an underlying stealth virus infection could create an in vivo environment of Borrelia growth promoting cells.

The rich variety of bacterial sources that have seemingly contributed to the prototypic viteria suggests a diverse array of bacterial-related antigens. Very few, if any, of the currently available commercial assays for Borrelia utilize strict criteria for specificity. At this time, it is considered more likely that within the broad scope of assimilated bacterial sequences, some would be found that simply cross react with the reagents commonly used in Borrelia testing.

Response to antibiotics

An argument advanced for Borrelia infection is that many patients show a partial clinical response following the administration of certain antibiotics. This conclusion discounts the various cellular activities mediated by antibiotics outside of their direct effect on bacteria. For example, erythromycin and clarithromycin (Biaxin) are known to suppress the cellular synthesis of chemokines. This observation is relevant, since at least some stealth viruses have multiple genes encoding both chemokine and chemokine receptors.

Significance

If the hypothesis of this paper is correct, patients mistakenly being diagnosed as having chronic Lyme disease should benefit from courses of therapy designed to suppress stealth viruses. Ongoing studies are evaluating the use of a combination of anti-oxidant, anti-rheumatic, antibiotics, and other drugs known to down regulate chemokine production. Patients should have their clinicians contact CCID for additional information.

References

Published articles on stealth viruses and viteria can be found on the web site www.ccid.org Any additional inquiries can be addressed to CCID at 1634 Spruce Street, South Pasadena CA 91030 USA or via email to wjohnmartin@ccid.org.

Encephalopathy Among Veterans: Post Traumatic Stress Disorder (PTSD): A Probable Stealth Virus Encephalopathy

The failure to recognize PTSD as a neurological rather than a psychological illness continues to plague military medicine and to under-serve many of its most troubled veterans. The role of brain-damaging stealth adapted viruses in individuals with PTSD needs to be explored.

/24-7PressRelease/ - BURBANK, CA, March 15, 2007 - A grossly inadequate understanding by military physicians of the condition labeled post traumatic stress disorder (PTSD) underscores much of the current controversy regarding sub-optional healthcare of veterans. As with Gulf War syndrome (GWS), the military hierarchy has resisted attempts to identify an infectious cause for PTSD. Subtle signs of organic brain disease, consistent with a viral encephalopathy have either not been sought or discarded when observed in PTSD patients. Instead, the cause of PTSD has been attributed to a maladaptive psychological response to the stress of war. All too often this reaction is suspiciously viewed as blending with a voluntary shirking of social responsibilities, which can lead the veteran to a seemingly unjustified expectation of lifelong disability. An adversarial challenge can ensure with the mistaken belief that any medical discussion of a potential physical disorder will likely prolong and intensify the disability claim.

Reports indicating that GWS is also a consequence of viral infection of the brain were ignored by the military because of the obvious inference that the disease might then be contagious and could potentially involve other family members. Even when veterans did report subsequent illnesses among family members, including children, the military was unwilling to perform virus cultures on the veteran or on close personal contacts. The Centers for Disease Control and Prevention (CDC) has also refused to culture blood samples from chronic fatigue syndrome (CFS) or GWS patients. Were they to do so, the majority of these patients would be shown to be actively infected with stealth adapted viruses. So too probably will the majority of veterans labeled as having PTSD.

Stealth-adapted viruses is simply a generic term used to describe cell damaging (cytopathic) viruses that lack the few critical components normally targeted by the cellular immune system. Stealth viruses do not evoke the typical inflammatory reaction expected with most conventional viruses. Still, signs of a brain infection can be seen especially in the subtle alterations of cognitive and emotional functions; typically accompanied by fatigue and insomnia. Stealth adapted viruses can be readily cultured in the laboratory using minor adaptations to existing established procedures.

The best characterized stealth adapted viruses originated from the simian cytomegalovirus of African green monkeys (SCMV). African green monkeys were used extensively to produce live polio virus vaccines. Indeed, the Food and Drug Administration (FDA) has confirmed the presence of SCMV DNA in several licensed polio vaccines approved for used during the 1970's. Again this finding has not been pursued for largely political and economic reasons. Consequently, the epidemic of stealth virus infections, both symptomatic and asymptomatic, has been allowed to proceed without regard to the medical consequences.

Stress is a known activator of cytomegalovirus and a clear association can exist between stress and stealth adapted virus induced brain damage. Few psychiatrists are aware of such an association and, therefore, do not address this issue as part of routine psychiatric care.

Veterans with PTSD should be tested for infection by stealth adapted viruses. If the culture is positive, their illnesses need to be further evaluated as a probable stealth virus encephalopathy and appropriate therapy instituted. Resolution of this issue may go a long way towards alleviating the frustration of many veterans and their family members.

Stealth Adapted Viruses, Alternative Cellular Energy (ACE) Pathway and Alzheimer's Disease

A contributing role for viruses in the pathogenesis of Alzheimer's disease is supported by substantial indirect evidence, especially in regards to the possible involvement of herpes viruses. The lack of a major inflammatory reaction associated with the disease has been a stumbling block in advancing the virus hypothesis. The lack of inflammation can, however, be due to an immune evasion process termed "stealth adaptation." This occurs when viruses lose or mutate the relatively few components normally targeted by the cellular immune system. Stealth adapted viruses have been cultured from patients with a wide range of non-inflammatory neuropsychiatric illnesses, including dementia. The most promising aspect of these studies is that both conventional and stealth adapted viruses can be effectively suppressed by a non-immunological defense mechanism, which involves the alternative cellular energy (ACE) pathway. Local activation of the ACE pathway in the vicinity of active herpes simplex virus (HSV) skin infections leads to expedited healing of the lesions. Recent studies have linked the ACE pathway to the intrinsic energy levels of the body's fluids. Moreover, the consumption and/or administration of energized fluids; ingestion of enerceutical™ foods and dietary supplements; and the application of certain energy delivering medical devices can enhance the ACE pathway; as shown by the reduced production of fluorescing ACE pigments. These same approaches to therapy warrant clinical trials in Alzheimer's patients. Even beyond the role of the ACE pathway in the suppression of virus infections, it can potentially compensate for deficiencies in the supply of adequate oxygen, blood and/or nutrients required for normal mitochondria mediated cellular metabolism. Studies will be presented on the results of monitoring and enhancing the ACE pathway in Alzheimer's disease patients. The major focus will be on evaluating the potential benefits of consuming ACE Water™ produced by a variety of inexpensive in-home methods. If successful, the studies will be followed by efforts at preventing or at least delaying the onset of Alzheimer's disease and other dementias.

The following article under the title "**Stealth Adapted Viruses – Possible Drivers of Major Neuropsychiatric Illnesses Including Alzheimer's Disease**" was published on 4/30/2014 in J Alzheimers Dis Parkinsonism 4: 144. doi: 10.4172/2161-0460.1000144

Abstract

Mainstream neurologists and psychiatrists have largely refrained from serious consideration of a virus cause of common brain diseases. This is mainly because of the general lack of any accompanying immune system stimulated inflammatory reaction within the brain. This article exposes a weakness in this argument by describing the process of "stealth adaptation" of viruses. Deletion or mutation of relatively few virus components can result in derivative viruses, which are no longer effectively recognized by the cellular immune system. Consequently, there is no triggering of the inflammatory response. Furthermore, the brain is uniquely susceptible to symptomatic illness caused by stealth adapted viruses. An understanding of stealth adaptation greatly expands the potential scope of viral illnesses. It also underscores the value of using virus cultures as a diagnostic tool and of taking appropriate measures to avoid transmission of infection. More importantly, therapeutic measures are available for suppressing both stealth adapted and conventional virus infections through enhancement of the alternative cellular energy (ACE) pathway. Such measures are available for clinical evaluation in treating many of the major illnesses affecting the brain, including Alzheimer's disease.

Introduction:

With the possible exception of cancer, diseases of the brain comprise most of the worst feared human aliments. Whether it is the increasing likelihood of autism occurring in children (1) or Alzheimer's disease occurring in the elderly (2); the current medical paradigms are failing to provide effective answers. Nor has significant progress been made in addressing many other tragic illnesses, such as schizophrenia, bipolar psychosis, Parkinson's disease and amyotrophic lateral sclerosis (ALS). Society is further challenged by a range of somewhat less severe but still disabling illnesses, including chronic fatigue syndrome (CFS), fibromyalgia, depression, drug addiction, criminal behaviors and intellectual impairments affecting both learning and work performances.

Various investigators have indeed suggested a possible infectious origin of several of the aforementioned neuropsychiatric illnesses (3-7). Because of the absence of noticeable inflammation, however, it is usually argued that if infectious agents are indeed involved, their effects must be indirect and quite possibly delayed. For example, common infections occurring during pregnancy can clearly lead to elevated levels of various cytokines as part of the immune response. There are supporting data that maternal cytokines may significantly inhibit normal fetal brain development, with potential later life consequences (8-10). Other researchers have suggested that certain infectious agents might potentially trigger the production of antibodies, which cross-react with neuronal components, thereby interfering with normal brain function. Such self-reacting antibodies may continue to form as part of an ongoing autoimmune process (11-13). None of these scenarios envisions viruses as the direct cause of ongoing cellular injury.

Evidence for a Role of Herpes Simplex Virus (HSV) in Alzheimer's Disease:

There are, nevertheless, data suggesting an active role of HSV in the ongoing pathogenesis of Alzheimer's disease (14-17). Additional indications include the following findings:

i)   HSV infected fibroblasts secrete both beta-amyloid and tau-proteins, which are the accepted diagnostic markers of Alzheimer's disease (18-19). Rather than being a primary cause of Alzheimer's disease, these markers may, therefore, be secondary phenomena occurring as a consequence of virus damage to neuronal cells. Moreover, these markers may simply accumulate more in the brains of older individuals because of an age-related decline in the normal clearance (removal) mechanism of degenerated cellular and extracellular materials (20).

ii)   Budding herpes viruses directly interact with amyloid precursor protein, affecting the normal transport and distribution of this protein (21-22)

iii)   HSV DNA is detected in association with the amyloid plaques and also with the neurofibrillary tangles that are mostly comprised of phosphorylated tau proteins (23).

iv)   Alzheimer's disease plaques and tangles also contain complement factors and HSV binding cellular proteins (24).

v)   The distribution of the Alzheimer's disease markers progressively involves increasing areas of the brain; consistent with the spreading of an infectious process (25).

vi)   Many of the gene alleles shown to significantly enhance susceptibility to Alzheimer's disease, e.g. ApoE-e4, are also known to promote the infectivity of cells by HSV (26-27).

vii) Anti-HSV antibody levels, including IgM, increase in conjunction with the onset of Alzheimer's disease (28); yet the avidity (binding capacity) of anti HSV antibody is lower in those with more severe disease (29), consistent with reduced antibody protection.

viii) Suggestive clinical improvements have occurred in Alzheimer's disease patients receiving therapies that could potentially inhibit HSV. Examples include intravenous gamma globulin (30) and statins (31).

HSV is an extremely common human infection and can be detected in brains of individuals with or without Alzheimer's disease (32). Similarly, some unaffected individuals can possess the same genetic markers as are present in a higher proportion of patients with Alzheimer's disease. Possibly, Alzheimer's disease results from infection with only certain variant HSV occurring in genetically prone individuals. It is noteworthy that studies to potentially distinguish HSV isolates from Alzheimer's disease patients and controls have not been pursued.

Lack of Disease Specificity of Identified Genetic Markers of Alzheimer's Disease:

Similar genetic markers operating in Alzheimer's disease are increasingly being found in association with other neuropsychiatric illnesses such as autism, bipolar psychosis and schizophrenia (26). Moreover, several of the identified genes not only enhance the infectivity of HSV but also of other viruses. Furthermore, many herpes viruses can induce similar types of acquired genetic changes as commonly seen in the genome of patients with various neuropsychiatric illnesses. These include shortened leukocyte telomeres (33) and altered copy numbers of certain gene sequences (34). These findings help deemphasize the focus on the existence of a specific Alzheimer's disease virus. Instead, they suggest the possibility of multiple types of pathogenic viruses, each of which can potentially cause differing diseases depending upon host genetic factors and on other variables. Still, the major stumbling block for many of those proposing an ongoing infectious cause of neuropsychiatric illnesses is the lack of an accompanying inflammatory response and the general inability to reliably culture viruses from the patients (3-4). In reality, these are not valid arguments since they do not exclude infections with stealth adapted viruses.

Detection of Stealth Adapted Viruses:

By adjusting the virus culturing technique, atypical viruses were detected in the vast majority of patients with symptomatic neurological and psychiatric illnesses (35-43). Some of the cultured viruses were molecularly characterized as derivatives of African green monkey simian cytomegalovirus (SCMV). Other cultured viruses showed differing patterns of molecular reactivity, consistent with the concept that stealth adaptation is a generic process, which may occur with all human and animal viruses. In the case of human cytomegalovirus (HCMV), an estimated 90% of the evoked cytotoxic T lymphocytes (CTL) are directed against only 3 of the over 200 virus coded components (44). Deletion and/or mutation of these 3 coding genes can result in a virus still capable of inducing illness, but without evoking any inflammation. Moreover, downsizing of the virus genome can help explain the apparent widening of the species susceptibility of certain stealth adapted viruses, including the possibility of transmission to bacteria (45).

Lack of Precision in Clinical Diagnoses of Neurodegenerative Illnesses:

A major overlap in clinical diagnoses exists in elderly patients between severe CFS and early Alzheimer's disease. Loss of short-term memory and other cognitive deficits are common to both illnesses. A 55 year-old physician maintained that he had CFS even though it became impossible for him to drive. Once he waited at a traffic light till another car came by because he had forgotten whether to proceed on the green or on the red signal. He was unable to take his patients' pulse rates since he could not recall both the counts and the elapsed time. An examining neurologist even contemplated a diagnosis of schizophrenia, when the patient spoke about "multiple little men in my brain not listening to each other." The neurologist categorized the patient as having Alzheimer's disease. Blood cultures and a culture of cerebrospinal fluid (CSF) yielded strong cytopathic effect (CPE) indicative of stealth adapted viruses.

In a family setting, a 65-year-old man showed progressive deterioration in his demeanor and personality. He began to express anger, complain of impaired memory and slept excessively during the day. He would sit idly with a blank stare. The illness forced his retirement as a bank manager. He was prescribed donepezil hydrochloride (Aricept) for Alzheimer's disease. Over the next several months his wife lost her capacity to provide adequate care as she too was becoming emotionally distant, fatigued and paranoid. The couples' daughter made arrangements to share her house with her parents. She soon developed CFS while her husband also began to lose some cognitive skills. He went on to develop amyotrophic lateral sclerosis (ALS). The health of their 4 children also began to deteriorate.

The eldest son noted a marked loss of short-term memory and diminishing muscle strength with frequent tingling. A 14-year-old daughter had a distinct mono-like illness with sore throat and fatigue that did not fully resolve. Her school andsporting performance changed from being a gifted student active on the softball team to barely being able to cope with her studies and relegated to a back-up cheerleading squad. Her mother withdrew her from school to try to provide home schooling to make up for the shortcomings in her learning capacity. She was prescribed Prozac for depression and also began to experience frequent migraines and to become somewhat obese. Two younger daughters also began to experience short-term memory loss and were soon unable to attend regular school because of an attention deficit disorder.

This striking family history is clearly consistent with an infectious process. In much the same way that CFS is not regarded as a contagious disease, it is possible that a strong bias exists against attributing any risk to caregivers of patients with Alzheimer's disease. Yet caregiver burnout is a recognized condition, which has not been thoroughly investigated (46-47). Many genetic and environmental factors can modify the course of an infection, but can also be mistaken as a primary cause of illness if the infectious agent is overlooked.

Animal Inoculation Studies:

Additional evidence for disease transmission by stealth adapted viruses was obtained in animal inoculation studies. Cultured stealth adapted viruses from both a CFS patient and a cognitively impaired patient with systemic lupus erythematosus (SLE) were inoculated into cats (48). The cats developed a severe acute encephalopathy with marked behavioral changes. Histological examination of the brain and other tissues showed foci of cells with cytoplasmic vacuolization in the absence of any inflammatory reaction. Nevertheless, the animals clinically recovered from the acute illness in the ensuring weeks; a finding consistent with the alternative cellular energy (ACE) repair process previously identified in cultures of stealth adapted viruses.

Alternative Cellular Energy (ACE) Pathway:

A cellular defense mechanism mediated by the ACE pathway can suppress the CPE caused by stealth adapted viruses. As reported elsewhere, the CPE in cultures of stealth adapted viruses will typically regress in infrequently re-fed cultures (49). Along with studies on human brain biopsies (50), mineral containing particulate materials accumulating in the infrequently re-fed cultures are effective in suppressing virus CPE. These materials, termed ACE pigments, have energy transducing (converting) properties and apparently maintain cellular viability in spite of marked disruption in the cells' mitochondria. ACE pigments are electrostatic; fluorescent in both ultraviolet (UV) and visible light, especially upon the addition of certain dyes, including neutral red and acridine orange; display kinetic movements well beyond Brownian motions; are occasionally ferromagnetic; and can act as electron donors (49). Further insight into the ACE pathway was provided by studies on the effectiveness of neutral red dye plus UV light in the expedited healing of skin lesions caused by HSV (51).

Methods are being developed to effectively monitor the ACE pathway in individual patients. Methods include the detection of fluorescing ACE pigments in dried perspiration and in saliva, with and without the use of a triggering agent, such as neutral red dye. The intrinsic energy level of urine and other bodily fluids can also be assessed. Even the prolongation of the duration of breadth holding may become a useful ongoing measure of beneficial enhancement of the ACE pathway.

The mode of action of ACE pigments in virus cultures has tentatively been identified as altering the kinetic activity of water molecules in the tissue culture medium. In turn, the activated water enhances the ability of the cells to reverse the virus induced CPE. The repair process is sustained by an ongoing water activation process and yet can be rapidly reversed by the replacement of ACE pigment-containing medium with fresh medium.

Enerceuticals™:

Natural products have been identified with ACE pigment-like, water-modifying activity and are termed enerceuticals™. Examples include fresh extracts of moringa oleifera leaves and certain preparations of humic/fulvic acids and zeolites. Plans are underway to assess the potential therapeutic benefits of these products in patients with Alzheimer's disease. Moreover, trials can also proceed assessing the potential benefit of consuming KELEA™ (kinetic energy limiting electrostatic attraction) activated water and/or using other direct means of enhancing the ACE pathway.

Additional Potential Benefits of Enhancing the ACE Pathway:

The ACE pathway can provide potential benefits to cells beyond enhancing the cells capacity to suppress virus induced CPE. Regular cellular metabolism requires adequate supplies of oxygen and nutrients along with efficient removal of carbon dioxide and metabolic waste products. Structural damage to the brain and its blood supply, as can occur in Alzheimer's disease patients, can lead to an insufficiency of cellular energy (ICE) in brain cells. A more effective ACE pathway may help restore impaired cellular function resulting from structural damage to the brain.

Conclusion:

A role for infectious agents in the pathogenesis of Alzheimer's disease deserves consideration. Stealth adapted viruses are overlooked by the cellular immune system. These viruses have also been largely overlooked in major studies on Alzheimer's disease. Instead the focus in Alzheimer's disease research has been the somewhat questionable assumption that beta amyloid and/or phosphorylated tau proteins are the primarily cause of brain damage. Additional animal inoculation studies using patient-derived stealth adapted viruses inoculated into older animals and into animals with an impaired ACE pathway will provide valuable testing of the underlying hypothesis of this paper. The opportunity also exists to proceed directly with clinical trials in Alzheimer's patients based on monitored enhancement of the ACE pathway. Beneficial effects will add credence to the stealth adapted virus hypothesis. Moreover, stabilization of illness or actual clinical improvements will support prevention strategies based on enhancing the ACE pathway throughout life. Prevention should also entail efforts aimed at minimizing the risks of ever becoming infected with stealth adapted viruses.

References

1. Centers for Disease Control and Prevention (2012) Prevalence of autism spectrum disorders--Autism and Developmental Disabilities Monitoring Network, 14 sites, United States, 2008. MMWR Surveill Summ. 61: 1-19.

2. Thies W, Bleiler L (2013) 2013 Alzheimer's disease facts and figures. Alzheimers Dement. 9: 208-45.

3. Ahokas A, Rimón R, Koskiniemi M, Vaheri A, Julkunen I, et al (1987) Viral antibodies and interferon in acute psychiatric disorders. J Clin Psychiatry 48: 194-6.

4. Taylor GR, Crow TJ, Ferrier IN, Johnstone EC, Parry RP, et al (1982) Virus-like agent in CSF in schizophrenia and some neurological disorders. Lancet 2: 1166-7.

5. Torrey EF (1988) Stalking the schizovirus. Schizophr Bull. 14: 223-9.

6. Pert CB, Knight JG, Laing P, Markwell MA (1988) Scenarios for a viral etiology of schizophrenia. Schizophr Bull. 14: 243-7.

7. Yolken R (2004) Viruses and schizophrenia: a focus on herpes simplex virus. Herpes 11 Suppl 2: 83A-88A.

8. Ratnayake U, Quinn T, Walker DW, Dickinson H (2013) Cytokines and the neurodevelopmental basis of mental illness. Front Neurosci. 7:180.

9. Buka SL, Cannon TD, Torrey EF, Yolken RH (2008) Collaborative study group on the perinatal origins of severe psychiatric disorders. Maternal exposure to herpes simplex virus and risk of psychosis among adult offspring. Biol Psychiatry 63: 809-15.

10. Brown AS, Patterson PH (2011) Maternal infection and schizophrenia: implications for prevention. Schizophr Bull. 37: 284-90.

11. Hornig M (2013) The role of microbes and autoimmunity in the pathogenesis of neuropsychiatric illness. Curr Opin Rheumatol. 25: 488-795.

12. Bach JF (2005) Infections and autoimmune diseases. J Autoimmun. 25 Suppl: 74-80.

13. Carter CJ (2011) Schizophrenia: A pathogenetic autoimmune disease caused by viruses and pathogens and dependent on genes. J Pathog. 2011: 128318.

14. Ball MJ (1982) Limbic predilection in Alzheimer's dementia: is reactivated herpes virus involved? Can J Neurol Sci. 9: 303–306.

15. Itzhaki RF, Lin WR, Shang D, Wilcock GK, Faragher B, et al (1997) Herpes simplex virus type 1 in brain and risk of Alzheimer's disease. Lancet 349: 241-4.

16. Itzhaki RF, Cosby SL, Wozniak MA (2008) Herpes simplex virus type 1 and Alzheimer's disease: the autophagy connection. J Neurovirol. 14: 1-4.

17. Wozniak MA, Itzhaki RF, Shipley SJ, Dobson CB (2007) Herpes simplex virus infection causes cellular beta-amyloid accumulation

and secretase upregulation. Neurosci Lett. 429: 95–100.

18. Wozniak MA, Frost AL, Itzhaki RF (2009) Alzheimer's disease-specific tau phosphorylation is induced by herpes simplex virus type 1. J Alzheimers Dis. 16: 341-50.

19. Shipley SJ, Parkin ET, Itzhaki RF, Dobson CB (2005) Herpes simplex virus interferes with amyloid precursor protein processing. BMC Microbiol. 5: 48.

20. Nakanishi H (2003) Neuronal and microglial cathepsins in aging and age-related diseases. Ageing Res Rev. 2: 367–81.

21. Cheng S-B, Ferland P, Webster P, Bearer EL (2001) Herpes simplex virus dances with amyloid precursor protein while exiting the cell. PloS One. 6: e17966

22. Wozniak MA, Mee AP, Itzhaki RF (2009) Herpes simplex virus type 1 DNA is located within Alzheimer's disease amyloid plaques. J Pathol. 217: 131-8.

23. Carter CJ (2011) Alzheimer's disease plaques and tangles: Cemeteries of a pyrrhic victory of the immune defence network against herpes simplex infection at the expense of complement and inflammation–mediated neuronal destruction. Neurochem Int. 58: 301-20

24. Ball MJ, Lukiw WJ, Kammerman EM, Hill JM (2013) Intracerebral propagation of Alzheimer's disease: strengthening evidence of a herpes simplex virus etiology. Alzheimer's Dement. 9: 169-75.

25. Lin WR, Shang D, Itzhaki RF (1996) Neurotropic viruses and Alzheimer disease. Interaction of herpes simplex type 1 virus and apolipoprotein E in the etiology of the disease. Mol Chem Neuropathol. 28: 135-41.

26. Carter CJ (2010) Alzheimer's disease: a pathogenetic autoimmune disorder caused by herpes simplex in a gene-dependent manner. Int J Alzheimers Dis. 2010: 140539.

27. Letenneur L, Pérès K, Fleury H, Garrigue I, Barberger-Gateau P, et al (2008) Seropositivity to herpes simplex virus antibodies and risk of Alzheimer's disease: a population-based cohort study. PLoS One 3: e3637.

28. Kobayashi N, Nagata T, Shinagawa S, Oka N, Shimada K, et al (2013) Increase in the IgG avidity index due to herpes simplex virus type 1 reactivation and its relationship with cognitive function in amnestic mild cognitive impairment and Alzheimer's disease. Biochem Biophys Res Commun. 430: 907-11.

29. Devi G, Schultz S, Khosrowshahi L, Agnew A, Olali E (2008) A retrospective chart review of the tolerability and efficacy of intravenous immunoglobulin in the treatment of Alzheimer's disease. J Am Geriatr Soc. 56: 772–4.

30. Wozniak MA, Itzhaki RF (2013) Intravenous immunoglobulin reduces β amyloid and abnormal tau formation caused by herpes simplex virus type 1. J Neuroimmunol. 257: 7-12.

31. Hill JM, Steiner I, Matthews KE, Trahan SG, Foster TP, et al (2005) Statins lower the risk of developing Alzheimer's disease by limiting lipid raft endocytosis and decreasing the neuronal spread of Herpes simplex virus type 1. Med Hypotheses. 64: 53–8.

32. Wozniak Shipley SJ, Combrinck M, Wilcock GK, Itzhaki RF (2005) Productive herpes simplex virus in brain of elderly normal subjects and Alzheimer's disease patients. J Med Virol. 75: 300–6.

33. Effros RB (2011) Telomere/telomerase dynamics within the human immune system: effect of chronic infection and stress. Exp Gerontol. 46:135-40.

34. Iourov IY, Vorsanova SG, Yurov YB (2013) Somatic cell genomics of brain disorders: a new opportunity to clarify genetic-environmental interactions. Cytogenet Genome Res.139: 181-8.

35. Martin WJ (1992) Detection of viral related sequences in CFS patients using the polymerase chain reaction. In: Hyde BM, editor. The Clinical and Scientific Basis of Myalgic Encephalomyelitis Chronic Fatigue Syndrome. Ottawa. Nightingale Research Foundation Press. pp 27883.

36. Martin WJ, Zeng LC, Ahmed K, Roy M (1994) Cytomegalovirus-related sequence in an atypical cytopathic virus repeatedly isolated from a patient with chronic fatigue syndrome. Am J Pathol. 145: 440-51.

37. Martin WJ (1994) Stealth viruses as neuropathogens. CAP Today 8: 67-70.

38. Martin WJ, Ahmed KN, Zeng LC, Olsen JC, Seward JG, et al (1995) African green monkey origin of the atypical cytopathic 'stealth virus' isolated from a patient with chronic fatigue syndrome. Clin Diagn Virol. 4: 93-103.

39. Martin WJ (1995) Stealth virus isolated from an autistic child. J Autism Dev Disord. 25: 223-24.

40. Martin WJ (1996) Severe stealth virus encephalopathy following chronic-fatigue-syndrome-like illness: clinical and histopathological features. Pathobiology 64:1-8.

41. Martin WJ (1996) Stealth viral encephalopathy: Report of a fatal case complicated by cerebral vasculitis. Pathobiology 64: 5963.

42. Martin WJ (1996) Simian cytomegalovirusrelated stealth virus isolated from the cerebrospinal fluid of a patient with bipolar psychosis and acute encephalopathy. Pathobiology 64: 646.

43. Martin WJ, Anderson D (1997) Stealth virus epidemic in the Mohave Valley. I. Initial report of virus isolation. Pathobiology 65: 51-6.

44. Wills MR1, Carmichael AJ, Mynard K, Jin X, Weekes MP, et al (1996) The human cytotoxic T-lymphocyte (CTL) response to cytomegalovirus is dominated by structural protein pp65: frequency, specificity, and T-cell receptor usage of pp65-specific CTL. J Virol. 70: 7569-79.

45. Martin WJ (2005) Alternative cellular energy pigments from bacteria of stealth virus infected individuals. Exp Mol Pathol. 78: 215-7.

46. Hubbell L1, Hubbell K (2002) The burnout risk for male caregivers in providing care to spouses afflicted with Alzheimer's disease. J Health Hum Serv Adm. 25: 115-32.

47. Takai M, Takahashi M, Iwamitsu Y, Ando N, Okazaki S, et al (2009) The experience of burnout among home caregivers of patients with dementia: relations to depression and quality of life. Arch Gerontol Geriatr. 49: e 1-5.

48. Martin WJ, Glass RT (1995) Acute encephalopathy induced in cats with a stealth virus isolated from a patient with chronic fatigue syndrome. Pathobiology 63: 115-8.

49. Martin WJ (2003) Stealth virus culture pigments: A potential source of cellular energy. Exp Mol Path. 74: 210-23.

50. Martin WJ (2003) Complex intracellular inclusions in the brain of a child with a stealth virus encephalopathy. Exp Mol Pathol. 74: 197-209.

51. Martin WJ, Stoneburner J (2005) Symptomatic relief of herpetic skin lesions utilizing an energy-based approach to healing. Exp Mol Pathol. 78: 131-4.

Pain Syndromes: Myofascial Disorders

Osteopathy and chiropractic disciplines focus on the physical interrelationships between all parts of the body; such that a disorder arising from one location can exert deleterious effects elsewhere in the body. Connective tissue provides a structural network, which operates throughout the body by way of fascia covering the organs, muscles, tendons, joints, nerves and blood vessels, and extending between and within such structures. This flexible, porous, web-like system comprises fibroblasts and extracellular collagen fibers, elastic fibers and cross-linking glycosaminoglycans. It provides sturdy, symmetrical mechanical support to the body's structures, while allowing for coordinated, purposeful muscle and joint movements. These normal movements can, in turn, assist in a continuing remodeling of the fascia by stimulating fibroblast-mediated replacement of degrading fibrous tissue, with suitably oriented newly synthesized fibers and cross-linking components. In some circumstances, atypical amounts of fibrous tissue can be formed, some of which may be unduly contracted or abnormally oriented. In other situations there can be a deficiency of fibrous support due to limited production or undue stretching.

Functional disruption of the fascia network can either be a cause or a consequence of adjacent illness, with further secondary mechanical effects potentially occurring elsewhere throughout the rest of the body. The term myofascia is commonly used to underscore the frequent symptomatic linkage of regions of persistently contracted muscles with accompanying mechanical strain extending into the adjoining fascia. Fibroblasts within the fascia can also convert into myofibroblasts with contractile properties. This occurs, for example, with inflammation leading to the term fibrositis. Fibroblasts can also transform to adipocytes when there is a need to store fat.

Pain is a dominant symptom of many myofascial disorders. It can either be constantly present or evoked by applied pressure, as in the testing for trigger points. These points can be focused on an easily palpable muscle or otherwise confined to a limited region or regions of the body. Muscle sensitivity can also be more generalized affecting entire muscle groups with radiating pain throughout much of the body. Pain is typically somewhat relieved by stretching of the involved muscle by osteopathic manipulative treatment (OMT) and chiropractic realignment. This induces relaxation of the locally contracted muscles, sometime accompanied by more widespread symptomatic relief. These treatments may also be acting by enhancing the blood and/or lymphatic circulation to an area or by stimulating the delivery of trophic factors via nerve cells.

Fatigue; cognitive impairment, such as memory loss and limited speaking vocabulary; emotional irritability and non-restful sleep are commonly associated symptom in some patients with recurrent, generalized myofacial pain. The combination of neurocognitive and myofascial symptoms comprises an illness called fibromyalgia (FM). Similar neurocognitive disorder, without marked myofacial pain, is more typical of the chronic fatigue syndrome (CFS). Many physicians and especially psychiatrists believe that most of these patients employ normal bodily stimulation as a psychological support mechanism to avoid dealing with reality. Terms such as somatizition and bodily distress disorder are used with the added reluctance of the physician to discuss any possible physical process. To do so, it is argued, is to fuel the patient's misconception of the illness by providing yet another attributable cause.

More objectively defined clinical conditions associated with myofascial pain include: lumbar and cervical spondylosis; piriformis (gluteal muscle) syndrome; tempromandibular joint and muscle disorder; carpel tunnel syndrome; vulvodynia in females and chronic pelvic pain syndrome in males. Altogether, myofasicial pain-related illnesses comprise many of the common medical conditions for which primary care is sought.

Less is known about the apparent association of myofasia disorder and organ dysfunction. Several potential mechanisms may be operating including: i) Unrelenting excess pressure on the parenchymal cells of the organ from contracted fascia covering and penetrating into the organ; ii) inadequate myofascial support of an organ during periods of body movements or even from the pull of gravity on an inadequately braced organ; iii) faulty delivery of nutrients and/or removal of excretory materials because the distorted fascia is impeding blood and/or lymphatic flow; and iv) aberrant neural activity in response to heightened or suppressed nerve cell activity caused directly by the distorted fascia surrounding the organ. Again, within this proposed association of primary myofascia induced organ dysfunction are a myriad

of many common clinical conditions, many of which are not primarily manifested by pain, but rather by impaired function. Based on apparent therapeutic success of OMD and other physical therapies, myofascial disorders are considered by some practitioners as a contributor to major diseases of the heart, lungs, liver, pancreas, kidney, brain, etc.

An interesting question is whether cancer can be reasonably included within the spectrum of potential cellular changes inducible by myofascial dysfunction. Anecdotal reports of regression of biopsy proven cancers following OMD suggest that the myofascia strain may possibly affect normal cellular maturation. Moreover, there are reports of an increased occurrence of myofacial pain in cancer patients.

Most medical practitioners are unskilled in applying physical massage as a means of relaxing unduly contracted muscles. Even if they were, remuneration at existing rates would be far below that of more routine patient visits. The typical medical practitioners would rather prescribe analgesics to help blunt the perception of pain. A more insightful approach is based on the understanding that muscles contract when there is an inadequate source of cellular energy. Thus, it is the muscle relaxation process, rather than contraction, which requires the expenditure of cellular energy. The widespread contraction of energy-starved muscles occurring after death is apparent in the process referred to as rigor mortis.

Normal individuals can readily adapt to the greatly increased energy needs of muscles during periods of strenuous exercise. It is noteworthy; therefore, that some people may have difficulty in even maintaining the energy needs of nonfunctioning muscles. Most researchers view the generation of adenosine triphosphate (ATP) by oxidative phosphorylation of food-derived metabolites as the only available source of cellular energy. This view has been challenged by the description of an alternative cellular energy (ACE) pathway. This pathway was initially identified in cells infected with stealth adapted viruses and in which major disruption of mitochondria, the predominant source of ATP, can occur. It is noteworthy; therefore, that stealth adapted virus infection is the cause of CFS and related illnesses. Could, there be a virus infection of the contracted muscles in fibromyalgia? Relatively few histological studies on affected muscle tissue of FM patients have been reported.

Illnesses such as irritable bowel syndrome, asthma and migraine are frequently seen in patients with multiple areas of localized muscle pain and tenderness. This association may be indirect, e.g. from aberrant neural activity coming to the brain or spinal cord from elsewhere in the body. It may also occasionally arise from genetically determined deficiencies in the actual composition of collagen fibers, such as in Ehlers Danlos Syndrome. More commonly, however, organ illness is seemingly occurring from a potentially reversible primary functional disruption in the myofascial network of the particular organ.

Several possible reasons exist for an insufficiency of cellular energy (ICE) in the myofascial network. Basically, these reasons can involve either normal cellular metabolism or the recently described alternative cellular energy (ACE) pathway. Defects in normal metabolism can occur from an impaired supply or utilization of nutrients and/or oxygen and the excess accumulation of metabolic products, again occurring from either faulty or excess metabolism or impaired circulation. Thus muscle contraction and myofascial distortion may be amenable to therapy via enhancement of the ACE pathway.

Malignancy: Positive Stealth Virus Cultures in Myeloma Patients: A Possible Explanation for Neuropsychiatric Co-Morbidity

Brian G.M. Durie, Russell A. Collins, W. John Martin Cancer Center, Cedars Sinai Medical Center, Los Angeles, CA, USA;
Center for Complex Infectious Diseases, Rosemead, CA, USA
(Presented at Am. Soc. Hematology Conf. Ab. 1953 and published in Blood 94: 10 Suppl. 1, 1999)

We have used viral culture techniques to screen patients with multiple myeloma for the presence of stealth-adapted viruses: a newly defined grouping of atypically-structured, poorly immunogenic viruses which induce a characteristic vacuolating cytopathic effect (CPE) in human and animal cell lines. Electron microscopy, serology and molecular-based assays have been used to further differentiate stealth-adapted viruses from conventional cytopathic viruses. Peripheral blood mononuclear cells from 20 patients with multiple myeloma were added to human MRC-5 fibroblasts. All cultures showed unequivocal, extensive foamy syncytial cell formation. Mononuclear cells from 10 patients were re-tested in a blinded fashion along with 10 samples obtained randomly from hospital outpatients. Nine of the 10 myeloma patient samples rapidly gave strong positives; the 10th became positive with serial observation, whereas no (zero) controls became positive. Positive cultures have also been obtained from bone marrow, CSF and pleural fluid of myeloma patients. Stealth viral infections have previously been linked to encephalopathy with complex and diverse neuropsychiatric manifestations. Detailed clinical review of the tested myeloma patients revealed neurologic abnormalities in 4 patients (brain and meningeal plasmacytomata, facial myoclonic seizures and nerve deafness), and prior neuropsychiatric abnormalities in a further 9 patients (ranging from emotional/cognitive difficulties to chronic fatigue syndrome). Since stealth virus replication can lead to varying re-combinations of mutated viral and cellular genetic sequences, virus assimilation and over-expression of genes coding myeloma growth factors could enable a stealth-adapted virus to promote the development of myeloma. Assessment of this will require sequence comparisons of stealth viruses from patients with and without myeloma. Our observations warrant these and other studies to clarify the significance of positive stealth virus cultures in myeloma patients.

Keywords: Multiple myeloma; Stealth virus

Drug Addiction: Relief from a Dreaded Sense of Mental and/or Physical Exhaustion

Why are so many individuals dependent upon drugs of addiction? What purpose do these drugs serve? A common answer turns out to be simply to help overcome a devastating sense of unexplained exhaustion. A methamphetamine addict expressed his response to abstinence by taking a bottle to bed each night because he could not muster the energy to pee in the bathroom. He would hear his dog whine to be let outside but in spite of knowing that he would be cleaning up dog feces the next day, he simply could not get out of bed! His mail would accumulate unopened and dishes would go unwashed. Who among us would not be tempted to seek a remedy, even if illegal, from such weariness and despair? With this image in mind, I was not surprised to obtain affirmation from someone addicted to Oxycontin. The pain of an earlier motorcycle accident was just an excuse that others could possibly understand. So too was their error in judging that he was self-pleasuring himself with an unearned euphoria. Unless he took his Oxy, he felt totally drained and both emotionally and physically exhausted.

Clearly, these individuals are lacking cellular energy. Their choice of medication has not been good since the chemicals they consume also affect specific biochemical pathways occurring in the brain. They deserve a trial of more selective products that can potentially provide energy through a recently described alternative (non-mitochondria) cellular energy pathway. These products have been termed enerceuticals™.

As opposed to pharmaceuticals, enerceuticals™ are not intended to treat a single disease. Indeed, they can potentially address many diseases and can work on plants and animals as much as on humans. The major distinction from pharmaceuticals is that rather than functioning in localized biochemical reactions, they generate biophysical energies that can operate widely throughout the body. A number of these products are being brought into clinical trials. It is time to test them as a possible relief from the exhaustion experienced by those trying to overcome drug addiction.

The format of these trials is to have selected individuals test various enerceuticals™ or inactive placebo controls. Providing the studies are done objectively, it should soon be apparent if individuals can differentiate between the products. Next will be to optimize the methods of delivery and to ensure safety over long term administration. Studies can be done overseas or, providing the Food and Drug Administration (FDA) has no objections, within the United States. Accordingly, a major study is being planned to be conducted among methamphetamine addicted prisoners in Thailand while a simple clinical trial is getting underway here in the United States. The major issues with the United States study are the limited choices of enerceuticals™ to be administration either as a skin or an intra-nasal spray, reliance upon anonymous self-reporting of the level of exhaustion and the possible continued dependence on either pharmaceutical or unapproved drugs. The United States study will, however, will be bolstered by parallel observations in athletes seeking to exceed usual performance levels while applying an enerceutical. For more information, contact the Institute of Progressive Medicine at 626-616-2868.

Suicide, SIDS, Anorexia Nervosa and Obsessive Compulsive Disorders

Dr. Jay Goldstein introduced me to the phrase "Betrayal by the Brain." It is particularly apt when applied to the tragedy of suicide. I recall a mother asking if I could help with the request of her daughter's poorly written suicide note that her body be used in CFS research. The patient had been labeled with the chronic fatigue syndrome (CFS) but was essentially depressed with unremitting pain. She took her life after being told by her clinician that he was no longer willing to prescribe narcotic pain medication. I reassured the mother that the research would continue without the need of her daughter's body and that she should proceed with normal funeral arrangements. Barely a week later, I spoke to a gentleman from New York whose son had died from suicide in Santa Barbara, California. The father explained that his son he had raised had essentially died several years ago when his personality had dramatically changed along with depression and fatigue. On yet another occasion, a coroner provided unfixed brain tissue from a recently deceased suicide victim. The tissue yielded a positive virus culture and was also a source of discrete mineral containing particles (as determined by EDX, see Chapter 2).

I was also once referred to a lady whose child had recently died from sudden infant death syndrome (SIDS). She told me how diligent she had been in not taking any medications during her pregnancy. When I ask why she had been prescribed the medications, she answered "for her CFS." I called the coroner, hoping to possibly obtain some tissue. Instead he boasted that SIDS was an easy diagnosis since the least you study the patient the more confident is your diagnosis. The mother had retained some breast milk and kindly provided some for cultures. As in an earlier cat study, it yielded a typical stealth adapted virus cytopathic effect (CPE).

Anorexia nervosa is a psychiatric illness with a significant mortality. It has been rationalized as a fear of being unattractively overweigh, From my experience, it has a far simpler explanation of food becoming abhorrent to the mind due to a faulty stimulus-response neuronal pathway. A failure of short-term memory (amnesia) can potentially explain repetitive behaviors shown by obsessively compulsive individuals. Memory loss can remove a frame of reference in the interpretation of many common happenings. This can lead to bewilderment and despair.

The opportunity exists to assess the workings of the ACE pathway in all such disorders and if found to be deficient to evaluate the usefulness of restoring energy to the pathway. To help prevent suicide and other fatal expressions of betrayal of the brain has been a motivating factor in pursuing research on stealth adapted viruses.

Stealth Adapted Virus Infections and Criminal Behavior

The repeated occurrence of bizarre criminal behaviors is consistent with an epidemic of psychiatric illnesses caused by brain-damaging stealth adapted viruses. It is urgent that Public Health officials address this crucial issue.

BURBANK, CA, December 21, 2012 /24-7PressRelease/ -- Tragic events, such as the mindless shooting in Newtown, reflect the failure of psychiatry to effectively prevent and treat mental illnesses. Psychiatrists are unaware that the brain can be damaged by stealth adapted viruses, which are essentially not recognized by the cellular immune system. Public Health authorities have resisted open discussion of stealth adaptation largely because some stealth adapted viruses unequivocally arose from contaminated vaccines. The lack of an accompanying inflammatory response and the molecular diversity of stealth adapted viruses have also limited interest in the pursuit of these viruses by the medical and scientific communities.

Infections with stealth adapted viruses can lead to marked behavioral changes and commonly cause mental illnesses. Infections transmitted from pregnant mothers to developing offspring explain the increasing prevalence of autism and attention deficit hyperactivity disorder (ADHD) in children. Infections transmitted among adults can also lead to a variety of dysfunctional brain illnesses, including the chronic fatigue syndrome (CFS). Stealth adapted viruses can be readily cultured from infected individuals and identified by a characteristic foamy, vacuolated cytopathic (cell damaging) effect.

Fortunately, the body can use an alternative cellular energy (ACE) pathway to defend against both stealth adapted and conventional viruses and to repair the cell damage caused by viruses. The ACE pathway is distinct from cellular energy derived from the metabolism of foods and involves the capturing of external energies through a biophysical process. Relatively simple, fluorescence-based methods have been developed for monitoring and enhancing the ACE pathway and validated in the prevention and therapy of oral and genital herpes virus infections. Limited studies in autistic children and in CFS patients have also yielded highly encouraging results.

Given the enormity of the problem of mental and other brain damaging illnesses, Public Health authorities have a responsibility to quickly learn more about stealth adapted viruses and therapeutic options. The authorities should certainly revamp the various advisory committees on mental illnesses, autism, CFS, etc., to include discussions on stealth adapted viruses and on the ACE pathway. Public health laboratories ought to be engaged in the culturing and further characterization of stealth adapted viruses, including those originating from African green monkey simian cytomegalovirus (SCMV). Lawyers could also usefully explore the issue of stealth adapted virus infection as a mitigating factor in criminal behavior. The inadvertent spreading of infections as a consequence of incarceration needs to be addressed.

For additional information please contact W. John Martin, MD, PhD. Medical Director, MI Hope Inc., a non-profit public charity, by phone at 626-616-2868 or by e-mail to wjohnmartin@hotmail.com

Is Obesity a Protective Response to Stealth Adapted Virus Induced Metabolic Dysfunction?

W. John Martin, M.D., Ph.D.
Institute of Progressive Medicine

"Why am I overweight?" a question being asked by an ever increasing number of individuals. They are commonly told they eat too much junk food; watch too much television; or are just plain lazy and undisciplined. But these explanations are not easily reconciled when they compare their life-styles with others who do not experience a weight problem.

As with any disease that is showing an alarming increasing prevalence, the first question should be whether an infectious process is involved. Another useful premise is that the body has a propensity towards repair. What is seen, therefore, as a manifestation of disease may, in fact, be a beneficial coping mechanism. In other words, the body may be making fat cells to protect itself from more serious damage potentially inflicted by an infectious agent. This paper is intended to help open discussion on how a stealth adapted virus infection could lead to obesity, and to its most serious accompanying complication, Type II diabetes.

Metabolic and endocrine studies on diabetes prone individuals have pointed to an early reduction in the ability of insulin to facilitate the movement of glucose from the blood to liver and muscle cells. Although less well studied, brain cells from pre-diabetic individuals may also require higher glucose levels to maintain normal intracellular glucose levels. More than all other tissues, the brain is particularly dependent on a continuing supply of glucose. Indeed, while the body can handle a ten fold increase in blood glucose levels, death from brain damage occurs if the normal fasting glucose level is reduced by just 50%.

Insulin stimulates the activation of glucose transporter molecules that reside at the cell surface. The ability to activate this receptor is impaired by elevated intracellular levels of lipids and in particular various fatty acids. The regulation of glucose transport, and in turn, blood glucose levels, is mainly controlled by the liver, muscle and possibly also the brain. If these organs are making too much lipids, including glucose transport inhibiting fatty acids, how is the body likely to respond? A reasonable possibility is for the body to expand upon an existing disposal system to store excess lipids. Obesity is, therefore, seen as a way to help redistribute toxic elevated levels of fat from critical tissues such as the liver, muscle and presumably brain, to a rather inert if unsightly depot. If successful, normal blood glucose levels can be maintained. If only partially successful, levels of blood glucose will rise to compensate for the impaired glucose transport from blood into critical tissues, including the brain.

What then accounts for the increased fatty acids in the liver, muscle and presumably brain? Based on a large amount of compelling data, viruses exist that essentially go unrecognized by the immune system because they lack the few critical components that are normally targeted by cytotoxic lymphocytes. A popular analogy is that of terrorists whom do not bear any insignia for easy recognition by homeland security. The terrorists can still, however, carry guns and bombs and inflict damage. The immune system evading viruses were grouped under the term stealth adapted. Another term appearing in the medical literature is virus escape mutants, although this term can also apply to viruses that lose sensitivity to anti-viral drugs. The three major hallmarks of tissues infected with stealth-adapted viruses are i) vacuolization with lipid accumulation; ii) disruption of mitochondria; and iii) appearance of complex structures termed alternative cellular energy pigments that provide a non-mitochondria source of cellular energy. The increased intracellular lipid needs to be tested for inhibitory activity on insulin mediated activation of the glucose transporter.

Lipid accumulation occurs when the lipids are either being over-produced or cannot be effectively shunted into and metabolized by mitochondria. Partial breakdown of lipids occurs outside of the mitochondria but the final steps are dependent upon normal mitochondria. From tissue culture studies, it appears that stealth adapted viruses can exert an increased demand upon the mitochondria, which eventually fail to provide adequate energy. Some of the energy needs can seemingly be met through the alternative cellular energy pathway. Indeed, this pathway can allow for long- term cell survival even in the absence of additional feeding.

Mitochondria failure in stealth adapted virus infected cells can lead to intracellular accumulation of lipid. In addition, some viruses, such as adenovirus 36, can directly stimulate lipid biosynthesis. With either mechanism, one can predict impairment of glucose transporter function and a response from the body to establish a preferable lipid storage mechanism.

Stealth adapted virus infected cell cultures are a rich source of lipids. An interesting observation led to the conclusion that excess lipids can induce the formation of fat cells, rather than the converse. HIV infected patients, especially when on prolonged therapy, can develop a fat wasting syndrome that can give an emaciated appearance, especially to the face. Attempts have been made to remove some fat from the abdomen and infuse it into the cheeks and other facial areas of the same patient. I was watching a plastic surgeon perform the very arduous task of liposuction from the abdomen of a patient. As the day was getting longer, I was wondering if there was going to be time to inject the fat that was being slowly collected. "No worry," was the answer from the plastic surgeon we can store the fat frozen and use it on another day. I took some of his frozen samples and easily confirmed that none of the fat cells were surviving the method of freezing. It basically meant that free fat was going to be injected and that it had the inherent capacity to stimulate the body's production of cells to contain the fat.

Another hint of this type of process has come from patients who develop painful collections of fatty tissues, similar to lipomas but far more painful. This condition is called Dercum's syndrome. Very similar to patients who experience outbreaks of herpes simplex and herpes zoster virus infections, Dercum patients can overnight develop a whole new set of these fatty lumps. Attacks are sometimes preceded by a day or two of tingling and occasionally the lumps correspond to a nerve pathway as is typically seen in shingles caused by herpes zoster virus reactivation. Of the several Dercum patients tested, all were positive by culture for stealth adapted viruses. Presumably, virus activation along a nerve pathway could account for the localized production of lipid and lead to the subsequent formation of cells to retain the fatty material. Consistent with a stealth adapted virus origin of Dercum disease, patients afflicted with this disorder share many of the brain disturbances seen in other stealth adapted virus infected patients. These conditions are variously classified as chronic fatigue syndrome, fibromyalgia, depression, etc.

Does this information provide any hints for the therapy of obesity and, in particular, Type II diabetes? The answer is a resounding yes. Viewing diabetes as an energy deprivation illness argues well for an energy enhancing approach to its control. There is still an unfortunate disconnect between the narrowly focused pharmaceutical approach and the non-validated, pseudo-science that permeates much of complementary-alternative medicine. As discussed elsewhere, the term Progressive Medicine is being employed to go beyond the notion that life is purely a series of biochemical reactions and yet demand the same quality of proof for efficacy of non-pharmaceutical therapies as required by the FDA for the pharmaceutical industry. A number of energy based products are currently available for formal testing in controlled clinical trials. Certainly, the preliminary data on some of these products, as reported to me by colleagues, are particularly encouraging. I have heard of patients who have exchanged a life of food-restricting starvation and fluid-restricting dehydration, for a return to completely normal dietary habits without insulin or other medications. Gone too is the obesity. The challenge now is to obtain participation of sufficient numbers of volunteers to critically validate these preliminary findings.

The other major Public Health challenge is to convince those who regulate the practice of medicine to examine patients for infection by stealth adapted viruses. Not to do so is potentially placing many of the world's population at an increasing risk, not only for obesity and diabetes, but also for the many other chronic illnesses attributable to persisting virus infections. Type I diabetes can result from direct virus damage to the insulin producing beta cells of the pancreas and/or from virus triggered autoimmune reactions to these cells. The possible role for stealth adapted viruses in some cases of Type I diabetes also needs to be explored.

Is Tissue Regeneration and Healing Without Scarring Mediated by the Alternative Cellular Energy (ACE) Pathway?

Two distinct healing processes can occur in response to tissue damage. The more usual response is inflammation followed by the predominant replacement of the damaged cells by fibroblasts in the form of a scar. A much preferred but far less commonly achieved outcome is the regeneration of tissue comprising the same specialized cell types as were lost in the tissue damaging event. This latter process is referred to as healing without scarring.

Inflicting tissue damage in fetuses of several species, including humans, does not lead to inflammation or to scarring if the damage occurs within the first trimester. On the other hand, scarring is especially pronounced in situations in which bacterial infections develop in the damaged tissue. These two observations are consistent with inflammation providing a major proliferative stimulus to fibroblasts. An independent driver of proliferation of more specialized cell types may be an energy requirement to allow for dedifferentiation of mature cells into cells that can replicate and/or to allow for the recruitment into the damaged tissue of fresh stem cells. These processes are still not well understood. Yet considerable evidence supports the proposal that the alternative cellular energy (ACE) pathway is a potent force allowing for tissue healing without scarring.

Dr. Stephen Palmer and I initially developed a copper-silver-citrate solution as an anti-bacterial product with preferential killing power for gram positive bacteria. The use of copper was based on the dearth of gram positive bacteria in the vicinity of abandoned copper mines with high levels of copper contamination. Copper is thought to compete with the cell wall receptor for iron in gram positive bacteria. As the solutions were being prepared using electrolysis, a considerable increase in volume was regularly observed. It was not attributed to temperature and was consistent with a water energizing process. Nuclear magnetic resonance (NMR) identified molecular complexes containing multiple copper and silver atoms bound to potassium citrate.

Dr. Palmer recounted rather remarkable observations concerning the healing properties of the solutions. His reports were confirmed by telephone interviews. The reports included the following:

Individual L.H. was using an electric screwdriver and inadvertently held his thumb in the path of the screw. The screw passed through both the front and the nail side of his thumb, leaving a gaping hole surrounded by flesh. It was apparent that the bone must have been penetrated. Two days elapsed with the thumb becoming more painful and beginning to throb. Because of concern about possible infection, the copper-silver-citrate solution was sprayed onto the wound. Apart from the stinging, there were no adverse effects. The individual was instructed to dip his thumb into a 2 oz container of the copper-silver-citrate solution twice daily for 7 days. The hole closed over in one day and the swelling had resolved. By day 3 the wound had healed without noticeable scabbing, discolorization or scarring. An ultrasound of his thumb at day 7 showed a fully intact terminal phalange (finger bone). His thumb became fully functional and unscarred with a normal appearing thumbprint.

Individual R.B. had a long-standing deep wound involving his left heel. His foot had been operated upon approximately a year earlier because of a methicillin resistant Staphylococcus aureus (MRSA) infection. The surgery included removal of a portion of the calcaneus (heel) bone. He was still left with an open, draining wound that required 9 daily dressings. He began to use the solution as a twice-daily spray and quickly noticed far less pain and drainage. When last examined, his wound had become noticeably smaller and seemingly the excised part of the calcaneus bone had begun to fill in. This allowed him to walk with only minimal pain. His physician is anticipating full recovery.

Individual J.B. had severe sunburn over his arms, neck and back with blister formation. He became pain free within an hour of his wife spraying the solution onto the affected areas. Interestingly, there was no subsequent peeling of the severely sun-burnt areas. The same individual developed a tooth abscess for which he sought medical and dental advice. His was able to see his physician but the dental appointment was scheduled for much later. His physician confirmed the abscess and rescheduled him to return in a few days. He took some of the copper-silver-citrate solution into his mouth in order to bathe the abscess for about 3 minutes. The pain ceased soon after the first treatment. He repeated the procedure twice daily and by the end of the second day, the swelling had gone. He cancelled his scheduled physician appointment and when he finally saw his dentist,

he was told his tooth was normal and there was no need for an extraction. Similarly, individual D.W. experienced complete recovery from a dental abscess by exposing his abscessed tooth to the solution for several days. He thereby avoided the need for a previously scheduled root canal surgery. What is most remarkable about these observations is the apparent healing of the underlying tooth pathology that led to the abscess formation. This is an extremely rare occurrence in routine dentistry.

The animal experience includes a dog that was treated for yeast and tropical Staphylococcus infection of its ears and abdomen. The skin in these areas was broken and inflamed. Three applications of the solution were made to the affected areas. Not only was there relief from the infection, as was expected from the product's antibacterial action, but the skin areas showed near complete restoration to normal appearance. Extensive infected and weeping areas on another dog cleared up without residual scarring within 6 days of daily applying the solution.

A goat breeder began to use the solution as an environmental disinfectant in the hope of controlling an outbreak of MRSA infections among her herd of goats. A veterinarian was prepared to perform euthanasia on a severely infected goat and also on an infected dog. Both animals were essentially cured the night before the planned euthanasia by a single application of the solution to the infected skin lesions. The veterinarian could find no signs of infection the next day in either animal. He requested some copper-silver-citrate solution for use as an internal irrigation disinfectant during a caesarian section to be performed on a pregnant goat. The operation was successful and seven offspring (kids) were delivered. The copper-silver-citrate solution was also applied twice daily to the site of surgical excision. Surprisingly, the mother goat was up and running by the next day. When examined by the veterinarian at 7 days post operatively, he concluded there were no internal adhesions, a common complication of caesarian section performed in goats. Even more remarkable, the surgical scar was quite unlike that expected a week after surgery. Instead, the veterinarian remarked that it looked years old. Two weeks later after three more applications of the solution, the scar was gone and fur had grown back over the incision area. As a measure of the health of this animal, all 7 of her kids survived, which is rather rare for caesarian section delivered goats.

In addition to these outstanding successes, several other individuals have used the spray on a variety of unhealed skin lesions, minor burns, allergy (poison oak) rash and even on skin blemishes, including an area of cellulite. Most individuals use diluted solutions. All of the individuals with painful skin lesions experienced pain relief soon after the application of the solution. A presumptive poison oak rash disappeared within an hour with no residual swelling or itchiness. The appearances of various skin blemishes, including the cellulite, have also been improved using daily application of the solution without any indications of toxicity.

A striking characteristic seen in these studies is the lack of scarring of the affected areas indicative of regeneration rather than simply of wound closing effect. While the healing could potentially be attributed to a biochemical action of the copper, silver or citrate ions, it is more likely a property of the activated water. Consistent with this conclusion is the apparent loss of healing capacity of solutions within weeks of being manufactured. Methods are being devised to reactivate the solution.

An argument for the solution working via the ACE pathway is the reports of healing occurring with other enerceutical™ products. Wound healing properties have also been reported with other enerceuticals™ products. These products include Enercel™, humic/fulvic acids, zeolites, colloidal silver, d-limonene and HB-101. These studies provide a firm foundation for directly testing current preparations of KELEA activated water for their healing and cosmetic properties.

Proposed Two-Way Relationship Between the Alternative Cellular Energy (ACE) Pathway and Cognitive Functioning

Abstract

The enormity of the problem of dementia, especially in the elderly, warrants consideration of fresh insights into possible causes and/or solutions, even before definitive proof is available. The physiological bases for many aspects of cognitive functioning, such as consciousness, empathy, mood and memory are still not well understood. Yet the value of positive mental attitudes and of avoiding stress in delaying the progression of many mental illnesses is widely acknowledged. These two themes may actually be related in that both could involve the alternative cellular energy (ACE) pathway. This brief paper outlines specific ways in which this hypothesis may be tested.

Introduction

Substantial progress has occurred in understanding the pathways and modes of signal transmission within the brain. The usual modeling of the brain is that it comprises innumerable circuits of electrically active neurons, intercommunicating via synapses, and organized according to specific functional needs of the body. The needs are conveyed to the brain via afferent signaling from within the body or from the environment and appropriate responses are relayed back to the body via efferent signaling. This model works well in terms of the motor, sensory and autonomic nervous systems. It is far less effective in explaining higher level cognitive functioning, much of which appears to originate from within the brain. The sense of self-awareness of one's own consciousness and that of fellow humans are attributes seemingly not present at birth but which develop during early childhood. Being able to comprehend events and establish memories for the events, such that the prior experience can be selectively recalled when a similar event recurs is an essential feature of normal brain functioning. Adequate recall of the prior event also helps in determining the quality and consistency of reactions. Memory recall is also a critical factor in interpersonal communications, including the appropriate choice of words to be spoken and comprehension of words being heard.

Higher level cognitive functioning is variously impaired in many of the major neurological and psychiatric illnesses affecting mankind. In some of these illnesses the impairments can be attributed to structural damage to neurons, as in certain genetic diseases and also following head trauma. For many illnesses, however, the impairments do not have a structural correlate. Moreover, the severity of the impairment can show wide fluctuations over time.

Stealth Adapted Viruses and the Alternative Cellular Energy (ACE) Pathway

As presented elsewhere in this book, many common neuropsychiatric illnesses are due to infection with viruses, which are not effectively recognized by the cellular immune system and do not, therefore, evoke an inflammatory response. The immune evasion mechanism is referred to as stealth adaptation. It can result from the deletion or mutation of the relatively few genes in conventional (intact) viruses that encode the major antigens recognized by the cellular immune system. These viruses can, nevertheless, be suppressed through a defense mechanism, which involves an energy source different from the calories provided by the metabolism of food. Both in virus cultures and in stealth adapted virus infected patients, electrostatic, mineral containing organic materials are formed, which can reverse the cytopathic effect (CPE) caused by stealth adapted viruses. The suggested mode of activation of these ACE pigments is that they can capture and transmit an external "kinetic energy limiting electrostatic attraction" (KELEA) force, which diminishes the level of hydrogen bonding between water molecules. This action appears to render the body's water more efficient in mediating biochemical reactions and may also help in maintaining electrical gradients across cellular membranes. The ACE pathway can be therapeutically enhanced by the consumption of various compounds, referred to as enerceuticals™, such as selected humic/fulvic acids, zeolites and herbal extracts. The ACE pathway can also be enhanced using external energy delivering devices, including the ultraviolet (UV) light illuminated of activated water containing neutral red dye. This latter phototherapy approach has been successfully used in a pilot investigative study treating children with autism, as described in Chapter 4. This Chapter included the following account:

Improved Cognition Following Activation of the ACE Pathway

A 14 year-old-girl received a home-based phototherapy session administered by her mother. Following the treatment, the mother was struck by seeing her daughter standing in front of a mirror and closely examining herself. It was as if her daughter was obtaining a fresh revelation of who she was as an individual. Best of all, she commented on being very pleased to be seeing herself. Prior to the treatment, the mother was especially sensitive to events that would invariably be distressful to her daughter. Yet these events would now pass without being emotionally upsetting to her daughter. When her mother commented upon how well she had behaved, her daughter simply answered that she remembered what her mother had said. The third and most striking post-treatment difference was the delight the daughter was showing in social interactions with her mother, even wanting to play silly games.

The criteria of self-awareness, emotional recall and empathy with others are proving to be useful markers of disease severity in other patients undergoing therapy for various neuropsychiatric illnesses, including the chronic fatigue syndrome (CFS). While the observed improvements could be explained solely on the basis of suppression of stealth adapted viruses, the findings are at least suggestive of the ACE pathway being directly able to enhance cognitive abilities.

Individual Able to Directly Activate Water

Various assays were established to distinguish regular water from KELEA activated water. The assays are based on the strength of intermolecular hydrogen bonding. It was, therefore, of interest that an individual came forth stating that she could activate water through her own will power. She felt that she was most effective when she mediated on the feelings of serenity, blissfulness and joy, yet maintained humility and gratitude. She would do so over several hours, surrounded by the water she wanted to activate. The level of activation she was able to achieve was more impressive than that accomplished using humic/fulvic acids or zeolites. Dr. Emoto is also renown for similar belief that humans can directly affect certain qualities of water.

Positive Mental Attitude and Healing

It has been a long held belief among many alternative medicine practitioners that for their therapies to work, their patients must maintain a positive, optimistic mental attitude. Indeed, some practitioners admit that this attitude is more valuable than any of the prescribed alternative modalities. Certainly, the opposite also holds, which is that stress can aggravate various illnesses. These aphorisms apply to a wide range of illnesses, well beyond those ascribed to stealth adapted viruses.

Dr. Jon Stoneburner made an intriguing observation in the performance of collaborative studies conducted with the author. He observed a branching pattern of deep skin fluorescence in the region of an outbreak of shingles occurring on the back of an elderly woman. The pattern was highly suggestive of a branching nerve, with the clear implication of ACE pigments being transmitted along nerves to an affected area. Particulate materials, which readily fluoresce with neutral red dye, have been isolated from suspension of post mortem brain material.

It will be of interest, therefore, to pursue the theme of a two-way positive feedback communication between cognitive functioning and the body's ACE pathway. A worthwhile goal is to essentially empower individuals to directly enhance their own ACE pathway through the power of cognition. The activated ACE pathway may, in turn, enhance the individual's power of cognition. Testing of this scenario could be extended to procedures such as laughing, qigong and joyfulness. The examination of water held in close proximity of the person may serve as a surrogate marker for intrinsic ACE pathway activating activity.

Several common medical conditions limit cellular energy formation via regular metabolism. For example, insufficient oxygen delivery to tissue from emphysema, anemia or impaired blood supply can lead to measurable cellular failure. At least theoretically, symptomatic relief from these conditions might be achievable by active cognitive self-enhancement of the ACE pathway.

While waiting for these provocative studies to take shape, it is decidedly worthwhile to directly proceed with well defined cognitive studies based simply on the consumption of KELEA activated water. So too can well conducted studies on the potential benefit of KELEA activated water in conditions, such as emphysema, anemia or impaired blood supply.

Can KELEA Activated Water Lead to Improved Quantity & Quality of Agricultural Produce?

This paper is intended to introduce a new paradigm regarding the role of water in supporting the alternative cellular energy (ACE) pathway in agricultural crops. It is proposed that water with ACE pathway enhancing activity can lead to higher crop yields and also improve the quality and the disease resistance of crops. Human and animal consumption of ACE enhanced foods (enerceuticals™) is proposed to offer clinical benefits beyond the supply of calories and conventional nutrients. Preliminary data supporting this paradigm are provided along with the request for participation by farming communities in further field trials.

It is widely acknowledged that water from different sources may vary in the functional capacity of promoting the growth of agricultural crops. The differences are typically attributed to natural variations in mineral content, hardness and/or pH of the water. Less accepted is the notion of possible changes in an intrinsic property of the actual water molecules, unrelated to known factors such as temperature and/or pH. Many proponents of structured or energized water or of water having a memory have made unsubstantiated claims, which are often biased by commercial interests. Still, it would be unfortunate if some measures of these statements were to be true and overlooked by the farming communities.

Based on an extensive review of many of the stated methods of energizing water, the author has confidently concluded that water can indeed be altered in the strength of the intermolecular bonding between water molecules. Energized or activated water processes greater kinetic activity with an enhanced tendency to vaporize and sustain an elevated vapor pressure in a closed container. Neutral red dye particles sprinkled onto activated water show rather dramatic linear dissolving patterns with to-and-fro movements. This pattern contrasts with the slowly and evenly expanding dye from stationary particles placed onto non-activated water. Another easily measurable difference is the increased rate of weight reduction in closed, but not completely sealed containers of activated water. A hypothesis was proposed to explain the water activation process in terms of an absorbed kinetic energy limiting electrostatic attraction (KELEA).

Among the various water activation methods examined, several comprised the use of heated ceramic materials. The heating process to 1,000°C or higher is likely to result in preferential hydrogen versus covalent intermolecular bonding. One such product is composed of volcanic rocks and called Kiko™ ceramics. The pebble-like stones are placed into either plastic or stainless steel perforated cartridges measuring 22mm x 160mm. The cartridges are placed either into the ground or within the water being provided for irrigation. This product is discussed because it has been evaluated in actual field trials of farming operations. Data from multiple types of crops have been reviewed and will be presented for peer-reviewed publication.

For the purpose of this report, two crops have been chosen, rice and sugarcane. These are two of the major staple foods consumed by more than half of the world's population.

Rice

Data from 5 studies were reviewed. Two were conducted in Thailand and three in the Philippines. Kiko cartridges were paced into the ground at approximately one per 40 square meters. Rice productivity and other observations in the Kiko-treated fields were compared with those in adjacent fields not containing Kiko cartridges.

A consistent finding was the greener hue of the Kiko-treated fields. At varying times, plants from the test and control fields were compared. These comparisons showed the Kiko-treated fields producing taller plants with more leaves on the tillers (branches). The root system was both denser and longer in the Kiko treated fields compared to control. The parlay (pre-husked rice) yield consistently exceeded the control by an estimated 35%. A similar differences was seen in the yield of rice grains, with those from the Kiko-treated fields being larger, more white in color and having up to 40% less broken forms.

Contributing to the greater yield of rice was a marked reduction in rodent (mice and rats) infestation. An invasion by Kuhol snails occurred in one of the sites but only involved the control field. Another farmer experienced an outbreak of tungro virus infection. It was far milder and of shorter duration in the Kiko-treated field. Two of the farmers commented upon the likely less need for both fertilizer and pesticides from using Kiko cartridges.

Sugarcane

An informative study was recently concluded in the Luzon region of the Philippines. Sugarcane points germinated in Kiko-treated water were subsequently grown in Kiko-treated soil. The efficiency of germination and the sugar yield were compared with the values obtained when regular water was used for germination and the sugarcane grown in non-treated soil. As detailed below, striking increases in the efficiency of germination and in subsequent sugar yield were recoded using the Kiko technology. Additional observations included markedly reduced infestation in the Kiko-treated crops.

The study was conducted on 3.3 hectares divided evenly into test and control areas. It was begun in January 2013 with harvesting in December 2013. Sugarcane points of the PHIL 74-64 variety were placed into either regular water or water treated with a single Kiko cartridge for varying periods prior to being planted. Kiko cartridges were inserted into the soil of the test field at 1 cartridge per 72 sq. meters.

Striking differences were noted in both the quantity and the quality of the resulting sugarcane. Specifically, there was earlier growth of plants (tillers) derived from the sugarcane points germinated in Kiko-treated compared with control water. Moreover, the stalks were of larger diameter and sturdier. At the time of the December harvest, the total yield in tons per hectare was 105 in the Kiko-treated area versus 93 in the control area. More impressive, the average sugar content expressed as piculs per ton (PS/TC) from the Kiko-treated area was 1.50, whereas the control was 1.35. This increase was similarly reflected in the increased number of 50Kg bags of sugar per ton (LKg/TC), being 1.90 in the Kiko-treated area versus 1.71 in the control area. The sugarcane from the Kiko-treated soil appeared to still be in the growing phase with greener stalks and longer intermodal lengths. Equally noteworthy was the absence of Downy mildew fungus, white grubs or rodent bites in the sugarcane from the Kiko-treated field.

Discussion

This preliminary report is provided to underscore the enormous impact on the world's food supply that is potentially achievable by inducing changes in the quality of ground water. The consistency of positive findings in multiple studies adds credence to the basic proposition that plants can respond differentially to variations in water, which are unrelated to mineral or other components. Similar provisional conclusions have been arrived at with many other approaches to enhancing water, although typically direct comparisons with controlled untreated areas have been lacking. One such approach involved the use of a terpene-rich extract from Japanese cedar, cypress and pine trees and plantains. The product is termed HB-101 and is typically used at 1:1,000 to 1:10,000 dilutions. Annual production of HB-101 was said to be in excess of 1,000 tons with usage by at least 20% of Japanese farming operations. It is unlikely that such extensive use would occur in the absence of conviction of its utility. Similar support is available for the use of other materials considered as beneficial as soil amendments. These include humic/fulvic acids, zeolites, shungite and many ceramics.

The lack of controlled studies has been a major impediment to the acceptance of water

having a variable intrinsic capacity to enhance the cultivation of crops. Even more limiting has been the lack of a perceived mechanism of action, whereby water molecules could be altered in a biologically relevant manner. As discussed in Chapter 6 of this book, very extensive data are available from numerous sources in support of water activation. Moreover, a testable hypothesis involving a kinetic energy limiting electrostatic attraction (KELEA) was proposed to explain the reduced level of hydrogen bonding between water molecules accompanied by increased kinetic activity of the activated water. This enhanced activity is regarded as a major component of an alternative cellular energy (ACE) pathway.

A common observation associated with the use of many of these products is the reduced infestation of treated crops. The cited examples with Kiko technology included rodents (rats and mice), insects (white grubs), snails, fungus and viruses. The farmers initially viewed the reduced infestation as being coincidental to the improved growth. It is more likely that both issues are related to enhancement of the ACE pathway.

Prior to Louis Pasteur's Germ Theory, the prevailing belief was that healthy tissues were naturally resistant to microbial contamination. While the extreme view that microbes could spontaneously arise from diseased tissue was clearly wrong, the basic belief that healthy plants as well as healthy animals were better able to resist infection holds true. This can lead to reduced requirements for toxic pesticides and/or the use of pesticide resistant genetically modified crops.

Medical science has largely equated the resistance mechanism in animals and humans solely to the immune system. The apparent benefits seen with the application of Kiko and other technologies to plants certainly support the ongoing studies of ACE enhancing procedures in the suppression of human infectious diseases. Indeed, there is an emerging synergy between ACE pathway studies in plants; animals and humans, with a clear focus on documenting benefits of KELEA activated water. An exciting challenge is to determine if major additional health benefits can be realized in humans and animals by consuming crops grown in KELEA activated water. Extracts of certain foods, e.g. moringa oleifera and ashitaba are already known to have substantial growth enhancing qualities when sprayed onto other plants and to have medical benefits when consumed by humans. The term enerceutical™ was introduced to describe these and similarly acting natural products. A determined effort is warranted to determine if many additional crops can be transformed into effective enerceuticals™ by being cultivated in KELEA activated water. The cooperation of farmers willing to provide test and control produce for rigorously controlled studies is being sought. In any event, all farming operations can presumably immediately benefit by enhancing the quality of the water being used for cultivation.

Acknowledgement: I wish to thank Mr. James T. Osugi, Founder Kiko Technology Ltd. for permission to review and report of the field trials. Additional input on pest control was also received from Mt. Geoff McMahon of Morningstar Minerals, producer of humic/fulvic acid soil amendments.

Topic 5: Progressive Medicine: Enhancing the Alternative Cellular Energy (ACE) Pathway

Rising cost of medical care in the absence of corresponding reductions in the prevalence of many chronic illnesses raises the concern that society is being misled by self-interests of the healthcare industry. The industry is engaged in two major pursuits; which are both destined to help ensure continued growth and financial gain. First is the major emphasis on identifying new clinical and laboratory-based criteria enabling patients to be further categorized into an ever-expanding array of diagnostic entities. Second is the parallel development of new pharmaceutical drugs, and to a lesser extent of novel dietary supplements, which are collectively promoted as necessary for optimal therapy of the ever-growing list of defined illnesses. At its extreme, these developments are ushering-in an era of personalized medicine, with highly sophisticated analysis of the patient's entire genome along with the characterization of innumerable cellular biochemical pathways (biochome). The data are used to design individually unique regimens of multi-drug therapies intended to specifically correct each and every aberration detected during the extended workup of the patient.

These costly expansions in medical services and provisions of pharmaceutical drugs run counter to the evolving proposition that many illnesses occur because of a common failure of a basic healing/regeneration process mediated by an alternative cellular energy (ACE) pathway. It is still widely believed that animals derive energy for cellular activities only through the metabolism of food, which largely involves cellular organelles called mitochondria. Cell survival, in spite of markedly damaged mitochondria, has been observed in tissues of humans and animals infected with stealth adapted viruses. These viruses fail to trigger a typical inflammatory reaction because of the deletion or mutation of the relatively few genes in most viruses, which actually code for antigens normally targeted by the cellular immune system. Infected cells can, nevertheless recover from the cell damaging effects caused by these viruses. They do so through the production of abundant mineral-bound organic materials, which are commonly pigmented and referred to as ACE pigments. These materials display a range of energy-related properties, including ultraviolet fluorescence (enhanced with certain dyes), selective electrostatic attraction, kinetic mobility, occasional ferromagnetism and formation of gas from water.

While ACE pigments were initially identified as a cellular response to stealth adapted viruses, their production is not restricted to this cause. Rather the ACE pathway appears to be a normal physiological process able to compliment normal food metabolism and possibly provide some additional unique functions. ACE pigments can be identified on the basis of the enhanced ultraviolet (UV) light fluorescence occurring in mixtures of ACE pigments and certain dyes, including neutral red. For instance, material made in response to herpes simplex virus (HSV), herpes zoster virus (HZV) and human papillomavirus (HPV) skin infections readily fluoresce when stained with neutral red dye solution (0.1-1.0 mg/ml) and illuminated with a simple, store-purchased ultraviolet-A (UV-A) light. Indeed, preliminary data obtained in collaboration with Dr. Jon Stoneburner, suggest that skin fluorescence in the region of recurrent HSV outbreaks is a potential surrogate marker for a pending outbreak or at least active virus shedding and, therefore, potential contagiousness.

Neutral red dye evoked fluorescence of saliva, urine and/or dried perspiration provides a potentially useful systemic marker to help identify individuals with an insufficiency of cellular energy (ICE). Urine and saliva can alternatively be tested for their ability to induce an increased rate of release of water molecules into the vapor phase using a simple weighing device. These tests can be easily performed by patients and by family members, as can the production of all but the most highly energized water and other liquids. In the laboratory setting blood smears can be examined for extracellular ACE pigments, although it needs to be noted that the granules within blood eosinophils will naturally fluoresce with neutral red dye.

ICE is expected to occur in a wide range of clinical circumstances, including i) excessive demands during cellular recovery from trauma and from infectious diseases; ii) impairments in oxygen delivery to tissues, as present in emphysema, anemia cardiovascular and cerebrovascular diseases; iii) metabolic disorders, such as diabetes; iv) cancer in which the tumor persistence may reflect a failure of energy-requiring growth-reversing processes, including cellular maturation and apoptosis; and v) ageing. A potential answer

to the correction of all of these conditions may lay in simple endeavors capable of enhancing the body's ACE pathway.

ACE pigments and other naturally occurring and synthetic counterparts are capable of increasing the energy absorbing capacity of liquids and foods. In turn, energized liquids, including water, can enhance the ACE pathway by simply being consumed. Terms such as "ACE Water" and/or "KELEA Activated Water" are being adopted to describe a range of highly energized products. These include solutions exceeding the kinetic performance levels of clinically proven homeopathic solutions. They also include more readily available water for regular human and animal consumption, as well as for widespread agricultural uses.

Community leaders can mainly assist by organizing trials among their memberships to determine the extent to which consumption of ACE or KELEA activated water or alcohol can improve the health of its members in rigorous Institutional Review Board (IRB) approved clinical studies. A useful parameter of such studies can be the diminish dependency on prescribed medications, such as anti-depressants, anti-hypertension and anti-diabetic therapies. So too can drug addicts potentially find an inexpensive alternative to their current mode of self-medication. Other very high priority illnesses include autism and the range of behavioral/learning disorders occurring in children, chronic fatigue syndrome (CFS) in adults and dementia in the elderly. Psychiatric patients also deserve special attention, especially if the ACE pathway is more critically involved in higher level brain functioning that is the conventional metabolism source of cellular energy.

The potential development of at least a universal partial remedy for many common, chronic illnesses will clearly be a challenge to the economic interests of the medical industrial complex. On the other hand, thoughtful healthcare providers, largely disenchanted by the relative ineffectiveness, yet not infrequent toxicity of many currently prescribed medicines, will see their own health and that of family members improve. Their professional time might also then be better spent exploring the many intriguing unanswered aspects of the biophysics of the ACE pathway and only having to attend to patients after correction of any underlying ICE using ACE Water.

Over 70% of fresh water consumption is used in agriculture and confirmation of the benefits of using KELEA activated water to increase crop production should lead to its universal adoption by all farmers. This is also likely to reduce the need for agricultural fertilizer and pesticides. Humans and animals should also subsequently benefit from consuming food with greater KELEA transferring capacity.

Progress in all of these areas will require concerted efforts in education and experimental validation. It will also benefit from the removal of political and business self-interest based barriers. Rather than a corporate model of incremental expansion based on return on investment, a more open, collaborative and far-reaching program is envisioned. Focusing on the ACE pathway will provide a unifying theme for these endeavors.

This book has offered insights into a new paradigm of healthcare. The foundation for the research was the discovery of stealth adapted viruses, originally from a patient with the chronic fatigue syndrome (CFS). Stealth adapted viruses differ from conventional viruses, from which they are derived, by the loss (or mutation) of the relatively few genes coding for antigens normally targeted by cellular immunity. More than sufficient data have been presented over the last two decades and summarized in Chapter 1, to prove the existence of stealth adapted viruses. The unwillingness of Public Health authorities to acknowledge or pursue these viruses reflects more than just intellectual limitation. It also involves opposition to the unequivocal fact that certain stealth adapted viruses arose from contamination of poorly regulated poliovirus vaccines. Moreover, contamination of the CHAT experimental poliovirus vaccine is the likely explanation for the origin of AIDS.

In retrospect, it was probably unfortunate that I initially linked stealth adapted viruses to CFS. The nature and cause of CFS is still an enigma to the medical profession. This disease also presents difficulties to employers and their insurance providers, which are generally reluctant to grant CFS disability. This is because the diagnosis of CFS is largely based on patients' self reported symptoms. Public Health officials have given support to insurance denials by emphasizing fatigue as the defining symptom and down playing the cognitive and other neurological manifestations shown by severely affected individuals. While the boundary between being lazy or ill is somewhat blurred, a definite divide has been placed between CFS and neuropsychiatric disorders. My early reports of stealth adapted viruses being present in CFS patients as well as in patients undergoing brain biopsies did not fit well into the mindset of many CFS patients or their clinicians. The reports, however, should, have been an alarm to Public Health officials, especially when I reported positive virus cultures in children with autism and, in conjunction with Dr. Tom Glass, also described severe illness occurring in virus-inoculated cats.

I understood the 1995 withdrawal of support from the University of Southern California (USC) School of Medicine. The Dean expressed his concern that controversial CFS research could split the alumnae into fractions and turn some away from making financial contributions. I also knew that the parent corporation of the manufacturer of live polio virus vaccines had financially supported the Chairman of the University Department in which I was working. Not surprisingly, in light of my being asked to do medico-legal testing, he closed my laboratory and confiscated what available funding I had. A major CFS patient organization was also discouraged since I had not delivered the test that would exclusively diagnose CFS patients. They too had, therefore, withdrawn their support. Yet during my leave from USC, I still managed to pursue the culturing and characterization of stealth adapted viruses. Indeed, the work was proceeding well till in 2002 when, on the directive of the Centers for Disease Control and Prevention (CDC), testing for stealth adapted viruses was deemed to be placing the

Nation's health in "Immediate Jeopardy" and no further clinical testing was allowed.

Again in retrospect, this was a useful turning point since it allowed more time to focus on how in the apparent absence of cellular immunity, the body was able to suppress stealth adapted viruses. I could readily observe suppression of virus activity in the virus cultures and had also witnessed striking recovery in the virus-inoculated cats. I closely examined intricately structured intracellular inclusions in brain biopsies from infected patients using electron microscopy. The cells were surviving in spite of marked disruption of the cells' mitochondria. Moreover, the structures were similar to those responsible for the repair of damaged cells in virus cultures. As I retained samples from earlier clinical testing, I noted yearlong survival of living cells in the absence of any additional feeding and in spite of the grossly deranged mitochondria. Even more remarkable, was the continuing production of lipids in cultures long after elimination of every remaining viable cell. Putting the pieces together lead to an understanding of the alternative cellular energy (ACE) pathway and of ACE pigments. This research is summarized, along with 35 Figures, in Chapter 2 of the book.

Dr Jon Stoneburner helped reinforce my understanding of the ACE pathway and its potential as therapy for both conventional and stealth adapted viruses. Genital herpes infections are especially common among African American adults with nearly 50% being antibody positive. Dr. Stoneburner had been pursuing a phototherapy approach to suppress herpes virus skin lesions using neutral red dye in conjunction with ultraviolet (UV) light. Although popular in the early 1970's, the use of phototherapy was largely abandoned because of negative findings in poorly thought out studies; one of which was published in 1975 in the New England Journal of Medicine. The successful use of phototherapy for herpes and how it relates to the ACE pathway is detailed in Chapter 3 of the book, along with updated modifications, which avoid any direct contact of neutral red dye with the herpes lesions.

The success with treating herpes virus infections quickly led to the testing of autistic children; a stealth adapted virus induced illness. It was extremely gratifying to see treated children proudly showing their newly acquired reading skills to their parents. It was also striking that ACE pathway therapy led to the immediate cessation of recurrent epilepsy in a 4-year-old autistic boy treated by his father. These studies are included in Chapter 4 of the book.

The suppression of stealth adapted viruses has been a continuing goal since they were initially cultured. I had tested various products used by farmers to enhance agricultural productivity, including humic/fulvic acids, zeolites and a Japanese tree/plant extract called HB-101. This latter product was particularly interesting since farmers were only required to add 5 drops per gallon (1:10,000 dilution) to the water used for irrigation. Such a low dilution was reminiscent of homeopathy. Indeed, I had previously shown suppression of reactivation of the cell damaging effect of stealth adapted viruses by including a highly diluted product originally called HANSI, in freshly added re-feeding culture medium. HANSI was also stated to have growth enhancing effects when used on plants, similar to the reports with HB-101, humic/fulvic acids and zeolites.

The stealth adapted virus culture results led to the renaming of HANSI to Enercel™ by its US manufacturer. When the opportunity came, I gladly offered to help evaluate Enercel™ in a clinical trial to confirm its efficacy in the therapy of children with tropical diarrhea (a major killer of children in developing countries). It was a real pleasure dealing with highly conscientious clinicians from El Salvador, who physically performed the trial. Two injections of Enercel™ significantly suppressed the severity of diarrhea. This study is described in Chapter 5 of the book and was presented at a National meeting.

I was greatly surprised and also intrigued by detecting Lidocaine in Enercel™ in personally performed GC-MS (gas chromatography-mass spectroscopy) studies. This was an undisclosed component in Enercel™ and alerted me, once again, how commercial considerations can negatively impact on the advancement of science. The finding of Lidocaine in the Enercel™ product also rekindled my interest in the work of Ana Aslan. She had previously popularized the rejuvenating benefits of procaine in the product Gerovital, but was largely stymied by political forces.

Profits, Political and Personality (P3) barriers have repeatedly suppressed the advancement of scientific discoveries. I undertook to objectively analyze the many reports that water and other fluids could potentially be activated (energized). This exercise is detailed in Chapter 6 of the book. It alludes to a series of remarkable individuals who have provided useful historical information, in spite of sometimes being elusive. The elusiveness was undoubtedly due mainly to scientific uncertainty but may also have reflected the protection of self-interests. Misleading information is especially apparent in the claim for symptom-specific therapies in homeopathy. Essentially, there are no data to support these claims. The alleged specificity of homeopathic products only serves to justify a continuing role of the homeopathic practitioner in formulating and/or selecting patient-tailored remedies. So too, the suppliers of various water energizing additives prefer to place the emphasis on their particular product rather then on a common outcome of using either their or a competitor's product.

The important first conclusion from Chapter 6 is that water and other fluids can indeed be activated. The Chapter then proceeds to intuitive thinking, backed by experimental observations, which led me to the confident suggestion that water and other materials could absorb a natural energy, which I refer to as KELEA (kinetic energy limiting electrostatic attraction). The energy can be displayed in water by its ability to alter the dissolving patterns of neutral red dye sprinkled onto the water. Documentation can also be easily obtained by increased vaporization, reflected in the rate of weight loss in capped (but not entirely sealed) containers and in the increased vapor pressure developing in sealed containers.

The working hypothesis is that KELEA is attracted to free electrical charges and serves to reduce the intermolecular hydrogen bonding between fluid molecules. Many dielectric substances, including humic/fulvic acids, zeolites, HB-101, Lidocaine, etc., can absorb, store and transmit the KELEA force to fluids, as can certain external devices. It is further possible that the brain itself may be able to directly receive and deliver KELEA. Moreover, this may potentially become a very valuable learned skill for everyone.

The science of KELEA is also interesting since even conventional cellular energy from photosynthesis and from food metabolism is fundamentally related to separating electrical charges. It is further possible that KELEA is related to a form of dark energy of physicists. Moreover, from the evidence of lipid synthesis in the absence of living cells, as shown in Chapter 2 of the book, KELEA activated water may help explain some reports of cold fusion. Its possible relationship to gravity has also been considered.

Two paths can now be followed: The first is to pursue the exciting and provocative physics of KELEA. This can best be done through collaboration with individuals and institutions with access to advance instrumentation to formally test the intuitive hypotheses. The other and more socially compelling task is the realization of potential benefits of this new scientific paradigm. The second Section of this book was primarily written to help in achieving this goal. It is intended to foster the organization of a multiplicity of ongoing human and animal studies, along with assessments of potential agricultural and industrial uses of KELEA-based technology.

Several of the medical aspects of this undertaking were anticipated in the "Goals of MI Hope Inc.," which I redefined in 2002. As stated then and still being stated, the goals are to:

1. Create a Movement to address the epidemic of brain damaging illnesses afflicting this country.

2. Attract bright individuals committed to "doing what is right," not merely what is expedient, convenient or financially rewarding.

3. Foster scientific, social and spiritual leadership, with effective spokespersons and media outlets.

4. Provide for the immediate social and spiritual care of individuals with brain damaging and other illnesses.

5. Investigate the role of vaccine-derived and other stealth-adapted viruses in either causing or contributing to many of these illnesses.

6. Pursue both basic and therapeutic-oriented research on alternative cellular energy pigments (ACE-pigments) with a focus on biophysics.

7. Define the pathways linking specific genetic abnormalities to the expression of various neurological and neuromuscular disorders.

8. Devise, develop, evaluate and market relatively inexpensive diagnostic and therapeutic procedures.

9. Resurrect and educate others on the principles and applications of suppressed medical and scientific technologies.

10. De-politicize Public Health.

I can now add to these goals, the activation of all water used in agriculture and possibly of all gasoline used in transportation. It may also be possible to facilitate the ease of electricity passage through wires (currently under investigation).

Along with the many medical applications, these additional goals are way beyond the efforts of a single institution. Nor should they be pursued in only one country with imposed patents or other restrictions.

The purpose behind the many commentaries included in the second Section of the book is to encourage the participation of many healthcare advocates and activists for many of the more common human illnesses. Along with agricultural and industrial proponents, numerous investigative teams could be assembled largely under their own guidance. A possible structuring of the different endeavors could be as follows:

Medical: A simple beginning will be to install KELEA delivering devices into regular water bottling plants to provide comparable batches of bottled water differing only in KELEA activity. Lead organizers from within various disease related patient support groups could help with the "blinded" distribution of coded water samples to participating volunteers from within their organizations. They could also assist in the documentation of individual patient responses prior to decoding. MI Hope Inc., (which also uses the name Institute of Progressive Medicine) can help in obtaining Institutional Review Board (IRB) and when necessary FDA approval for the studies performed in the United States. A central clearinghouse of data could be established as an ongoing resource to compile data and help design future modifications in the protocols.

The range of illnesses to be studied extends way beyond the daunting epidemic of stealth adapted virus infections (CFS, autism, childhood learning disorders, post traumatic stress disorder (PTSD) and probably many of the commonly defined neuropsychiatric illnesses). It should include illnesses caused by conventional viruses, including herpes, HIV, hepatitis viruses, etc. So too should bacterial infections be studied. Tuberculosis has been shown to respond to Enercel™ as has gastrointestinal diarrhea. A major topic for study should also be cancer in the expectation of achieving energy based remissions. There are also supportive data for the potential value of enhancing the ACE pathway in a myriad of illnesses with a known insufficiency of cellular energy (ICE). These include conditions with impaired tissue delivery of oxygen, as in emphysema and anemia; of blood, as in cardiovascular and cerebrovascular diseases; and of metabolic factors, as in nutritional deficiencies. The potential achievement of non-scarring wound healing; including trauma and in both stroke and heart attack patients, can be explored. So too is the possibly of delaying ageing. Again lead organizers for each of the various medical conditions could help in the recruitment and evaluation of efficacy in the particular disease for which they are offering assistance.

Laboratory personnel could also become involved in the testing of KELEA activated water in animal models for various human illnesses. Again, the comparisons can simply be between animals consuming KELEA activated versus regular water. The laboratory studies can extend to tissue culture observations on cells and to the assessment of possible changes in biochemical reactivities. Major virology laboratories should undertake the culturing of stealth adapted viruses so that they can be further characterized. The cultures can also be a source of ACE pigments. The culturing technology is not difficult and can be easily explained and demonstrated. The possible passage of stealth adapted viruses through bacteria warrants serious consideration.

Even beyond the strictly medical application of KELEA activated water, understanding the new paradigm of healthcare can potentially elevate the spirit of all mankind. Indeed, fascination with the wonderment of Nature can inspire a deeper respect for all things living and may help minimize human-to-human and human-to-environment atrocities. Even knowing that impaired brain functioning may soon be treatable can alleviate some of the current despair, which characterizes much of mental illness. An understanding that stealth adapted viruses may help explain the unfortunate predicament that many people are facing may also help minimize the patients' sense of personal failure and may be used to mitigate criminal sentencing. Disseminating this type of information will entail a major educational program with all the skills of social and other media. This task will become particularly valuable when the information includes actual results of clinical trials.

Agricultural: Approximately, 70% of all fresh water being used is in agriculture. Controlled studies can directly compare different means of KELEA delivery to crops in terms of the extent of improved productivity, longer shelf life and reduced use of pesticides. In addition to helping quantify the results, lead organizers for farming applications could undertake studies to determine if consumers of food grown with KELEA activated water will experience health benefits beyond those consuming food grown with regular water. The term Enerceutical™ is used to describe foods which seemingly can deliver KELEA upon consumption. Prominent contenders are moringa oleifera and ashitaba; already well known for their health benefits. Interested lead farmers could also help with studies intended to assess KELEA activated water for its beneficial effects on the health, vitality and productivity of their farm animals.

Industrial. Lead investigators in the industrial arena can help in the evaluation of KELEA activated water as being less corrosive than regular water. This can greatly improve the performance of heat exchange water-cooling plants. Activated water may also allow for more refined mixing of cement and other components, etc. Interestingly, the Aquapor device installed in some old European building to help dissipate moisture from the house foundations, may simply be working via KELEA. This could be easily tested. Preliminary studies have been performed on improving the combustion of KELEA energized gasoline and this topic should be further pursued. Although, still very preliminary, KELEA induced reduction in electrostatic attraction may also potentially lead to a slight reduction in the resistance to flowing electricity, thereby saving in the cost of electrical transmission along wires. Based on studies with Brown's gas, KELEA may also be able to reduce the rate or manner of radioactivity decay. Collectively, all of these possible endeavors will require major task forces, yet not to proceed as quickly as possible would be a disservice to mankind.

Role of Philanthropic and Other Organizations: I would hope that philanthropic organizations will quickly embrace the opportunity of helping move these goals forward. They could provide substantial financial support for lead investigators and initially underwrite the cost of production and distribution of KELEA activated water. They could also help provide the communication avenues for many in the world to understand and appreciate that a dynamic quality can, indeed, be imparted to water and to many other substances. I realize that the proposed scope of activities is enormous, but still well within the organizational skills of philanthropic organizations. Religious organizations could very much help in the recruitment of volunteers for clinical studies.

While, I am undoubtedly biased, I fully anticipate that natural enhancements of the ACE pathway will potentially greatly reduce the scourge of chronic illnesses, especially those that impair the brain. I can also foresee fundamental changes in agriculture. Most importantly, I am pleased to share in the excitement that Nature had one more gift to humanity. Let us use it for the betterment of mankind.

For more information or answers to specific questions, please feel free to contact the author at 626-616-2868 or by e-mail to either wjohnmartin@hotmail.com or wjohnmartin@ccid.org

Printed in the United States
By Bookmasters